# Fundamentals of
# Low Gravity
# Fluid Dynamics
# and Heat Transfer

**Basil N. Antar**
University of Tennessee Space Institute
Tullahoma, Tennessee

and

**Vappu S. Nuotio–Antar**

# Fundamentals of Low Gravity Fluid Dynamics and Heat Transfer

**CRC Press**
Boca Raton   Ann Arbor   London   Tokyo

**Library of Congress Cataloging-in-Publication Data**

Antar, B. N.
    Fundamentals of low gravity fluid dynamics and heat transfer / Basil N. Antar, Vappu
S. Nuotio-Antar.
       p.  cm.
    Includes bibliographical references and index.
    ISBN 0-8493-8913-5
    1. Liquids–Effect of reduced gravity on. 2. Fluid
mechanics. 3. Heat–Transmission. I. Nuotio-Antar, Vappu S. II. Title.
   TA357.5.R44A58  1993
   620.1'064–dc20                           93-11613
                                                     CIP

© 1993 by CRC Press, Inc.

No claim to original U.S. government works

International Standard Book Number 0-8493-8913-5

Library of Congress Card Number 93-11613

Printed in the United States of America   1 2 3 4 5 6 7 8 9 0

Printed on acid-free paper

# *Preface*

*T*his book deals with only one application of fluid dynamics and heat transfer, namely that of fluid response to the diminution or elimination of the gravity body force, such as in space environment.

Human presence in space manifests itself in two forms: space exploration and space utilization. Missions such as the Viking, Voyager, Galileo, and others similar fall under the first group. The recent Hubble telescope and the forthcoming large X-ray facility, the AXAF telescope, also count as exploratory. By space utilization, on the other hand, is meant those endeavors where the results are more directly beneficial to the earthbound mankind; all commercial, communication, navigation, military, and weather satellite systems belong to this group. The ongoing Earth Observing Satellite (EOS) system, part of the larger Mission to Planet Earth program, the various Space Shuttle flights (STS), as well as the upcoming Space Station Freedom, are also part of the latter category, as they provide platforms for experimentation and data gathering.

All unmanned missions involve complicated machinery for operational purposes as well as unique propulsion systems. The manned ones, such as the Space Shuttle and the Space Station, require additional habitable environments. Most of the above operate with fluids, whether in the form of gases or liquids or both. Furthermore, a sizable portion of the Space Shuttle program includes performing laboratory experiments involving fluids such as in the materials and biological processing modules. Space vehicles and satellites also have extensive thermal requirements in the form of cooling, refrigeration, and heat dumping. It is therefore clear that a thorough understanding of both fluid behavior and heat transfer in low gravity is of fundamental importance.

Our present knowledge of [general] fluid dynamics and heat transfer is quite extensive due, in large part, to the tremendous experimental and theoretical effort already expended toward this purpose. Deterministic thermohydraulic processes are well understood and can be predicted. Examples of these include single-phase laminar flows, climatic flows, and large-scale planetary flows. More complex

v

situations, including turbulent and multiphase flows, can only be handled by utilizing the large amount of empirical data available. In this category fit almost all engineering applications, short-range weather forecasting, and many others.

For the deterministic thermohydraulic processes, the diminution or absence of gravity does not pose much added complexity over terrestrial counterparts. Adequate mathematical description is possible thus making the forecasting of fluid behavior in low gravity fairly straightforward with a reasonable assurance of success. Examples of these include buoyancy-driven fluid convection, stationary stable liquid interfaces, and gentle liquid sloshing. The low-gravity behavior can be estimated through solutions to the governing equations coupled with validation experiments conducted in the Earth's environment. All of these topics are thoroughly examined in this book, allowing the reader to develop his or her own predicting capabilities.

For the more complex systems, forecasting the behavior of low-gravity thermohydraulic processes becomes more involved. Examples include two-phase and high-speed flows, boiling and combustion. These are not well understood under terrestrial conditions, and it turns out to be very difficult to isolate the effects of gravity in the theoretical treatment. Traditionally, scientists and engineers have relied heavily on empirical data extracted from intensive experimentation. The present cost of low-gravity testing, however, makes the development of similar databases prohibitively expensive or, in realistic terms, impossible. The following procedure is therefore recommended to ease the situation: First, a thorough investigation of the specific problem must be made. Second, a few key subprocesses must be identified. Third, low-gravity experiments must be conducted to further the understanding of the effects of gravity on these key elements. Fourth, the data from such experiments must be used as the foundation for developing an empirical basis for predicting capabilities on the total system.

This book is the result of a number of courses presented by the authors to inform and educate persons not traditionally trained in fluid mechanics and heat transfer on the effects of low gravity on these processes. The contents of the book represent the present day understanding of the topic. The first part puts forward the mathematical foundation needed for this; the subject area of fluid mechanics and heat transfer being vast, no pretense is made to cover more than the essentials. The second part presents some of the major low-gravity applications in which fluids play a crucial role. A sizable amount of the known data for these cases is covered. It is hoped that this part will be of substantial benefit and a shortcut for readers who want to perform research in this area with little or no previous background.

The intent of the book is both to provide classroom instructional material as well as be a guide for practicing scientists and engineers involved in low-gravity design. For the former, we recommend following the contents of the book in its present sequence.

The first chapter discusses the meaning of low-gravity environment and the modes of its availability. Chapter 2 summarizes the theoretical basis of fluid mechanics and heat transfer and can be omitted by readers with a background in it, although it may be instructive to glance through this chapter for notational consistency. In Chapter 3 we deal with the simplest fluid configurations, namely stationary with hydrostatic balance and slowly moving inviscid flows. Both of these are treated without the complicating factor of thermal environment. In Chapter 4, the topic of free convection, involving slow motion of fluids under the effects of gravity and surface tension, is discussed. In these two cases the thermal environment is fundamental to fluid motion.

The second part of the book discusses a number of low-gravity applications for which there exists some understanding and in which thermohydraulic processes play a crucial role. Known data for each case are presented. Chapter 5 presents the effects of fluid dynamics and heat transfer on materials processing and the advantages gained in a low-gravity environment. The subject of freely suspended drops and bubbles has fascinated scientists for a very long time. They were the first low-gravity experiments conducted involving fluid dynamics and are treated in Chapter 6. Chapter 7, on two-phase liquid vapor flows, analyzes fundamental thermohydraulic processes encountered in many engineering low-gravity mechanisms. This chapter describes a fluid process that is not well understood in a terrestrial environment and for which a low-gravity data base is sorely needed. Combustion is another complex thermohydraulic process that is not well understood on the Earth and for which a low-gravity data base is also missing. This is the topic of Chapter 8. Chapter 9, on fluid management in space, represents an example of how to develop operating design criteria on complex thermohydraulic systems with intuition, analysis, and experimentation.

Many individuals, including colleagues and students, have greatly contributed to our knowledge of the subject matter of this book, and we are deeply indebted to them. In particular, we owe a great deal to Frank Collins, George Fichtl, William Fowlis, Ziad Saghir, and Harvey Willenberg. This book is a direct outgrowth of a series of lectures presented at the Boeing Company, Huntsville, during the summer of 1992 and later repeated at the Canadian Space Agency. The support of these organizations is gratefully acknowledged.

We dedicate this book to our daughters Alli Martina and Annukka Aida Rose, in the hope that they be inspired to exceed our expectations of them.

May 1993

Tullahoma, Tennessee
Huntsville, Alabama

# Contents

## Chapter 4   Free Convection   87

## Chapter 5   Materials Processing   141

## Chapter 6   Drops and Bubbles   171

# Chapter *1*

---

# *Low-Gravity Environment*

*T*his book is about the behavior of fluids subject to low or zero gravity force. We therefore begin with a description and a definition of a low-gravity environment.

Whether onboard satellites or interplanetary vehicles, most space flight environments are characterized by low-gravity conditions. Therefore, space experiments and hardware have to be designed as well as tested with this in mind before they are placed in orbit. At the present time, however, even the most economical space flights are prohibitively expensive for large-scale research efforts, which is why several terrestrial systems have been developed in which low gravity can be simulated at a reasonable cost. These include drop towers and airplanes flying a special trajectory. Although these simulations are constrained by short time durations, making them of limited value, their availability has resulted in accumulating a large knowledge base on some problems.

## *1.1* DYNAMIC WEIGHTLESSNESS

The condition of dynamic weightlessness or the apparent lack of gravity occurs in all free-falling objects and space vehicles. This is not so much due to their increased distance from the Earth but rather to their free-falling nature, as we shall see in the following.

According to Newton's law of gravitation, any two objects have a gravitational attraction to each other in direct proportion to their masses and in inverse proportion to the square of the distance between their centers of masses. Thus, the gravitational acceleration $g(R)$ for a spacecraft at an altitude $h = R - R_0$ above the Earth's surface is calculated from Newton's law to be

$$g(R) = [R_E/(R_E + h)]^2 g_0. \qquad (1.1.1)$$

In (1.1.1), $R_E = 6371$ km is the mean equatorial Earth radius, and $g_0 = 9.81$ m/sec$^2$ is the gravitational acceleration at sea level. Relation (1.1.1) shows that for a spacecraft at an altitude of 1600 km above the surface of the Earth, the gravitational acceleration is only 64% of its ground-level value. This means that under these conditions alone, $g/g_0 \approx 0.64$.

Galileo was the first to define the state of dynamic weightlessness in his book *Mathematic Proof Concerning Two Sciences.* In it Galileo correctly interpreted that the weight of a load is felt only because we try to prevent it from falling down. If, however, we were moving downward at the same velocity as the load we are supporting, that load will appear to be weightless. This dynamic state is analogous to trying to strike with a spear somebody who is running ahead at the same velocity as that of the spear. The spear will never strike its target. Such an interpretation of weightlessness may be experienced at a very low level daily when riding an elevator.

Any body that is in a state of free fall, be it falling toward the Earth, around the Earth, or around the Sun, is in a state of dynamic weightlessness. This state is equivalent to the state of no gravitational force, i.e., the state of zero gravity.

The state of dynamic weightlessness is experienced by all Earth-orbiting vehicles. When a space vehicle is injected into its designated orbit, brought up to the orbital velocity, and made to move without propulsive assistance, it is in a state of free fall toward the Earth. Everything in it is also in a state of free fall and does not exert forces on any support. This dynamic state will be referred to as *low gravity* or *zero gravity*. When we refer to a low-gravity environment we really mean that the statics or dynamics of that system, relative to the traveling vehicle, can be treated as though it were in fact in a low-gravity field.

Consider, for example, a spacecraft describing a circular orbit of radius $R$ around the Earth as shown in Figure 1.1. Newton's second law of motion requires that the gravitational force acting at the center of mass of the spacecraft be equal to the product of its mass, $M$, and its centripetal acceleration, $a_s$,

$$g(R)M = a_s M. \tag{1.1.2}$$

The centripetal acceleration, given by $a_s = \omega^2 R$, is directed toward the Earth's center, $\omega$ is the angular velocity of the spacecraft and $g(R)$ is the local gravitational acceleration of the spacecraft given by equation (1.1.1).

Newton's second law of motion, applied to a particle of mass $m$ moving inside the spacecraft with acceleration $a_p$ in an inertial reference frame, can be written as

$$F_{\text{rel}} + W = m a_p, \tag{1.1.3}$$

where $W = m g_p$ is the weight of the particle and $F_{\text{rel}}$ is the force on the particle relative to the center of mass of the spacecraft. The particle acceleration, $a_p$, is given by the sum of its relative acceleration with respect to the center of mass of the spacecraft, $a_{\text{rel}}$, and the acceleration of the spacecraft, i.e.,

$$a_p = a_s + a_{\text{rel}}. \tag{1.1.4}$$

Substituting expression (1.1.4) into equation (1.1.3) we obtain

$$F_{\text{rel}} + m(g_p - a_s) = m a_{\text{rel}}. \tag{1.1.5}$$

Since the position vector of the particle with respect to the center of mass of the spacecraft, $r$, is much smaller than the orbital radius of the satellite, i.e.,

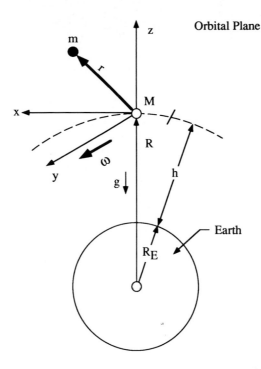

**Figure 1.1**

$r \ll R$ in Figure 1.1, then $g_p \approx g(R)$. With this assumption equation (1.1.5) reduces to the following:

$$F_{\text{rel}} = ma_{\text{rel}}. \tag{1.1.6}$$

This is the dynamical equation for a particle of mass $m$ inside the spacecraft, and it is identical to the equation for this particle in a true zero-gravity environment. This happens only when the spacecraft accelerates at exactly the same rate as the particle if it were free falling in the gravitational field, i.e., $g_p = a_s$.

The condition $g_p = a_s$ is exactly satisfied only at the center of mass of the spacecraft and only when there are no other forces acting on the vehicle. It is for this reason that the condition of zero-gravity environment is never quite achieved in reality, making the term *zero-g* a misnomer. Even for the most advanced solar system missions one can envision, gravitational forces are never really absent. That is why the term *low gravity* will be used throughout the text.

For realistic spacecraft systems there exist several forces that could induce small, gravity-like disturbances on objects inside a spacecraft. These forces may be grouped into one of three main categories depending on the nature of the forces acting on the system: The first includes forces whose origin is due to the physical displacement of the specific object inside the spacecraft from the center of mass of the spacecraft. Such forces are commonly known as *tidal forces* and *gravity gradient forces*. The second group includes forces which act on the spacecraft itself and which also affect all objects within the spacecraft. These include drag due to the Earth's atmosphere, and solar radiation pressure. Thruster firings for attitude control and orbital maneuvers also contribute to the total acceleration environment of objects within the spacecraft.

Third, forces due to the mass redistribution inside the spacecraft, that are transient in nature, also contribute to the total acceleration environment of specific bodies within the spacecraft. Prime examples of these are crew activities and the motion of mechanical parts. These forces do not involve any momentum changes of the spacecraft's center of mass. Impulses caused by internal forces are always compensated for by equal and opposite impulses. All transient forces therefore result in an excitation of the flexibility modes of the spacecraft. The induced transient accelerations, called the *g-jitter*, are characterized by a broad frequency spectrum. Even though g-jitter may reach high peak values, the resulting displacements of particles with respect to the center of mass are small because of its compensated and random nature.

## 1.2 *TIDAL EFFECTS*

The racks containing the experiments and the various operating mechanisms within a spacecraft cannot all be placed at exactly the center of mass of the vehicle. As a result, a number of kinetic effects associated with the actual spacecraft environment produce artificial, gravity-like forces. For example, any unconstrained object inside a spacecraft will be moving in its own orbit around the Earth. If the spacecraft is held in an inertial orientation—that is, at a constant orientation relative to the fixed stars—an object released motionless relative to the spacecraft at some distance from the center of mass along the flight path will have an orbit whose shape is identical to that of the spacecraft. Such an orbit will always either lead or lag the position of the center of mass of the spacecraft. Due to the inertial orientation of the craft, this object will move full circle around the center of mass during the course of one orbit.

An object released motionless relative to the spacecraft at some distance from the center of mass along the radial vector from the center of the Earth will have an orbit slightly different from that of the spacecraft. Since the velocity required to maintain a circular orbit varies inversely with the square root of the distance from the center of the Earth, this object will have a slightly different velocity which will put it into an elliptical orbit with a different period. This trajectory will cause the object to slowly drift away from its initial position as the spacecraft describes its orbit around the Earth. The accelerations required to continuously alter the trajectories of such interior objects to keep them in the same relative configuration are of the order $10^{-7}g_0$ for each meter of radial displacement from the spacecraft's center of mass.

The physical separation of a body from the spacecraft's center of mass gives rise to what is known as the *tidal force*. This has an acceleration, $a_t$, given by

$$a_t = g_0(R_E/R)(r/R) \qquad (1.2.1)$$

where $r$ is the distance between the object and the spacecraft's center of mass.

The actual value for this residual acceleration, $a_t$, depends on the parameters of the orbit and the spacecraft size. For a displacement of 1 m from the center of mass of a spacecraft at an altitude of 300 km, the component of the acceleration along the tangent to the orbit is a minimum, being of the order of $10^{-10}g_0$, while the component along the axis in the orbital radius direction is greater, being of the order of $10^{-6}g_0$ to $10^{-7}g_0$ (see [1]). This component of acceleration is commonly called the *gravity gradient* acceleration. For larger orbiting structures, such as the Space Station, the tidal forces may become noticeable.

Objects inside a spacecraft that maintain a fixed orientation with respect to the Earth will have yet another residual force acting on them. When released motionless relative to the spacecraft they experience forces due to the rotation of the spacecraft. These forces tend to modify the orbits of the objects, resulting again in motions relative to the spacecraft. This inertial force $F_i$ acting on a body of mass $m$ can be written in terms of a coordinate system fixed at the center of mass of the spacecraft as:

$$F_i = m\omega^2 R, \tag{1.2.2}$$

where $\omega$ is the angular velocity of the spacecraft and $R$ is the radius of the vehicle's orbit. Reference [1] shows that this inertial force gives rise to a corresponding residual acceleration, $a_i$, acting on the body, in the amount of

$$a_i = \omega^2 R. \tag{1.2.3}$$

## *1.3* ATMOSPHERIC DRAG AND SOLAR RADIATION PRESSURE

The atmosphere at altitudes corresponding to low Earth orbits is considered composed of rarefied gas having mean molecular thermal velocities ranging from $U_m = 750$ m/s at 150 km to $U_m = 1200$ m/s at 450 km altitude. To remain in a circular orbit at 400 km, for instance, a spacecraft must travel at a speed given approximately by $U_s = 7850$ m/s. It is very clear that the orbital velocity of a spacecraft $U_s$ is always much larger than $U_m$. This disparity in the velocities thus results in a highly directional momentum transfer to the spacecraft due to the collision of the gas molecules with the vehicle.

The induced drag force, $F_D$, on the spacecraft due to this momentum transfer can be calculated as

$$F_D = \rho C_D U_s^2 A_p/2 \tag{1.3.1}$$

where $C_D$ is the drag coefficient whose value depends on the shape of the spacecraft and the type of molecular reflectance, $\rho$ is the mean atmospheric density, and $A_p$ is the spacecraft cross-sectional area projected onto the plane normal to the flight direction.

The drag coefficient $C_D$ is ordinarily calculated using aerodynamic theory in which the body shape and the fluid state play important roles. Such calculations usually result in values of $C_D$ ranging from 2 to 3.5. The lower limit is used for a diffusely reflecting sphere while the upper limit is used for a specularly reflecting flat plate with flow normal to its surface.

The deceleration due to atmospheric drag, $a_D$, is calculated from equation (1.3.1) to yield

$$a_D = \rho C_D U_s^2 A_p/(2M), \tag{1.3.2}$$

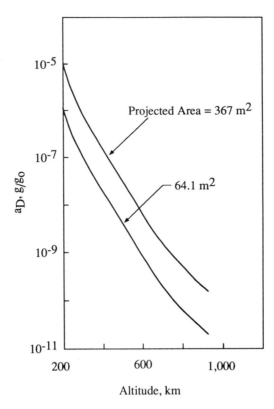

**Figure 1.2** Calculated variation of atmospheric drag with altitude for the Space Shuttle according to [4].

where $M$ is the spacecraft mass. The mean atmosphere density, $\rho$, varies diurnally for a given altitude as well as with the solar activity for longer time scales.

The vehicle cross-sectional area $A_p$ may vary due to changes in the vehicle's attitude or due to the rotating sun-tracking solar array. Thus there is an inherent variation of $a_D$ of twice the orbital frequency, with the amplitude depending on the orbital parameters, type and dimension of solar arrays, etc. The average value for the amplitude of $a_D$ is approximately $3 \times 10^{-7} g_0$. The direction of $a_D$ is constant for a vehicle in a gravity-gradient mode. In an inertial orientation, $a_D$ rotates in the orbital plane around the center of mass with the orbital period.

Figure 1.2 shows calculated values of $a_D$ as a function of altitude for the Space Shuttle according to reference [4]. These curves were constructed with values of $C_D$ set at $C_D = 2$ and for a spacecraft of mass $M \approx 9.1 \times 10^4$ kg. At an altitude of 200 km, $a_D \leq 10^{-5} g_0$. Figure 1.2 shows that a change of altitude from 200 km to 300 km reduces $a_D$ by one order of magnitude.

The other parameter affecting $a_D$ is the ratio of the projected area of the spacecraft to its mass, $A_p/M$, which is small for the Space Shuttle due to its large mass. For free-flying experiment platforms containing large solar arrays, the value of $A_p/M$ could be much larger. Initial design criteria for the Space Station could yield values of $A_p/M$ of the order of $10^{-2}$ m$^2$/kg. Assuming an

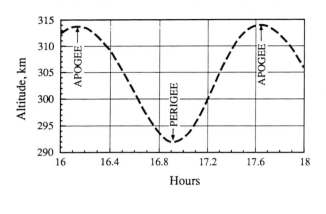

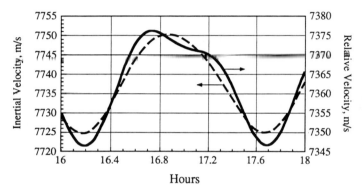

**Figure 1.3**   STS-40 altitude and velocity profiles for one orbital period during MET day 7 from [2].

altitude of 500 km, the drag deceleration $a_D$ for the Space Station is expected to be in the neighborhood of $4 \times 10^{-7} g_0$.

Reference [2] presents results from instrument recordings for the actual residual values due to atmospheric drag, $a_D$, for one orbit of the Space Shuttle flight STS-40. Figure 1.3 shows the exact altitude and the vehicle velocity for one orbital period of MET day 7. Figure 1.4 shows a comparison between the calculated values due to drag for the same period and the measured accelerations. Figure 1.4 shows all three $x$-, $y$-, and $z$-components of the acceleration due to atmospheric drag, $a_D$. The calculated values in Figure 1.4 are based on the actual shuttle data for both attitude and altitude.  The actual measured data recordings are from an accelerometer placed at some distance from the center of mass of the Space Shuttle. In Figure 1.4 the values due to the gravity gradient and those due to the rotation of the Space Shuttle have been subtracted from the total acceleration field recorded by the instrument to yield the deceleration due to atmospheric drag.  The results of [2] confirm that the acceleration due to atmospheric drag, $a_D$, can be predicted well from known atmospheric data once the attitude and altitude of the spacecraft are known.

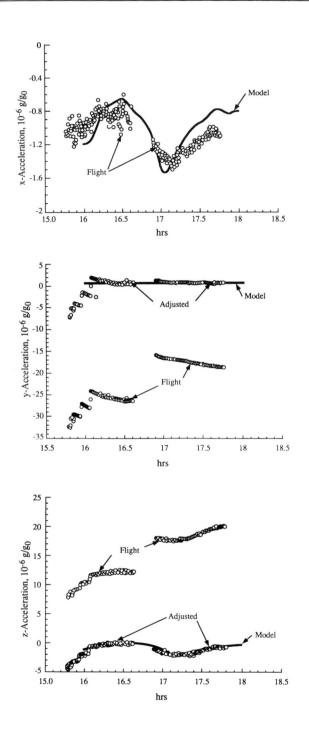

**Figure 1.4**   The three acceleration components of $a_D$ for one orbital period of STS-40 for MET day 7 from [2].

Atmospheric drag forces can be canceled by employing an active control system consisting of a small free-floating mass in a cavity at the center of mass and a thruster system to fly the spacecraft centered to this reference. This *drag-free satellite* concept has been realized with the U.S. experimental Navigational Satellite TRIAD I. Residual accelerations due to impulses of the thrusters have been found to be less than $10^{-11}g_0$.

Solar radiation can cause a pressure force directed away from the sun that is due to photon momentum transfer. The magnitude of the solar radiation pressure, $p_{sr}$, is calculated by [4] as

$$p_{sr} = \frac{1 + \beta}{c} E, \qquad (1.3.3)$$

where $0 \leq \beta \leq 1$ is the surface reflectivity, $c$ is the speed of light, and $E = 1.4$ joule/(m$^2$s) is the solar constant. For a perfectly reflecting surface, $\beta = 1$, the solar radiation pressure is $p_{sr} = 9.33 \times 10^{-6}$ Pa, while for a black surface, $\beta = 0$, $p_{sr} = 4.67 \times 10^{-6}$ Pa.

The residual acceleration induced by solar radiation pressure can be calculated from equation (1.3.3) as

$$a_{sr} = p_{sr} A_p / M, \qquad (1.3.4)$$

where $A_p$ is the vehicle's projected cross-sectional area and $M$ its mass. Normally accelerations due to solar radiation pressure are several orders of magnitude lower than the accelerations caused by atmospheric drag.

## *1.4* OSCILLATORY ACCELERATIONS: G-JITTER

The accelerations experienced by bodies within the spacecraft discussed in the previous two sections are characterized by long time scales that are of the order of one orbital period. For this reason these accelerations are usually referred to as *steady* or *quasi-steady*. Normally, there are also disturbances within the spacecraft that give rise to accelerations that are characterized by shorter time scales. Such disturbances, commonly referred to as *g-jitter*, are inherently local and stochastic in nature. These include crew movement and spacecraft operations. The difference between the long and short time-scale accelerations justifies superimposing the orbital disturbances and the g-jitter, the latter being the irregularly fluctuating contribution to the local acceleration field. The specific amplitude and frequency of the g-jitter accelerations depend on the dynamic behavior of the spacecraft structure, the location of the body, and the type and location of the sources generating the contributing forces. These may be divided into external and internal sources.

Transient external accelerations are primarily induced by pulse-like forces and torques due to thruster firings for orbital maneuvers and attitude control, docking maneuvers, extravehicular activities, waste and water dumps, etc. The Space Shuttle, for instance, is equipped with thrusters for pitch and roll control and these firings cause uncompensated transient accelerations. In the case of

**Table 1.1** Typical acceleration levels and frequencies caused by various spacecraft activities.

| Activity | Acceleration | Frequency |
|---|---|---|
| Water dump[a] | $8.0 \times 10^{-6} \, g_0$ | |
| Thruster firings[a] | $6 \times 10^{-5} - 6 \times 10^{-4} \, g_0$ | |
| Breathing[a] | $10^{-5} - 10^{-4} \, g_0$ | 0.3–2 Hz |
| Coughing[a] | $5 \times 10^{-5} - 2 \times 10^{-4} \, g_0$ | 0.3–2 Hz |
| Sneezing[a] | $2 \times 10^{-5} - 3 \times 10^{-4} \, g_0$ | 0.3–2 Hz |
| Console operation[a] | $10^{-5} - 3 \times 10^{-5} \, g_0$ | 0.3–2 Hz |
| Body bending[a] | $9 \times 10^{-5} - 3 \times 10^{-4} \, g_0$ | 0.3–2 Hz |
| Arm rotation (90°)[a] | $4 \times 10^{-5} - 2 \times 10^{-4} \, g_0$ | 0.3–2 Hz |
| Leg rotation (45°)[a] | $7 \times 10^{-5} - 2 \times 10^{-4} \, g_0$ | 0.3–2 Hz |
| Crouch and stand[a] | $3 \times 10^{-4} - 5 \times 10^{-4} \, g_0$ | 0.3–2 Hz |
| Treadmill jog[b] | $2.1 \times 10^{-3} \, g_0$ | 2.81 Hz |
| EVA tethered soar[b] | $1.7 \times 10^{-4} \, g_0$ | 0.34 Hz |
| EVA bay soar[b] | $9.3 \times 10^{-5} \, g_0$ | 0.28 Hz |
| Treadmill walk[b] | $8.2 \times 10^{-5} \, g_0$ | — |
| IVA hab soar[b] | $5.5 \times 10^{-5} \, g_0$ | 0.28 Hz |
| MT translation[b] | $2.3 \times 10^{-5} \, g_0$ | 0.28 Hz |
| Console ops[b] | $1.5 \times 10^{-5} \, g_0$ | 0.34 Hz |
| Centrifuge[b] | $3.0 \times 10^{-6} \, g_0$ | 0.86 Hz |

[a]Data from [3].
[b]Data from [8].

Skylab, on the other hand, attitude control was generally accomplished by means of control moment gyros which did not impart external forces to the spacecraft. In that case it was necessary to periodically dump the angular momentum accumulated from the integrated gravity-gradient torques by means of cold-gas thrusters. Similar control systems are being studied for the Space Station. Disturbances by docking and berthing maneuvers can occur during operation of the Space Station. An additional acceleration effect can arise from the shift of the total center of mass during docking maneuvers.

Compensated transient accelerations generated by internal activities can be pulse-like for crew motion or continuous for machinery operations. These activities are usually of very short duration accompanied by large acceleration magnitudes. Table 1.1 shows typical acceleration levels calculated for the various internal and external activities with their approximate frequency ranges.

Transient accelerations can be reduced by isolating either the vibration source or the payload element in question from the support structure, or by a damping mechanism (passive elements, e.g., springs or dampers, or active systems). Passive damping is an effective means to dampen vibrations and reduce propagation of shocks. It can be achieved by the use of rubber pads or other viscoelastic materials. In general, high-frequency vibrations can be shifted to lower frequencies with such techniques. Most of the scientific experiments show, however, an increased sensitivity to lower frequencies. Hence, care must be taken not only to reduce the vibrational frequencies but also to substantially lower the amplitudes of the accelerations; otherwise this approach can be counterproductive.

## 1.5 TOTAL ORBITAL ACCELERATION ENVIRONMENT

The total acceleration environment experienced by a body placed inside a spacecraft is made up of the sum of the various acceleration components discussed in the previous sections and is composed of the sum of the following three components: quasi-steady, transient, and oscillatory accelerations, in terms of the acceleration time scales. The *quasi-steady* accelerations are due to Earth's gravity gradient, spacecraft altitude and attitude, atmospheric drag, and solar radiation pressure. These accelerations have amplitudes of the order of $10^{-9}g_0$ to $10^{-6}g_0$, with frequencies of the order of the orbital frequency, which is approximately $10^{-4}$ Hz. Reference [7] shows recordings of quasi-steady accelerations of magnitude $10^{-6}g_0$ that were made during some Spacelab missions on board the Space Shuttle. Table 1.2 shows calculated estimates of the various quasi-steady accelerations that are always present and contribute to the total acceleration level for two different orbital radii.

*Transient* accelerations can have magnitudes as large as $10^{-2}g_0$ and tend to vary considerably in orientation. The disturbances giving rise to them are rarely sustained for more than a fraction of a second. These accelerations are caused by crew activities and thruster firings. Figures 1.5 and 1.6 show typical accelerometer recordings from Spacelab 3 that may be attributed to these activities.

*Oscillatory* accelerations can have magnitudes comparable to the transient acceleration levels or of the order of $10^{-5}g_0$ to $10^{-3}g_0$. These accelerations also fluctuate rapidly in orientation and contain a broad range of frequencies but act over relatively long periods of time. Recorded oscillatory accelerations are usually related to machinery rotation and vibrations and also to the structural modes of the spacecraft that are excited by both transient and oscillatory sources. Frequency domain analysis from one-second to fifteen-minute intervals of Spacelab 3 acceleration data show transient and oscillatory accelerations between 0.005 Hz and 50 Hz possessing amplitude spectra with a maximum amplitude no greater than $10^{-3}g_0$ as seen from Figures 1.5 and 1.6.

There have been many accelerometer measurements of the total acceleration environment aboard the Space Shuttle and other spacecrafts. Figure 1.7 shows a typical accelerometer recording for a 220-second period during a high-$g$ interval

**Table 1.2** Typical quasi-steady acceleration levels.

| Source | Altitude of circular orbit [km] | |
|---|---|---|
| | 240 | 1610 |
| Aerodynamic origin (maximum solar activity) | $7.0 \times 10^{-6}$ | $4.6 \times 10^{-12}$ |
| Geomagnetism | $9.5 \times 10^{-12}$ | $5.1 \times 10^{-13}$ |
| Light pressure | $3.1 \times 10^{-9}$ | $3.1 \times 10^{-9}$ |
| Internal gravitation | $3.3 \times 10^{-8}$ | $3.3 \times 10^{-8}$ |
| In-flight orientation control | $4.3 \times 10^{-7}$ | $2.4 \times 10^{-7}$ |
| External gravitation (nonuniformity of Earth's gravitational field) | $4.3 \times 10^{-7}$ | $2.4 \times 10^{-7}$ |

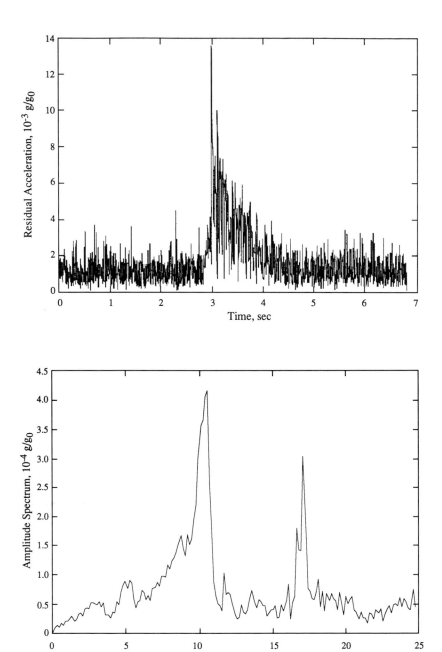

**Figure 1.5** Accelerometer data and corresponding spectral content from Spacelab 3 according to [7].

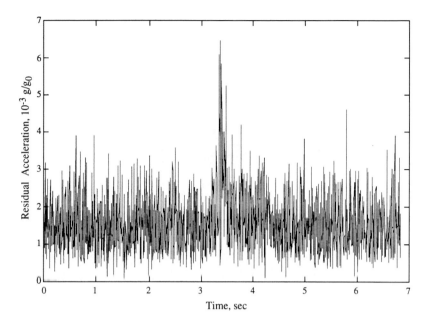

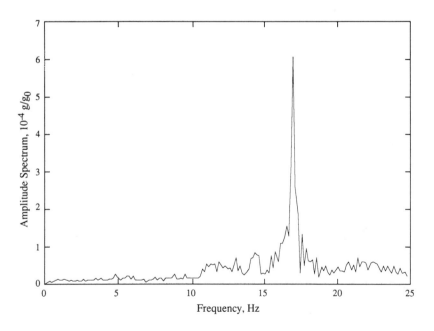

**Figure 1.6**  Accelerometer data and corresponding spectral content from Spacelab 3 according to [7].

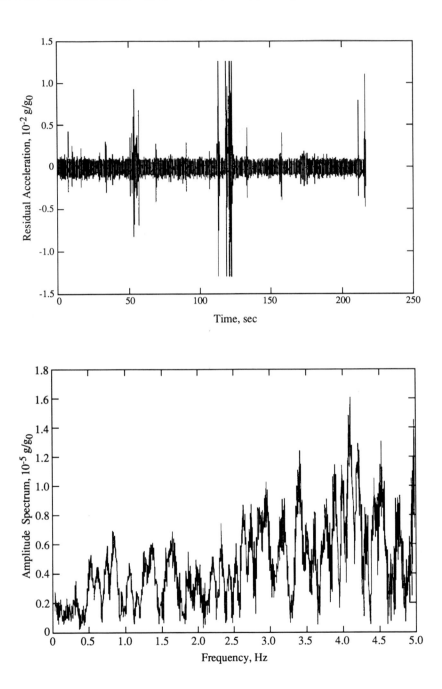

**Figure 1.7**   Accelerometer data for the total acceleration level of Spacelab 3 according to [7].

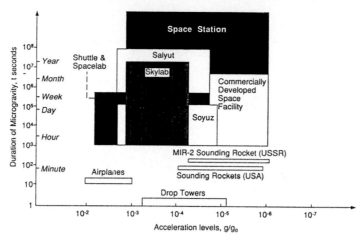

**Figure 1.8**   Quasi-steady low-gravity environment of major test facilities, from [8].

together with a spectral analysis for the data. These data are from the Spacelab 3 flight of the Space Shuttle. The data show peaks of $10^{-2}g_0$ for extremely short periods with the highest content of $10^{-5}g_0$ for up to 5.0 Hz.


## *1.6* GROUND-BASED LOW-GRAVITY ENVIRONMENT SIMULATION

The interest in low-gravity research has lead to the development of ground-based simulations as cheaper alternatives to orbital flights. The three main categories of alternatives are sounding rockets, aircrafts, and drop towers. All three suffer from the same disadvantage, providing only short time durations of low-gravity environment. These decrease from several minutes for the sounding rockets to several seconds for the drop towers. Figure 1.8 summarizes all commonly available simulation systems with their respective approximate low-gravity time durations.

The cheapest alternative to orbital flight that provides excellent low-gravity environment is the *sounding rocket*. A typical sounding rocket profile is shown in Figure 1.9. During the ascent phase, the rocket is stabilized by a spin about the longitudinal axis, which is induced by slanted external fins. The coasting phase is reached after motor separation and despin. At an altitude of 90 km, where aerodynamic forces are sufficiently low, the quasi-steady residual acceleration is usually below $10^{-4}g_0$. The duration of the free-fall phase ranges anywhere from 4 to 15 minutes, depending on the altitude and the type of rocket used. The residual accelerations which are usually composed of atmospheric drag, residual spin, rate control thruster firings and operational disturbances compare favorably with that of manned orbital vehicles. During re-entry into the atmosphere, drag forces increase gradually and degrade the free-fall condition until at 90-km altitude it is no longer applicable. At an altitude of 5 km, parachutes are deployed to return the payload without mechanical damage. The maximum acceleration levels that are possible with this system can reach up to $30g_0$ during the re-entry phase and $12g_0$ during liftoff.

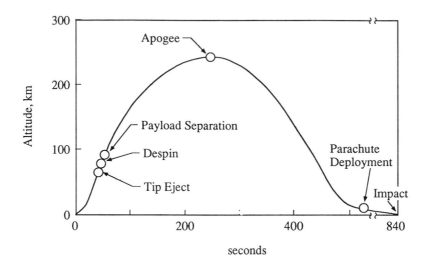

**Figure 1.9**   Sounding rocket flight profile from [4].

Another alternative to spaceflight for low-gravity research is offered by specially equipped aircrafts able to fly *parabolic trajectories* for simulating weightlessness. In this flight mode the plane climbs rapidly at a 45° angle in a pull-up maneuver, slows down as it traces a parabola, and then descends at a 45° angle in the pull-out maneuver. The acceleration and deceleration forces produce approximately twice the normal gravity, $2g_0$, during the pull-up and pull-out legs of the flight, while the pushover at the top of the parabola produces a quasi-steady acceleration of less than $10^{-2}g_0$ for a 30-second time period. During a period of 5 to 15 seconds the residual quasi-steady acceleration can be smaller than $10^{-3}g_0$. Figure 1.10 shows typical flight attitude and corresponding acceleration levels for one parabola for the NASA KC-135 airplane.

Normally the parabolic flight trajectory is repeated several times during a single flight, up to 40 times for the NASA KC-135 airplane. Aerodynamic noise and vibrations of the outer skin of the aircraft as well as engine vibrations can produce oscillatory dynamic disturbance levels of the order of $10^{-5}g_0$. The accelerations from such disturbances can be lowered in a free-floating state during which there is no contact between the body and the aircraft. Figure 1.11 shows accelerometer recorded values for a sequence of 20 parabolas executed by the KC-135 during a single flight.

*Drop tubes* and *drop towers* allow for simulation of weightlessness on Earth. Although the duration of the free fall is of the order of only a few seconds, several tests can be performed per day depending on the complexity of the facility and the particular experiment. Drop tubes normally accommodate small experimental packages in which the free-fall period lasts from 1.6 to 4.6 seconds depending on the tower height. Drop towers normally accommodate larger experimental packages and generally employ a drop shield or a containment-like projectile that reduces the aerodynamic drag during the free fall. Drop tubes are usually evacuated to reduce the aerodynamic drag. One drop tube in operation is the Japanese facility JAMIC, for which the free fall height is approximately 490 m, achieving a

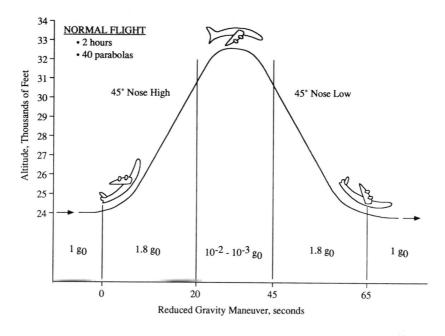

**Figure 1.10**   NASA's KC-135 aircraft typical flight trajectory.

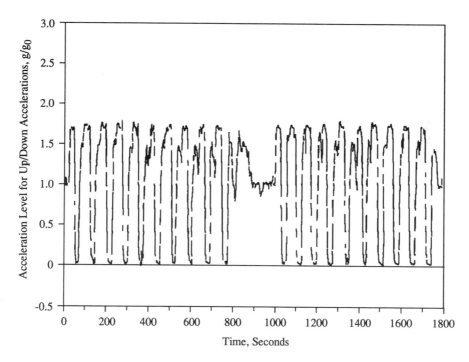

**Figure 1.11**   Typical accelerometer recording of the up/down accelerations on board the NASA KC-135 airplane.

quasi-steady gravity level of $\approx 10^{-4} g_0$ for a period of up to 10 seconds. This facility is the longest drop tube in existence to date. It can accommodate payloads of approximately 1 $m^3$ in volume with a maximum weight of 1000 kg. The maximum deceleration rate at breaking time for it is approximately 10 $g_0$. Further details regarding this facility can be found in [5].

# Chapter *2*

# *Equations Governing Fluid Flow and Heat Transfer*

T he study of low-gravity fluid mechanics and heat transfer evolves from a fundamental understanding of fluid behavior and the consequent transfer of energy from the fluid to its surroundings under terrestrial conditions. When the terms involving the gravity component of the body force are reduced or eliminated, the effects of low gravity on the dynamics of fluid flow become evident. In energy transport itself, conduction and radiation remain the same, in principle, but convection is indirectly affected by the gravitational field, as it is changed due to the changing dynamics of the fluid motion. The total energy transport will therefore have to be looked upon from a global point of view, taking into account the total mass and energy transport of the system as a whole. If, for example, buoyancy-driven flows do not take place, then natural convection of heat cannot be relied on as a mode for transferring heat. The only other modes available for energy transport in this case are conduction and radiation. Measurements of the total energy flux in this case show that it is significantly modified by the changes in the gravity field.

Many areas of engineering applications, including the design and analysis of mechanisms, developed traditionally from the empirical point of view. A design was first conceived through insight, then the inventor built, tested, and modified it. The associated physics developed either simultaneously or subsequently, through analysis and theoretical reasoning.

Designing mechanisms for space operations is very different from the conventional method. There is no laboratory for conveniently building and testing them. There was not even much of an insight about how things work in low or zero gravity. This makes analytical models very attractive and useful as viable alternatives for conceiving and testing space designs, at least in a qualitative sense.

Analytical model building has become possible only recently, due largely to substantial advances made in the areas of the foundations of fluid dynamics and heat transfer. The prevalent practice of computational fluid dynamics (CFD) is one source, and an example. It is in this spirit that any space design involving fluid mechanics and heat transfer should be approached, by first relying on analytical modeling, using the few low-gravity experimental data available, before proceeding to the final validation by testing in a true low-gravity environment.

In this chapter the equations governing the motion of fluids are derived, and the terms containing the effects of gravity are identified. Then specific equations applicable to circumstances that arise in low-gravity fluid dynamics are inferred and discussed.

## 2.1 *EQUATIONS OF MOTION FOR FLUID DYNAMICS*

The motion of fluids is governed by the basic conservation principles for Newtonian mechanics in the continuum frame of reference. These laws are the conservation of mass, momentum, and energy. In order for these conservation principles to be useful as predictive tools, they must be translated into equivalent mathematical equations whose solutions provide the required answers. It should be mentioned here that the fluids being dealt with satisfy the continuum assumption, i.e., the assumption that the mass, velocity, temperature, and all other pertinent variables are well defined at every point in space.

The first principle to be discussed is the *conservation of mass*, which can be stated as follows: The amount of matter in any material region must remain constant. An equivalent statement is that the time rate of change of the mass, $M$, of a material region is zero. Mathematically this can be written as

$$\frac{d}{dt}M_{MR} = \frac{d}{dt}\int_{MR} \rho \, dV = 0, \qquad (2.1.1)$$

where $\rho$ is the fluid density and $V$ is the volume of the material region.

At this point a property relating the differentiation of a material element to the rate of change of surface and volume integrals over a material region must be used (see [9]). This property can be stated in the following form for any function $f$:

$$\frac{d}{dt}\int_{MR} f \, dV = \int_{MR} \frac{\partial f}{\partial t} \, dV + \int_{MR} n_k w_k f \, dS, \qquad (2.1.2)$$

where $f$ can be any scalar, vector, or tensor function; $S$ is the surface boundary for the material volume $V$ described by the normal vector $n_i$; and $w_i$ is the velocity of each element of the surface. Substituting the fluid density, $\rho$, for $f$ in equation (2.1.2), the conservation of mass equation becomes

$$\int_{MR} \frac{\partial \rho}{\partial t} \, dV + \int_{MR} n_i v_i \rho \, dS = 0, \qquad (2.1.3)$$

where the fluid velocity $v_i$ is substituted for $w_i$ as the material region is assumed to be totally enclosed within the fluid.

Using Gauss' theorem which relates volume integrals of any tensor function $T_{ij}$ to surface integrals (see [11]),

$$\int_V \frac{\partial T_{ij}}{\partial x_i} \, dV = \int_S n_i T_{ij} \, dS, \qquad (2.1.4)$$

the following equation for the principle of mass conservation is obtained:

$$\int_{MR} \left[ \frac{\partial \rho}{\partial t} + \frac{\partial \rho v_i}{\partial x_i} \right] dV = 0. \tag{2.1.5}$$

This is the *integral form* of the equation of mass conservation. The *differential form* for this equation is obtained by noting that, due to the arbitrariness of the material region over which the above integration is performed, the integrand must be equal to zero (see [11]), i.e.,

$$\frac{\partial \rho}{\partial t} + \frac{\partial \rho v_i}{\partial x_i} = 0. \tag{2.1.6}$$

This form for the conservation of mass equation is commonly known as the *continuity equation*. If the *material derivative* is defined as

$$D/Dt = \partial/\partial t + v_i \partial/\partial x_i, \tag{2.1.7}$$

then the continuity equation takes the following form:

$$\frac{D\rho}{Dt} = -\rho \frac{\partial v_i}{\partial x_i}. \tag{2.1.8}$$

In vector notation, the above equation can be written as

$$\frac{D\rho}{Dt} = -\rho \nabla \cdot \mathbf{v}, \tag{2.1.9}$$

where $\mathbf{v}$ is the velocity vector of every fluid element.

Equation (2.1.9) can be physically interpreted, as *the volume of a material element may change only due to changes in its density* (see [11]). Such interpretation is possible since $(\nabla \cdot \mathbf{v})$ represents the rate of expansion of a material region.

The second conservation principle used to describe fluid dynamics is the continuum form of Newton's second law of motion, which is also known as the principle of *conservation of linear momentum*. It states that the time rate of change of the momentum of a material volume must be equal to the sum of the forces acting on the fluid elements in this volume:

$$\frac{d}{dt} \int_{MR} \rho v_i \, dV = \sum_i \mathbf{F_i}. \tag{2.1.10}$$

Normally there are two types of forces that act on bulk matter. The first are the long-range forces, commonly known as the *body forces*, which act equally on all the material within the volume and decrease with increasing distance from the volume. Examples include the gravity force, the centrifugal force, and the electromagnetic force. These forces will be denoted here as $\mathbf{F_B}$.

The second type of force acting on a fluid element is a short-range force originating from the molecular forces, which decrease very rapidly with increasing distance from the boundaries. Such forces are due primarily to the momentum exchange across boundaries such as gas-liquid or liquid-solid. The value of the force acting on a fluid element is proportional to the surface area of the element regardless of the volume of the element. These short-range forces are commonly known as the *surface forces*, designated here as $\mathbf{F}_S$. Substituting the body and surface forces on the right-hand side of equation (2.1.10), Newton's second law of motion can be written as

$$\frac{d}{dt} \int_{MR} \rho v_i \, dV = \int_{MR} \rho F_{Bi} \, dV + \int_{MR} F_{Si} \, dS. \qquad (2.1.11)$$

It can be shown (see, for example, reference [9]) that, for all of the applications discussed in this course, the surface forces $F_{Si}$ can be represented by

$$F_{Si} = \tau_{ij} n_j, \qquad (2.1.12)$$

where $\tau_{ij}$ is the stress tensor representing the *i*th component of the force per unit area exerted over a plane surface element normal to the *j*th direction at a specific position in space and time. $n_j$ is the unit normal of the surface element. Equation (2.1.11) takes the following form, after using Gauss' theorem and the derivative for a material element:

$$\int_{MR} \left[ \frac{\partial \rho v_i}{\partial t} + \frac{\partial \rho v_i v_j}{\partial x_i} \right] dV = \int_{MR} \rho F_{Bi} \, dV + \int_{MR} \tau_{ij} n_j \, dS. \qquad (2.1.13)$$

By using Gauss' theorem again and after dropping the subscript *B* from the body force term, equation (2.1.13) becomes

$$\int_{MR} \left[ \frac{\partial \rho v_i}{\partial t} + \frac{\partial \rho v_i v_j}{\partial x_i} - \rho F_i - \frac{\partial \tau_{ij}}{\partial x_j} \right] = 0. \qquad (2.1.14)$$

Basically, this is the continuum form of *Newton's second law of motion* as it applies to a material fluid region. Due to the arbitrariness of the region it can be shown that the integrand must also be equal to zero, which leads to the differential form for the momentum equation, namely:

$$\frac{\partial \rho v_i}{\partial t} + \frac{\partial \rho v_i v_j}{\partial x_i} - \rho F_i - \frac{\partial \tau_{ij}}{\partial x_j} = 0. \qquad (2.1.15)$$

Equation (2.1.15) can be written in general vector form as

$$\frac{\partial \rho \mathbf{v}}{\partial t} + \nabla{:}\rho \mathbf{v}\mathbf{v} = \rho \mathbf{F} + \nabla{:}\tau. \qquad (2.1.16)$$

The left-hand side of the above equation is the time rate of change of momentum of a fluid element, while the right-hand side represents the sum of all the forces acting on that element. The above equations cannot be integrated unless the necessary constitutive relations for the stress tensor, $\tau_{ij}$, are given in terms of the flow variables, such as the velocity vector in this case.

It can be shown, for a homogeneous and isotropic fluid and due to the symmetry of the stress tensor and under the assumption of a Hookian material, that the stress tensor may take the following form (see [9]):

$$\tau_{ij} = -p\delta_{ij} + \lambda e_{kk}\delta_{ij} + 2\mu e_{ij}, \qquad (2.1.17)$$

where $\mu$ is the viscosity of the fluid, $\lambda$ is the second coefficient of viscosity, $p$ is its pressure, and $e_{ij}$ is the rate of strain tensor defined by

$$e_{ij} = \frac{1}{2}\left(\frac{\partial v_i}{\partial x_j} + \frac{\partial v_j}{\partial x_i}\right). \qquad (2.1.18)$$

Invoking Stokes's assumption, which states that the thermodynamic and mechanical pressures for normal fluids differ by only a small amount, leads to the following relation (see [9]):

$$\lambda = -(2/3)\mu.$$

Substituting these assumptions into the momentum equation gives us

$$\frac{D\rho v_i}{Dt} = \rho F_i - \frac{\partial p}{\partial x_i} + \frac{\partial}{\partial x_j}\left[2\mu\left(e_{ij} - \frac{1}{3}e_{kk}\delta_{ij}\right)\right], \qquad (2.1.19)$$

where $\delta_{ij}$ is the Kronecker delta. This equation is usually known as the *Navier–Stokes equation*.

The third conservation principle used in analyzing fluid dynamics is the *conservation of energy*. This principle states that the rate of change of energy within a material region is equivalent to the rate of energy received by the region due to work done on the region and the flux of heat transferred to the region, i.e.,

$$\frac{d}{dt}\int_{MR}\rho\left(E + \frac{1}{2}v^2\right)dV = -\int_{MR}n_iq_i\,dS + \int_{MR}n_i\tau_{ij}v_j\,dS$$
$$+ \int_{MR}\rho F_iv_i\,dV, \qquad (2.1.20)$$

where $E$ is the internal molecular energy, $v^2 = v_iv_i$, and $q_i$ is the heat flux. Using Gauss' theorem and the expression for the derivative of a material region and the arbitrariness of the region, the following differential equation is obtained for the conservation of energy principle (see [9]):

$$\frac{\partial}{\partial t}\left[\rho\left(E + \frac{1}{2}v^2\right)\right] + \frac{\partial}{\partial x_i}\left[\rho v_i\left(E + \frac{1}{2}v^2\right)\right]$$
$$= -\frac{\partial}{\partial x_i}q_i + \frac{\partial}{\partial x_i}(\tau_{ij}v_j) + \rho F_iv_i. \qquad (2.1.21)$$

It can be shown that the kinetic energy ($\frac{1}{2}\rho v^2$) may be eliminated from the energy equation by using the momentum equations, giving the energy equation in terms of the internal energy only, as

$$\frac{\partial}{\partial t}(\rho E) + \frac{\partial}{\partial x_i}\rho Ev_i = -pe_{kk} + 2\mu\left(e_{ij}e_{ij} - \frac{1}{3}e_{kk}e_{kk}\right) - \frac{\partial}{\partial x_i}q_i. \qquad (2.1.22)$$

In vector notation, the above equation takes the following form:

$$\frac{\partial}{\partial t}(\rho E) + \nabla \cdot (\mathbf{v}\rho E) = -p\nabla \cdot \mathbf{v} + 2\mu\left[\mathbf{e}:\mathbf{e} - \frac{1}{3}(\nabla \cdot \mathbf{v})^2\right] - \nabla \cdot \mathbf{q}. \qquad (2.1.23)$$

Normally, the energy equation can be written in terms of the fluid temperature, instead of the internal energy, by using the various thermodynamic equalities. The heat flux can be defined using Fourier law for heat conduction, i.e.,

$$\mathbf{q} = -k\nabla T$$

where $k$ is the coefficient of thermal conduction of the fluid and $T$ is its temperature. Using the thermodynamic definitions for the internal energy, it can be shown that the energy equation takes the following form:

$$\rho c_v \frac{DT}{Dt} = -T \frac{\partial p}{\partial t}\bigg|_\rho + \Phi + \frac{1}{\rho}\frac{\partial}{\partial x_i}\left(k\frac{\partial T}{\partial x_i}\right) \qquad (2.1.24)$$

where $c_v$ is the specific heat at constant volume and $\Phi$ is the commonly known dissipation function defined by

$$\Phi = 2\mu\left(e_{ij}e_{ij} - \frac{1}{3}e_{kk}e_{kk}\right).$$

## 2.2 INCOMPRESSIBLE FLOW

There is a large area of applications of fluid flows in which the density of the fluid at any specific position does not vary with time. This condition can be interpreted mathematically as

$$\frac{D\rho}{Dt} = 0. \qquad (2.2.1)$$

The equations for fluid motion under such constraint take a special form known as the equations of motion for *incompressible flow*. Specifically, under the incompressibility condition the equation of state will reduce to

$$\frac{D\rho}{Dt} = 0. \qquad (2.2.2)$$

The conservation of mass equation (continuity equation) will reduce to

$$\frac{\partial v_i}{\partial x_i} = 0. \qquad (2.2.3)$$

The momentum equation will reduce to

$$\rho\frac{Dv_i}{Dt} = -\frac{\partial p}{\partial x_i} + \rho F_i + \mu\frac{\partial}{\partial x_j}\frac{\partial v_i}{\partial x_j}. \qquad (2.2.4)$$

Finally, the energy equation will reduce to the following form:

$$\rho c_p \frac{DT}{Dt} = k\frac{\partial^2 T}{\partial x_i \partial x_i} + \Phi + \beta T\frac{Dp}{Dt} \qquad (2.2.5)$$

where $c_p$ is the specific heat at constant pressure and $\beta$ is the coefficient of thermal expansion.

The velocity, pressure, and temperature fields for incompressible flow are governed by the system of equations (2.2.3)–(2.2.5).

## *2.3* INVISCID FLOW

The above equations may be simplified a great deal if the fluid viscosity is set equal to zero everywhere within the fluid. It should be cautioned that such a condition is totally unrealistic physically and will render the mathematical description of the problem ill posed (see [9]). Nevertheless, inviscid flows have been treated extensively in the past and under certain conditions have been found useful, for example in aerodynamics.

When the fluid viscosity is set equal to zero, i.e. $\mu = 0$, then the incompressible flow equations will reduce to:

$$\frac{\partial v_i}{\partial x_i} = 0 \qquad \text{or} \qquad \nabla \cdot \mathbf{v} = 0 \tag{2.3.1}$$

for the conservation of mass (continuity) and

$$\rho \frac{Dv_i}{Dt} = -\frac{\partial p}{\partial x_i} + \rho F_i, \qquad \rho \frac{D\mathbf{v}}{Dt} = -\nabla p + \rho \mathbf{F}, \tag{2.3.2}$$

for the momentum equations.

Equations (2.3.2) are known as *Euler's equations.* If, in addition, the flow is assumed to be irrotational—that is, the vorticity vector, $\boldsymbol{\omega} = \nabla \times \mathbf{v}$, is zero everywhere—then it is possible to define a velocity potential function such that

$$\mathbf{v} = -\nabla \phi. \tag{2.3.3}$$

Substituting this condition into the continuity equation, the velocity potential is satisfied by Laplace's equation:

$$\nabla^2 \phi = 0 \tag{2.3.4}$$

which for rectangular Cartesian $(x, y, z)$ coordinates take the following form:

$$\frac{\partial^2 \phi}{\partial x^2} + \frac{\partial^2 \phi}{\partial y^2} + \frac{\partial^2 \phi}{\partial z^2} = 0. \tag{2.3.5}$$

When the velocity potential is substituted into the momentum equation, equation (2.3.2) will take the following form:

$$-\frac{\partial}{\partial t}(\nabla \phi) + \nabla \mathbf{v} \cdot \mathbf{v} = -\frac{1}{\rho} \nabla p + \rho \mathbf{F}. \tag{2.3.6}$$

If the last term in the above equation, the body force term, is also derivable from a potential function, i.e.,

$$\mathbf{F} = -\nabla \Pi, \tag{2.3.7}$$

then substituting this representation for the body force into the Euler equation and integrating once, results in the following equation:

$$-\frac{\partial \phi}{\partial t} + \frac{p}{\rho} + \frac{1}{2}\mathbf{v} \cdot \mathbf{v} + \Pi = C(t), \tag{2.3.8}$$

where $C(t)$ is the constant of integration and is a function of time only. Equation (2.3.8) is known as the *Bernoulli's equation.* Normally, inviscid and irrotational flows are completely described by first solving Laplace's equation for the velocity potential to determine the velocity field. Subsequently, the pressure is determined by solving the algebraic Bernoulli's equation. Note that the gravity force is included within the potential function $\Pi$, in Bernoulli's equation.

## *2.4* ROTATING FLOWS

The equations of motion above were derived in a stationary, inertial reference frame. There are, however, a great deal of fluid flows that are either rotating or enclosed within a container that is rotating with respect to a laboratory reference frame. Examples of these flows are planetary and stellar flows and flows in rotating turbines and pumps.

In order to accurately describe such flows using the basic conservation laws, it is necessary to derive the fluid dynamic equations in a rotating reference frame. The equations of motion in a moving reference frame are identical in form to the equations in a stationary frame when a fictitious body force is added to the real body and surface forces acting on the fluid element. Such a force takes the form

$$-\mathbf{f_0} - 2\mathbf{\Omega} \times \mathbf{v} - \frac{d\mathbf{\Omega}}{dt} \times \mathbf{r} - \mathbf{\Omega} \times (\mathbf{\Omega} \times \mathbf{r}),$$

where $\mathbf{f_0}$ is the acceleration vector of a fluid particle at a distance $\mathbf{r}$ relative to a stationary reference frame, and $\mathbf{\Omega}$ is the rotation vector of that particle. If the fluid particle is rotating steadily and its reference frame is not accelerating, then the equations of motion in a rotating reference frame will take the following form for an incompressible flow (see [10]):

$$\frac{D\mathbf{v}}{Dt} + 2\mathbf{\Omega} \times \mathbf{v} = -\frac{1}{\rho}\nabla p + \mathbf{F} - \mathbf{\Omega} \times (\mathbf{\Omega} \times \mathbf{r}) + \nu\nabla^2\mathbf{v} \qquad (2.4.1)$$

where $\nu = \mu/\rho$ is the kinematic viscosity.

The centrifugal term, $\mathbf{\Omega}\times(\mathbf{\Omega}\times\mathbf{r})$, may also be written through vector identities as the gradient of a potential function in the form

$$\nabla(\mathbf{\Omega} \times \mathbf{r})^2.$$

Also if the only body force acting on the system is due to gravity, then body force is derivable from a potential, i.e.,

$$\mathbf{F} = \nabla(\mathbf{g} \cdot \mathbf{r}).$$

Under these conditions the equations of motion for an incompressible, rotating fluid may be written as

$$\frac{D\mathbf{v}}{Dt} + 2\mathbf{\Omega} \times \mathbf{v} = -\frac{1}{\rho}\nabla P + \nu\nabla^2\mathbf{v} \qquad (2.4.2)$$

in which $P$ is commonly known as the modified pressure and is defined as

$$P = p + \rho\mathbf{g} \cdot \mathbf{r} + \rho(\mathbf{\Omega} \times \mathbf{r})^2. \qquad (2.4.3)$$

In this form neither the gravity nor the centrifugal forces appear explicitly in the equations of motion, as the effect of these forces is lumped into the modified pressure. In this way it is possible to solve the equations of motion without the complications introduced by the gravity and centrifugal forces. The conservation of mass and internal energy equations retain the same form as for the nonrotating reference frame.

## 2.5 THE HYDROSTATIC BALANCE

If there is no fluid motion, the velocity field will be zero everywhere, i.e., $\mathbf{v} = 0$, and the equations of motion reduce to the following form:

$$\nabla p = -\rho \mathbf{F} + \rho \mathbf{\Omega} \times (\mathbf{\Omega} \times \mathbf{r}). \tag{2.5.1}$$

This equation shows that in the absence of fluid flow, the pressure force is identically balanced by the sum of the gravity force and the centrifugal force. This balance of forces in the absence of fluid motion is commonly known as the *hydrostatic balance*.

## 2.6 THE NONDIMENSIONAL FORM

It is customary in fluid dynamics to nondimensionalize the equations of motion. In fact, in many cases the effects of low gravity on fluid dynamics can be much better appreciated in terms of nondimensional numbers than in dimensional form. In order to proceed with the nondimensionalization process, first appropriate scales must be chosen for all of the variables involved. From [11], the following scales are chosen for time, velocity, length, temperature, and pressure, respectively:

$$L/U, \quad U, \quad L, \quad \mu_0 U/k_0, \quad \rho_0 U^2,$$

where the subscript $_0$ identifies reference values. Using these scales, all the flow variables can be nondimensionalized as:

$$x_i = x_i^*/L, \quad t = t^*/(L/U), \quad v_i = v_i^*/U, \quad \rho = \rho^*/\rho_0,$$

$$p = (p^* - p_0)/(\rho_0 U^2), \quad T = (T^* - T_0)/(\mu_0 U/k_0). \tag{2.6.1}$$

Note that dimensional variables are identified by an asterisk in the above scalings.

Substituting these variables into the governing equations, equations in terms of the nondimensional variables result with coefficients composed of specific groupings for the chosen scales, known as *nondimensional numbers*. The equations for incompressible flow in a nondimensional form can now be written as follows:

$$\frac{D\rho}{Dt} = 0 \tag{2.6.2}$$

for the equation of state,

$$\frac{\partial v_i}{\partial x_i} = 0 \tag{2.6.3}$$

for the conservation of mass (continuity),

$$\frac{Dv_i}{Dt} = -\frac{\partial p}{\partial x_i} + \frac{1}{Fr}F_i + \frac{1}{Re}\left(\frac{\partial}{\partial x_j}\frac{\partial v_i}{\partial x_j}\right) \tag{2.6.4}$$

for the momentum equation, and

$$\frac{DT}{Dt} = \frac{1}{Re}\left[\frac{1}{Pr}\frac{\partial^2 T}{\partial x_i \partial x_i} + e_{ij}e_{ij}\right] + \frac{B}{Pr}\left(\frac{Dp}{Dt}\right) \tag{2.6.5}$$

for the conservation of energy.

The nondimensionalization process resulted in a set of nondimensional numbers appearing as coefficients to some of the terms in the equations. Specifically, these numbers are the *Reynolds number*, the *Prandtl number*, and the *Froude number*, which are defined as:

$$Re = \frac{UL\rho_0}{\mu_0} = \frac{\text{Inertia Force}}{\text{Viscous Force}} = \frac{\rho Dv_i/Dt}{\mu\nabla^2 v_i}, \tag{2.6.6}$$

$$Pr = \frac{\mu_0 c_{p0}}{k_0} = \frac{\text{Viscous Diffusion}}{\text{Thermal Diffusion}}, \tag{2.6.7}$$

$$Fr = \frac{U^2}{g_i L} = \frac{\text{Inertia Force}}{\text{Buoyancy Force}}. \tag{2.6.8}$$

The above nondimensional numbers are extremely important and significant in fluid mechanics. The Reynolds number, $Re$, which expresses the ratio of the convective motion to the diffusion of momentum through viscosity, normally determines whether the flow is laminar or turbulent. Turbulent flows are very complex and difficult to analyze. The Prandtl number, $Pr$, which expresses the ratio of the diffusion of momentum to the diffusion of energy, is a function of the fluid material and can be determined a priori. The Froude number, $Fr$, represents the ratio of the inertia force to the buoyancy force due to the gravitational acceleration which is extremely significant for low gravity flows as will be shown later. Depending on the form of the nondimensionalization procedure and the specific flow conditions, many more nondimensional numbers can be obtained from the equations of motion and energy and will be introduced throughout the text. However, the three nondimensional numbers discussed above are the most frequently encountered.

## 2.7 TURBULENT FLOW

Perhaps the most significant nondimensional number in all of fluid mechanics is the Reynolds number, named after the British scientist Osborne Reynolds (1842–1912). Depending on the value of this number, the flow field for any specific set of conditions may be classified as laminar or turbulent. Laminar flows always occur when the Reynolds number is small (the value of a small Reynolds number is different for different flows). As the Reynolds number is increased for any specific situation, the flow is said to go through a transition process to turbulent flow. The equations enumerated above are sufficient for describing laminar and transitional flows. However, these equations, in the specific form in which they are written, cannot at the present time be used to describe turbulent flows.

Unfortunately, almost all flows encountered in nature or resulting from laboratory experiments are classified as turbulent flows. And although scientists and engineers have dealt with turbulent flows for at least one hundred years, still to date they do not completely understand them. The most salient feature of turbulence is its non-repeatability, or apparent randomness. One of the best ways of dealing with turbulent flows has been through the *Reynolds decomposition process* in which all field variables are divided into an averaged mean value and a random fluctuation; see for instance [12]:

$$g_i(x_i, t) = G_i(x_i) + g_i'(x_i, t), \qquad (2.7.1)$$

where $G_i$ is the time-averaged value of the field variable, defined as

$$G_i(x_i) = \lim_{t \to \infty} \frac{1}{t} \int_{t_0}^{t_0 + t} g_i \, dt, \qquad (2.7.2)$$

and $t$ is a very large time interval.

For incompressible flows, decompostion of the velocity, pressure, and temperature fields and considerable algebra result into the following equations for the conservation of mass, momentum, and energy, respectively, in terms of the time-averaged variables:

$$\frac{\partial}{\partial x_i} U_i = 0, \qquad (2.7.3)$$

$$U_j \frac{\partial}{\partial x_j} U_i = -\frac{1}{\rho} \frac{\partial P}{\partial x_i} + \frac{1}{\rho} \frac{\partial}{\partial x_j} \left[ \mu \frac{\partial}{\partial x_j} U_i - \rho \overline{u_i' u_j'} \right], \qquad (2.7.4)$$

$$U_j \frac{\partial}{\partial x_j} T_i = \frac{\partial}{\partial x_j} \left[ \kappa \frac{\partial}{\partial x_j} T_i - \overline{T' u_j'} \right], \qquad (2.7.5)$$

where $\kappa = \rho c_p / k$ is the thermal diffusivity.

It can be seen that the variables which the equations attempt to solve with this method are the averaged values and not the instantaneous functions. However, the Reynolds stress terms $\overline{u_i' u_j'}$ and the temperature-velocity fluctuation correlation functions $\overline{T' u_j'}$ are additional unknowns brought about by using the decomposition method. The addition of these unknowns to the original variables introduces the *closure problem*, which may be stated as follows: the turbulence equations for the averaged variables, equations (2.7.3)–(2.7.5), always possess more unknowns than equations. In order to bridge this gap *turbulence models* must be defined in which the Reynolds stress terms are written in terms of the averaged field variables. Almost all of turbulence research is devoted to defining such models that are rational and hopefully universal.

*Chapter 3*

# Liquid-Gas Capillary Surfaces

*I*n a low- or zero-gravity environment, the driving forces are the surface, rotational, and other forces that remain the same regardless of the value of gravity. They give rise to a different kind of fluid dynamics. In many terrestrial applications, the force of gravity is by far the largest and, consequently, the smaller effects are neglected. The absence of gravity therefore means that the new fluid dynamics is mainly shaped by the secondary effects. Surface tension, which occurs along the boundaries between two different phases, such as liquid-vapor or liquid-solid, is expected to be a dominant force.

In this chapter we first investigate fluid response to all the above forces, including gravitation, and then drop the gravity-affected terms to find the free-surface response to surface and body forces. The shape of the free surface is altered according to the variation of the forces acting on the fluid.

First we investigate the meaning of surface tension force and its ratio to the force of gravity. The equations governing the shape of static axisymmetric free surfaces are introduced next. Finally, the small amplitude sloshing problem is outlined in detail.

## 3.1 *SURFACE TENSION FORCE*

Surface tension forces result from interfacial molecular activity at the boundary between two media of different materials. The difference in intermolecular forces between the phases in contact gives rise to excess energy at the interface known as the surface energy. In the simplest case, in which liquid is surrounded by its own saturated vapor, the density of the vapor is normally only a fraction of that of the liquid. The molecules inside the liquid bulk are subjected to attractive forces of neighboring molecules while those near the boundary with the vapor experience an unbalanced cohesive force directed away from the interface. This imbalance implies that, in an isothermal process, some work must be done in order to move molecules from the bulk of the liquid to the surface. The increase in the number of molecules in the surface layer causes it to extend, and this occurs by doing some work against the unbalanced forces at the interface. In other words, work in the amount $dW$ is done when the surface area changes by a small amount, $dS$, in the following way:

$$dW = -\sigma dS, \tag{3.1.1}$$

where $\sigma$ is the surface tension and has the dimensions of $J/m^2$, or $N/m$.

Since the surface area in this case changes at a constant temperature, the work done is equal to the change in the *free energy*, or that part of the internal energy of the liquid that is capable of doing mechanical work, namely

$$dW = -d\Phi, \tag{3.1.2}$$

where $\Phi$ is the free energy of the surface. Equating expressions (3.1.1) and (3.1.2) results in a definition for the *coefficient of surface tension* as the free energy of a unit surface area of the interface, i.e.,

$$\sigma = \Phi/S. \tag{3.1.3}$$

It can be seen from the above expression that the surface of the liquid at a vapor-liquid interface has an excess of free energy over the bulk of the liquid.

A state of equilibrium is defined as that state of a system where its energy is at a minimum. For a free surface of liquid this implies a spherical form, and we see now why very small drops and bubbles normally assume this shape. The forces tending to contract the surface area of the liquid are surface tension forces that act in the tangential direction of the surface.

The coefficient of surface tension is a function of the thermodynamic state of the materials. Its magnitude depends on the properties of the different media in direct contact. On the free surface of a liquid, $\sigma$ depends on the temperature and the pressure in both the vapor and the liquid. As these parameters increase, $\sigma$ decreases, becoming equal to zero at the critical temperature, $T_{cr}$. The critical conditions imply that there is no physical distinction between the liquid and the vapor. The dependence of $\sigma$ on $T$ can be accurately described by the following empirical relation:

$$\sigma = C(T - T_{cr}), \tag{3.1.4}$$

where $C$ is an experimentally determined constant that depends on the type of the liquid. Table 3.1 shows some measured values of the surface tension for a number of "liquid-gas" and "liquid-liquid" pairs.

The value of $\sigma$ at a liquid-fluid interface in equilibrium may be significantly affected by the presence of surface active (or adsorbed) materials at the surface of the liquid. For instance, a drop of lubricating oil placed on the free surface of water spreads out into a very thin sheet covering the whole surface. The small amounts of oil and grease and some other contaminating substances that are present in water under normal conditions spread out over any free surface and have a large effect on the surface tension. Normally, the effect of adsorbed contaminant molecules at free water surface is to decrease the surface tension by an amount that increases with the surface concentration of the adsorbed material.

The surface tension force can contribute significantly to the motion of fluids, especially at the interface between liquid bulks and gas bulks. Under terrestrial

**Table 3.1** Surface tension for some "liquid-gas" and "liquid-liquid" pairs.

| Substance | Medium in contact | Temperature (°C) | Surface Tension, $\sigma$ (dyne/cm) |
|---|---|---|---|
| Hydrogen | Vapor | − 252.0 | 2.0 |
| Nitrogen | Vapor | − 195.9 | 8.3 |
| Nitrogen | Vapor | − 183.0 | 6.2 |
| Oxygen | Vapor | − 182.7 | 13.0 |
| Ethyl alcohol | Vapor | 20.0 | 22.0 |
| Mercury | Vapor | 0 | 513.0 |
| Mercury | Vapor | 20.0 | 475.0 |
| Mercury | Chloroform | 20.0 | 357 |
| Water | Olive oil | 20.0 | 20.0 |
| Water | Air | 20.0 | 72.75 |
| Water | Air | 100 | 58.8 |
| Gold | Vapor | 1130 | 1102 |
| Gold | Vapor | 1070 | 612 |
| Silver | Vapor | 1060 | 750 |

conditions, the surface tension force is normally included in the interface conditions and most often is neglected except in a very small number of specific applications, since its value is much smaller than other body forces, including gravity. Such applications include flows in small fluid bulks such as thin liquid films, drops, and bubbles. The body force in these cases is small because the magnitude of the body force is proportional to the small mass of the fluid bulk. In an environment where the force of gravity in the form of body force is negligible, the effects of surface tension forces would appear significant compared to the gravity body force and would have a substantial influence on fluid motion. This is exactly the situation in low-gravity environment, in which the value of the gravity force is small, making the effects of the body force very small. Thus in low-gravity environment the surface tension force usually becomes the more dominant of the two forces.

It is possible to estimate the conditions under which surface forces become dominant over gravity forces by calculating the ratio of the values of the surface forces to body forces. The surface tension force, $F_\sigma$, may be approximated by

$$F_\sigma \approx \sigma L, \tag{3.1.5}$$

where $L$ is a characteristic length scale. On the other hand, the body force due to gravity, $F_g$, is represented by the product of the mass of the system times the acceleration due to gravity, which can be written as

$$F_g = \rho L^3 g. \tag{3.1.6}$$

Upon taking the ratio of these two forces, the following expression results:

$$F_g/F_\sigma = \rho g L^2 / \sigma. \tag{3.1.7}$$

This ratio is commonly known as the *Bond number.*

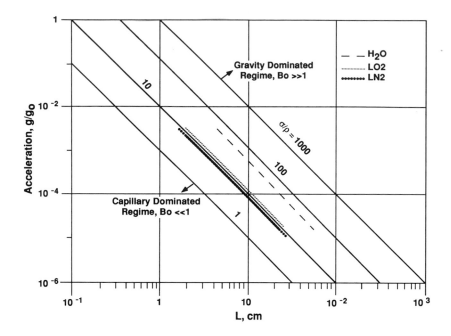

**Figure 3.1**   Regions of applicability for large and small Bond numbers for different fluids.

As an example, for water at 20°C, for which $\rho = 1$ gm/cm$^2$ and $\sigma = 72.75$ dyne/cm, and with an assumed volume of 1 m$^3$ where $L = 100$ cm, the ratio of the two forces is calculated to be $F_g/F_\sigma = 1.35 \times 10^5$. For this system the surface tension and gravity forces become comparable when the value of $g$ is approximately $7 \times 10^{-6}$. Available data show that the drag force experienced by an earth-orbiting vehicle is approximately of that magnitude at an altitude of 240 km above the surface of Earth. At higher altitudes it is thus possible to expect the surface tension force to become dominant. Figure 3.1 shows the two regimes in which either the body force (gravity) or the surface force (surface tension) dominate as a function of the fluid properties, the normalized values of the gravity, $g/g_0$, and typical fluid system length scales, $L$.

When a liquid comes in contact with a solid flat surface, the three phases of solid, liquid, and saturated vapor are in contact around the circumference bounding the liquid, as shown in Figure 3.2. An element of length $dl$ of the circumference is acted upon, in this situation, by the following three forces

$$F_{ls} = \sigma_{ls}\,dl, \qquad F_{vs} = \sigma_{vs}\,dl, \qquad F_{lv} = \sigma_{lv}\,dl \qquad (3.1.8)$$

at the interfaces between the liquid and the solid, the solid and the vapor, and the liquid and the vapor, respectively. If gravity is neglected and the interface is in equilibrium, then these forces must be in a static balance, as shown by the following equation:

$$F_{vs} = F_{lv}\cos\alpha + F_{ls}. \qquad (3.1.9)$$

As shown in Figure 3.2, $\alpha$ is the angle between the tangent to the liquid surface at the point of contact and the plane surface of the solid. This angle is commonly

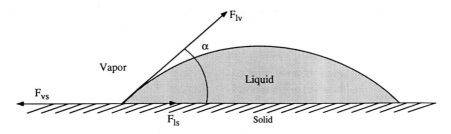

**Figure 3.2**   Representation of the forces acting at a solid-liquid-gas triple point.

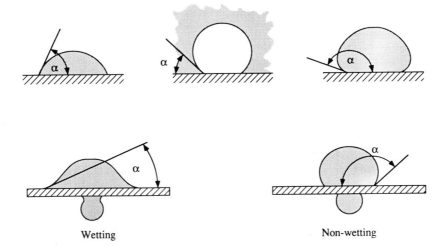

Wetting

Non-wetting

**Figure 3.3**   A sketch of the different wetting conditions.

called the *contact-*, or *wetting-angle.* Substituting the values of these forces into equation (3.1.9), we get

$$\cos \alpha = (\sigma_{vs} - \sigma_{ls})/\sigma_{lv}. \qquad (3.1.10)$$

If, on the other hand, the values of the various surface tension coefficients are such that

$$\sigma_{vs} - \sigma_{ls} \geq \sigma_{lv} \qquad (3.1.11)$$

then the value of $\alpha$ must be 0, in which case the liquid spreads to form a thin film on the solid surface. This condition is called *total wetting*, and it typically occurs for cryogenic liquids such as liquid hydrogen, liquid nitrogen, and liquid oxygen. Otherwise, if

$$\sigma_{vs} + \sigma_{lv} = \sigma_{ls} \qquad (3.1.12)$$

then $\alpha = \pi$ and complete *non-wetting* is observed. As an example, $\alpha$ for water on paraffin ranges from 106° to 109°, and for water on Teflon, $\alpha = 112°$. In the intermediate cases for which $\alpha < \pi/2$, *partial wetting* occurs, which is the more frequently encountered situation, or $\alpha > \pi/2$, and nonwetting takes place. Figure 3.3 shows examples of the different wetting conditions that may arise. A similar situation exists at the interface between two immiscible liquids.

**Table 3.2**   Values for the contact angle between different solid-liquid pairs in air, according to [17].

| Solid | Liquid | Contact angle, deg |
|-------|--------|--------------------|
| Glass | Water | 0 |
| Glass | Mercury | 128–148 |
| Glass | Hydrogen | 0 |
| Glass | Nitrogen | 0 |
| Glass | Oxygen | 0 |
| Steel | Water | 70–90 |
| Steel | Hydrogen | 0 |
| Steel | Nitrogen | 0 |
| Steel | Oxygen | 0 |
| Paraffin | Hydrogen | 106 |
| Aluminum | Nitrogen | 7 |
| Platinum | Oxygen | 1.5 |

In all capillary problems and problems concerning free surfaces, the values of both the contact angle and the surface tension are assumed to be known experimental data. At the present time there is no physical theory that allows us to calculate the values of the surface tension and the contact angle or even to predict the existence of a completely definite contact angle for a given media in contact. The values of surface tension are obtained by empirical methods or determined experimentally together with the contact angle by measuring the height to which liquid rises in a capillary tube or solid fiber. Another method for determining both values is measuring the shapes of drops and bubbles. Table 3.2 gives contact angle values for some liquid-solid pairs in air.

## 3.2 *EQUILIBRIUM SHAPES OF FREE SURFACES*

In this section a theory is developed for evaluating the equilibrium shapes of the surface of a capillary liquid contained in a vessel that is subjected to a specific body force field, $\mathbf{F}_B$. The body force can be in the form of the force of gravity, a centrifugal pseudoforce due to a rotating fluid mass, both, or neither. In this section we assume that the liquid, with density $\rho$, is incompressible and homogeneous, and that it and its vapor, or another gas, completely fill the vessel. The container walls are assumed to be absolutely rigid (i.e., nondeformable surface). The equations governing the equilibrium shape of the free liquid surface are derived when there is no motion in either the liquid or the vapor adjacent to it.

The pressure distribution in the liquid can be determined from the momentum equations of fluid dynamics, which for an incompressible fluid take the following form:

$$\rho \frac{D\mathbf{v}}{Dt} = -\nabla p + \rho \mathbf{F}_B + \mu \nabla^2 \mathbf{v}, \tag{3.2.1}$$

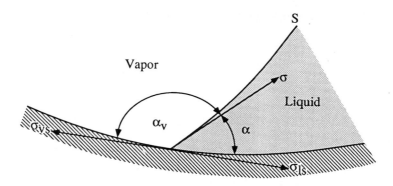

**Figure 3.4**

where $\mu$ is the liquid viscosity and $p$ its pressure. Since the liquid and the vapor are assumed to be stationary, the velocity field **v** in the momentum equations must be set equal to zero everywhere. The resulting equations of motion will reduce to the hydrostatic balance:

$$\nabla p = \rho \mathbf{F}_B. \qquad (3.2.2)$$

Equation (3.2.2) is commonly known as *Euler's condition* for free-surface analysis. The pressure can be determined by integrating equation (3.2.2), if the body force is known and integrable. If it is further assumed that the body force is derivable from a potential function $\Pi$, i.e.,

$$\mathbf{F}_B = -\nabla\Pi, \qquad (3.2.3)$$

then equation (3.2.2) can be readily integrated. Not all body forces can be written in terms of the gradient of a potential function. Electromagnetic forces, for example, cannot. However, both of the forces treated here, namely the gravity force and the centrifugal force, are derivable from a potential function.

Laplace has shown that the curvature of a capillary surface is determined by the pressure difference across that surface and the value of the surface tension in the following manner:

$$p_v - p = \sigma \left( \frac{1}{R_1} + \frac{1}{R_2} \right), \qquad (3.2.4)$$

where $p_v$ is the pressure in the gas, and $R_1$ and $R_2$ are the two principal radii of curvature for the capillary surface separating the liquid and the gas. Equation (3.2.4) is commonly known as *Laplace's condition.*

In addition to equations (3.2.3) and (3.2.4), equation (3.1.10) for the *Dupre-Young* condition on the solid, liquid, and gas contact line must hold, namely

$$\sigma \cos \alpha = \sigma_{vs} - \sigma_{ls} \qquad (3.2.5)$$

where $\sigma$ is the surface tension between the liquid and vapor, and $\alpha$ is the contact angle as shown in Figure 3.4. An alternative form of the Dupre-Young condition in terms of the contact angle between the vapor and the solid, $\alpha_v$, as shown in Figure 3.4, is

$$\sigma \cos \alpha_v = \sigma_{ls} - \sigma_{vs}. \tag{3.2.6}$$

Equation (3.2.6) specifies that the contact line of a liquid and a gas can be stable only if

$$|\sigma_{vs} - \sigma_{ls}| \leq \sigma. \tag{3.2.7}$$

For all of the cases discussed in this chapter it will be assumed that this inequality is always satisfied.

Since all the media in contact—namely, the solid, the liquid, and the gas—are assumed to be homogeneous and at adiabatic conditions, then the values of all the various $\sigma$'s are also assumed to be known constants. The values for both $\alpha$'s also must be constants for the same reason since they are determined in terms of the $\sigma$'s.

For the case in which the body force is given in terms of a gradient of a potential, Euler's equation becomes

$$\nabla p = -\rho \nabla \Pi, \tag{3.2.8}$$

which can be integrated to yield

$$p = -\rho \Pi + c, \tag{3.2.9}$$

where $c$ is an arbitrary constant of integration.

Substituting the value of $p$ into Laplace's equation for the free-surface curvature, the following equation for the hydrostatic equilibrium condition of a free-surface results:

$$\sigma \left( \frac{1}{R_1} + \frac{1}{R_2} \right) = \rho \Pi + c. \tag{3.2.10}$$

In equation (3.2.10) the pressure in the vapor phase $p_v$ has been absorbed into the constant of integration $c$ without loss of generality. Equation (3.2.10) constitutes a *necessary and sufficient condition for the existence of a free liquid surface in an equilibrium state.* In addition, the Dupre-Young condition must be satisfied on the equilibrium contact line.

Physically, equation (3.2.10) states that the equilibrium shape of a free surface of a liquid is primarily determined by the type of body and surface forces acting on the liquid. This is the equation that will be used to determine the shapes of equilibrium free capillary surfaces. In the analysis of this chapter, equation (3.2.10) will be the governing differential equation for the free-surface shape while the Dupre-Young condition, given by equation (3.2.5), will be used as the boundary condition for determining the solution to the free-surface equation. In addition to these equations, other constraints can be used; for example, a condition will be imposed requiring that the total volume be given for both the liquid and the vapor.

The shape of any surface can be described by the spatial distribution of the normal and tangential vectors to that surface. For a Cartesian coordinate system, reference [17] shows that the sum of the radii of curvature for any surface shape function, $S = f(x, y)$, is given by

$$\frac{1}{R_1} + \frac{1}{R_2} = \pm \nabla \cdot \frac{\nabla f}{\sqrt{1 + (\nabla f)^2}} \tag{3.2.11}$$

where $\nabla f$ is the gradient of the function $f$ and $(\nabla \cdot)$ is the divergence of a vector function, both defined in this case by

$$\nabla f = (\partial f / \partial x, \partial f / \partial y), \qquad \nabla \cdot = (\partial / \partial x + \partial / \partial y).$$

The respective sign in equation (3.2.11) is chosen in such a way that the unit normal vector to the free surface is directed away from the liquid and toward the vapor. Upon substituting the definition for the curvature function into the free-surface equation, we obtain:

$$\pm \nabla \cdot \frac{\nabla f}{\sqrt{1 + (\nabla f)^2}} = \frac{\rho}{\sigma} \Pi + c / \sigma. \qquad (3.2.12)$$

An essential boundary condition, necessary for determining a solution for the equilibrium free surface of the liquid, is the shape of the vessel wall, $\Psi$, containing both the liquid and the vapor. $\Psi$ is defined as the vessel surface shape and $l$ is the *contact line* of the liquid free surface with the vessel wall. Let us assume that the equation for the vessel surface $\Psi$ is given in the form

$$\Psi(x, y, z) = 0, \qquad (3.2.13)$$

where $\Psi$ is a scalar function such that $\Psi > 0$ inside the vessel, and $|\nabla \Psi| \neq 0$ on the vessel wall. Since the required equilibrium surface of the liquid must lie inside the vessel, then we must have

$$\Psi(x, y, f(x, y)) > 0 \text{ on the free surface } S, \qquad (3.2.14)$$

with the requirement that

$$\Psi(x, y, f(x, y)) = 0 \text{ on the contact line } l. \qquad (3.2.15)$$

The boundary condition (3.2.15) is derived using the fact that the cosine of the contact angle, $\alpha$, is given by the inner product of the normal to the free surface, $\mathbf{n}_s$, and the normal to the vessel wall, $\mathbf{n}_w$:

$$\mathbf{n}_s \cdot \mathbf{n}_w = \cos \alpha. \qquad (3.2.16)$$

In a Cartesian coordinate system, condition (3.2.16) may be expressed in the following form:

$$\cos \alpha = \pm \frac{-(\partial f / \partial x)(\partial \Psi / \partial x) - (\partial f / \partial y)(\partial \Psi / \partial y) + \partial \Psi / \partial f}{\sqrt{1 + (\nabla f)^2} |\nabla \Psi|}, \qquad (3.2.17)$$

on the contact line $l$. The upper and lower signs are chosen depending on whether the liquid lies below or above the free surface $S$ with respect to the $z$-axis.

The above equations and boundary conditions are very general; nevertheless, the complexity of the free-surface problem should be obvious. The governing equation is nonlinear in the free-surface function $f(x, y)$, with at least one boundary condition also being nonlinear. This difficulty will become clearer with the specific solutions derived below.

The potential function for the body force $\Pi$, in the equations describing the shape of the equilibrium free surface, is assumed to be a known function of space. In most applications relevant to low-gravity fluid dynamics, the force of gravity and the centrifugal force on a fluid particle in a rotating reference frame are the two most relevant body forces that can be defined by the gradient of a potential function. Thus in all of the analysis performed in this section we will consider the case in which the liquid is acted upon by a homogeneous gravitational field of intensity $g$ directed along the $z$-axis, and also by a centrifugal force field due to the homogeneous rotation of a mechanical system about the $z$-axis with angular velocity, $\Omega$. In this case the mass density of the potential function takes the following form:

$$\Pi = gz - \frac{1}{2}\Omega^2 r^2, \tag{3.2.18}$$

where $r = \sqrt{x^2 + y^2}$ is the radial distance from the axis of rotation. The sign convention adopted here is the same as in [17], where positive loading means $g > 0$, and negative loading means $g < 0$. The case $g = 0$ corresponds to zero-gravity conditions.

Upon substituting the value for the potential function, $\Pi$, given by equation (3.2.18), into the free-surface equation (3.2.12), the following equation results:

$$\pm\nabla\cdot\frac{\nabla f}{\sqrt{1 + (\nabla f)^2}} = \left(\frac{\rho g}{\sigma}\right)z - \left(\frac{\rho\Omega^2}{2\sigma}\right)r^2 + \frac{c}{\sigma}. \tag{3.2.19}$$

Note that the coefficient of the first term on the right-hand side of equation (3.2.19) has the dimension (length)$^{-2}$, while the second term has the dimension (length)$^{-3}$.

Equation (3.2.19) can be nondimensionalized by choosing a suitable length scale, $L$, to yield a nondimensional equation for the free-surface shape in the following form:

$$\pm\nabla^*\cdot\frac{\nabla^* f^*}{\sqrt{1 + (\nabla^* f^*)^2}} = Bo z^* - Ro r^{*2} + c^*. \tag{3.2.20}$$

The quantities with an asterisk indicate nondimensional variables, $Bo$ is the *Bond number* defined by

$$Bo = \rho g L^2/\sigma,$$

and $Ro$ is the *rotation number* defined by

$$Ro = \rho\Omega^2 L^3/(2\sigma).$$

The Bond number may be interpreted as the ratio of the gravitation force to the capillary force, while the rotation number is the ratio of the centrifugal force to the capillary force. The existence of these two nondimensional numbers in the free-surface equation leads to the similarity law articulated in [16]: *Whenever a*

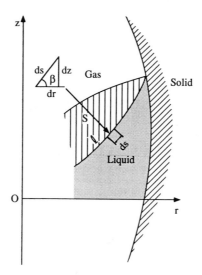

**Figure 3.5**

*unique solution exists for any two capillary surfaces possessing the same values for the Bond number, the rotation number, and the contact angle, over geometrically similar domains, the capillary surfaces themselves will be geometrically similar.* Thus the nondimensional equation groups the solutions to the free-surface problem into equivalence classes of geometrically similar surfaces that cannot be distinguished from each other.

## *3.3* AXISYMMETRIC FREE SURFACES

One of the simplest geometries for which the above formulation can be applied is that which possesses symmetry about one of the coordinate axes. There are basically two simple, commonly used geometries for which such symmetry exists; these are the axisymmetric and the planar two-dimensional geometries. The axisymmetric geometry is important in many practical situations, especially in low-gravity applications, since containers like cylinders or spheres are commonly used in experimental investigations and in technology applications. Also, symmetric problems usually exist whenever the container shape and the external body force field have a common axis of symmetry. Such configurations lead to equilibrium free surfaces, which turn out to be symmetric with respect to that axis.

It is convenient when dealing with axially symmetric problems to work with a cylindrical coordinate system represented by $(r, \theta, z)$, in which the $z$-axis is the axis of symmetry. The axisymmetric equilibrium free surface may be defined by the curve $l$ formed by the intersection of the free surface $S$ with the plane $\theta = const$, as shown in Figure 3.5. The curve $l$ will be called the *equilibrium curve*. Let $s$ be any parameter measured along $l$; one such parameter is the arc length of $l$ measured from a specific point. In axisymmetric geometry, the sum of the principal radii of curvature $(1/R_1 + 1/R_2)$ can be represented in the following form (see [17]):

$$\pm\frac{1}{r\,dr/ds}\frac{d}{ds}\left(\frac{r\,dz/ds}{\sqrt{(dr/ds)^2 + (dz/ds)^2}}\right). \qquad (3.3.1)$$

In equation (3.3.1), the upper (lower) sign should be used in the case for which the liquid remains to the right (left) of an observer moving along $l$ in the direction of increasing $s$.

The coordinates defining the curve $l$ for the axisymmetric case can be written in terms of the parameter $s$ as $r = r(s)$ and $z = z(s)$. If $\beta = \beta(s)$ is the angle between the $r$-axis and an element of the arc length $ds$ as $s$ increases, then according to Figure 3.5,

$$\sin\beta = dz/ds, \qquad \cos\beta = dr/ds, \qquad (3.3.2)$$

which upon differentiation yields the following relationship:

$$\frac{d^2r}{ds^2} = -\frac{d\beta}{ds}\frac{dz}{ds}, \qquad \frac{d^2z}{ds^2} = \frac{d\beta}{ds}\frac{dr}{ds}. \qquad (3.3.3)$$

The sum of the principal radii of curvature can be expressed in the following form, when the definitions of (3.3.3) are used:

$$\frac{1}{R_1} + \frac{1}{R_2} = \pm\left(\frac{d\beta}{ds} + \frac{\sin\beta}{r}\right) = \pm\left(\frac{d\beta}{ds} + \frac{1}{r}\frac{dz}{dr}\right). \qquad (3.3.4)$$

The free-surface equation for the axisymmetric geometry takes the following form after substituting the values for the curvature of the surface into equation (3.2.10):

$$\pm\left(\frac{d\beta}{ds} + \frac{1}{r}\frac{dz}{dr}\right) = \frac{\rho}{\sigma}\Pi(r,z) + c/\sigma. \qquad (3.3.5)$$

Equation (3.3.5) can be further split into two equations—one for $r$ and another for $z$—in terms of the arc length $s$ in the following manner: Upon substituting the first part of (3.3.3) into equation (3.3.5), an equation for $r$ in terms of $s$ results in

$$\frac{d^2r}{ds^2} = -\frac{dz}{ds}\left\{\pm\left[\frac{\rho}{\sigma}\Pi(r,z) + c\right] - \frac{1}{r}\frac{dz}{ds}\right\}. \qquad (3.3.6)$$

On the other hand, when the second part of (3.3.3) is substituted into equation (3.3.5), an equation for $z$ in terms of $s$ results in

$$\frac{d^2z}{ds^2} = \frac{dr}{ds}\left\{\pm\left[\frac{\rho}{\sigma}\Pi(r,z) + c\right] - \frac{1}{r}\frac{dz}{ds}\right\}. \qquad (3.3.7)$$

Normally, the two equations (3.3.6) and (3.3.7) must be solved simultaneously to determine the coordinates of the curve $l$, which represents the equilibrium surface as a function of both $r$ and $z$. In addition, Figure 3.5 shows that the following geometric constraint must hold:

$$\left(\frac{dr}{ds}\right)^2 + \left(\frac{dz}{ds}\right)^2 = 1. \qquad (3.3.8)$$

The equations for the coordinates of the equilibrium free surface in axisymmetric geometry is obtained by substituting the definition for the body force potential, $\Pi$, given by (3.2.18), into equation (3.3.5) to yield

$$\frac{1}{r}\frac{d}{dr}\left(\frac{r\,dz/dr}{\sqrt{1+(dz/dr)^2}}\right) = bz - wr^2 + k, \qquad (3.3.9)$$

where

$$b = \pm\frac{\rho g}{\sigma}, \qquad w = \pm\frac{\rho\Omega^2}{2\sigma}, \qquad k = \pm\frac{c}{\sigma}.$$

Equation (3.3.9) can in turn be split into two equations, one for $r$ and another for $z$, in terms of the arc length parameter, $s$, in the following manner:

$$\frac{d^2r}{ds^2} = -\frac{dz}{ds}\left(bz - wr^2 + k - \frac{1}{r}\frac{dz}{ds}\right), \qquad (3.3.10)$$

$$\frac{d^2z}{ds^2} = \frac{dr}{ds}\left(bz - wr^2 + k - \frac{1}{r}\frac{dz}{ds}\right). \qquad (3.3.11)$$

The signs of the parameters $b$ and $w$ in equations (3.3.10) and (3.3.11), for a given value of $g$, depend on whether the liquid is above or below the free surface in the vicinity of the point $s = 0$. For the solutions that follow, it will always be assumed that the liquid is below the equilibrium surface $S$ with respect to the direction of the $z$-axis. This is the same convention adopted in [17]. In this case the plus sign must be used in the definition for both $b$ and $w$, and thus the signs of $b$ and $g$ will be the same.

Equations (3.3.10) and (3.3.11) are sufficient to describe the equilibrium free surface when solved simultaneously, subject to the imposed constraints. These equations comprise an initial value problem requiring four initial conditions for their solution. In the case of a simply connected free surface, the arc length $s$ along the surface curve $l$ is usually measured from a specific point on the $z$-axis. One such point, for example, is $r = 0$.

The equations for the equilibrium line will have a unique solution (see [17]) whenever the following initial conditions are imposed:

$$r(0) = \frac{dz}{ds}(0) = 0, \qquad z(0) = z_0, \qquad \frac{dr}{ds}(0) = 1. \qquad (3.3.12)$$

Normally, the height of the free surface at the axis of symmetry, $z_0$, is either arbitrary or defined by other means, e.g., the vessel fill level. There is no loss of generality in translating the origin of the coordinate system by an amount of $z_0$ along the $z$-axis. Such a translation will modify the initial conditions in (3.3.12) to the following form:

$$r(0) = \frac{dz}{ds}(0) = 0, \qquad z(0) = 0, \qquad \frac{dr}{ds}(0) = 1. \qquad (3.3.13)$$

Notice, that since

$$\frac{1}{R_1} + \frac{1}{R_2} = bz - wr^2 + k$$

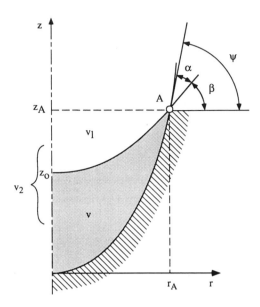

**Figure 3.6**

for the axisymmetric problem, then at the point $r = 0$ the surface *mean curvature* $k$ is given by

$$k = \frac{1}{R_1} + \frac{1}{R_2} = \frac{2}{R(0)},$$ (3.3.14)

$$R(0) = 2/k,$$

where $R(0)$ is the radius of curvature of the free surface at the axis of symmetry $r = 0$.

The complete derivation of the free surface problem requires a number of physical constraints in addition to equations (3.3.10), (3.3.11), and (3.3.13). These are the shape of the vessel and the liquid fill level. For the axisymmetric geometry under consideration here, the volume generated by rotating any curve in the $(r, z)$ plane about the axis of symmetry $z$ is given by

$$\pi \int r^2 \, dz$$

times a constant. The volume enclosed by rotating the curve describing the equilibrium free surface about the $z$-axis and extending from $z = 0$ up to point $A$ on the vessel wall, as shown in Figure 3.6, is given by:

$$v_1 = \pi \int_O^A r^2 \, dz$$ (3.3.15)

while the total volume enclosed by rotating the meridian of the vessel surface wall about the axis of symmetry is given by

$$v_2 = \pi \int_O^A R^2 \, dz,$$ (3.3.16)

where the vessel wall meridian coordinates are given by $(R, z)$ as shown in Figure 3.6.

If $\psi$ is the angle between the $r$-axis and the tangent to the vessel wall at the point of contact $A$, then the values of $v_1$, $v_2$, and $\psi$ are all functions of the coordinates of the contact point $A$. If the contact angle $\alpha$ is known as well as the volume of the liquid, then the problem of finding the free-surface equilibrium curve $l$ must be solved subject to the following geometric constraints:

$$\beta = \psi - \alpha, \qquad v_1 = v_2 - v. \qquad (3.3.17a, b)$$

These conditions are illustrated in Figure 3.6.

When equation (3.3.11) is multiplied by $r$ and integrated between any two points, $s_1$ and $s_2$, on the equilibrium free-surface curve, the following formula results:

$$\left( 2r\frac{dz}{ds} - kr^2 - br^2z + \frac{\omega}{2}r^4 \right) \bigg|_{s=s_1}^{s=s_2} = -b \int_{s_1}^{s_2} r^2 \frac{dz}{ds}\, ds. \qquad (3.3.18)$$

This equation allows for the use of the vessel shape constraints given in equations (3.3.17) in determining the solution for the free-surface equation.

The shape of the equilibrium free surface can be determined by integrating the governing equations (3.3.10) and (3.3.11) simultaneously, subject to the initial conditions and the boundary constraints given in (3.3.13), (3.3.17), and (3.3.18). The resulting differential equations are nonlinear, for which constructing an analytical solution is impossible except in a few specific situations. One such case involves a nonrotating vessel in a zero-gravity environment. Otherwise, the solution for the free-surface problem for any given rotation rate and gravity level can only be determined numerically. The specific integration methods used for arriving at a solution to this problem will be outlined in subsequent sections. Note that the governing equations contain the constant $k$, which must be determined as part of the solution by using the vessel constraints given in conditions (3.3.17).

## 3.4 AXISYMMETRIC FREE SURFACE IN ZERO GRAVITY

The simplest solution that can be constructed for the free-surface equations and conditions is for the case of *zero gravity*, which results from setting $b = 0$ in the equations for the equilibrium curve. If in addition to zero gravity it is assumed that the system is not rotating, by setting $\omega = 0$, there exists a simple analytical solution to the equilibrium curve problem. The solution for this case is given in [17] and is outlined here.

For $b = 0 = \omega$, equations (3.3.10) and (3.3.11) can be simplified to the following form:

$$r\frac{dr}{ds} = -kr\frac{dz}{ds} + \left(\frac{dz}{ds}\right)^2 \qquad (3.4.1)$$

$$\frac{d}{ds}\left( r\frac{dz}{ds} - \frac{1}{2}kr^2 \right) = 0. \qquad (3.4.2)$$

Integrating equation (3.4.2) once and applying the initial conditions (3.13) results in

$$\frac{dz}{ds} = \frac{1}{2}kr. \tag{3.4.3}$$

When the value for $dz/ds$ is substituted into equation (3.4.1), the following equation results for the coordinate $r(s)$ in terms of $s$:

$$\frac{d^2r}{ds^2} + \frac{1}{4}k^2r = 0. \tag{3.4.4}$$

Equation (3.4.4) may easily be integrated to yield the following solution for $r(s)$ after applying the initial conditions given in (3.13):

$$r = \frac{2}{k}\sin\left(\frac{ks}{2}\right). \tag{3.4.5}$$

Similarly, the solution for $z(s)$ is obtained by integrating equation (3.4.3) after substituting the value for $r$ from (3.4.5) and applying the initial conditions. These operations give the following solution:

$$z = \frac{2}{k}\left(1 - \cos\frac{ks}{2}\right). \tag{3.4.6}$$

For $r > 0$, the solution derived in (3.4.5) and (3.4.6) for the free surface represents a semicircle of radius $2/|k|$ whose center is located at the point $(0, 2/k)$. In addition, reference [17] shows that the equilibrium free-surface curve represented by this solution is stable, indicating that such a solution is physically meaningful. Thus for the zero-gravity case all nonrotating free surfaces in axisymmetric geometry take circular shapes in the neighborhood of the axis of symmetry.

Using these solutions, the intersection point of the free surface with the vessel wall can also be determined from analytical expressions for the functions $\beta(k, s)$ and $v_1(k, s)$, which take the following form for the zero-gravity case:

$$\sin\beta = dz/ds = \frac{1}{2}kr \text{ and} \tag{3.4.7}$$

$$v_1 = \frac{8\pi}{3k^3}\left[1 - \cos\left(\frac{ks}{2}\right)\right]^2\left[2 + \cos\left(\frac{ks}{2}\right)\right]. \tag{3.4.8}$$

The equilibrium shapes for specific values of the contact angle $\alpha$ and liquid volume $v$ can finally be determined through constraints (3.13.7a) and (3.13.7b) using the above expression for $\beta$ and $v_1$.

These solutions for the zero-gravity case can be illustrated by using a particular shape for the vessel wall in the form of $R = R(z)$ which is being wetted by a liquid. Taking the origin at the bottom of the vessel in this case, the values for $\beta$ and $v_1$ then can be evaluated by the following expressions:

$$\left.\frac{dR}{dz}\right|_{z=z_A} = \cot(\beta_A + \alpha), \tag{3.4.9}$$

$$\frac{\pi}{3}\frac{R^3(z_A)}{\sin^3\beta_A}(1 - \cos\beta_A)^2(2 + \cos\beta_A) = \pi\int_0^{z_A} R^2(z)\,dz - v, \tag{3.4.10}$$

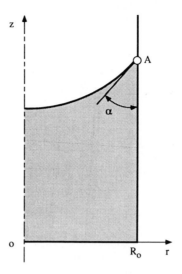

**Figure 3.7**

where $A$ is the point of contact of the liquid surface with the vessel wall. These two equations may be solved simultaneously to determine the values of $\beta_A$ and $z_A$ once the vessel shape $R(z)$, the contact angle $\alpha$, and the volume of the liquid $v$ are all known. The values of $r_A = R(z_A)$ and $k_l = (2 \sin \beta_A)/r_A$ also can be determined from these equations. The following specific examples, from [17], illustrate the utility of the solutions derived above:

1. For a cylindrical vessel with circular cross section of radius $R_0$, $R$ in this case is given by $R = const. = R_0$, as shown in Figure 3.7:

$$\beta_A = \frac{\pi}{2} - \alpha, \tag{3.4.11}$$

$$z_A = \frac{v}{\pi R_0^3} + \frac{R_0}{3 \cos^3 \alpha}(1 - \sin \alpha)^2(2 + \sin \alpha). \tag{3.4.12}$$

2. For a conical shape vessel for which $R = Cz$ (see Figure 3.8):

$$\beta_A = \psi - \alpha, \qquad \cot \psi = C, \tag{3.4.13}$$

$$z_A = \left( \frac{3v}{\pi \cot^2 \psi} \right)^{1/3}$$

$$\times \left\{ 1 - \frac{\cot \psi}{\sin^3 (\psi - \alpha)}[1 - \cos(\psi - \alpha)]^2[2 + \cos(\psi - \alpha)] \right\}^{1/3}. \tag{3.4.14}$$

3. For a liquid drop on a plane (see Figure 3.9):

$$\beta_A = -\alpha, \tag{3.4.15}$$

$$k = -2 \left[ \frac{\pi(1 - \cos \alpha)^2(2 + \cos \alpha)}{3v} \right]^{1/3}. \tag{3.4.16}$$

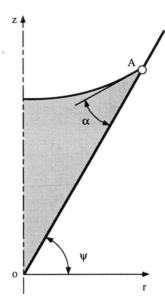

**Figure 3.8**

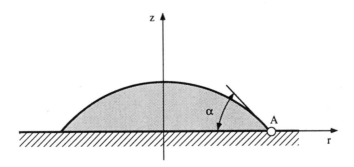

**Figure 3.9**

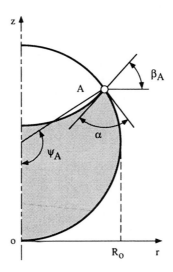

**Figure 3.10**

4. For a sphere of radius $R_0$, the following equations must be solved to determine the values for $\beta_A$ and $\psi_A$ (see Figure 3.10):

$$\beta = \psi - \alpha, \tag{3.4.17}$$

$$\frac{\sin^3 \psi}{\sin^3 \beta}(1 - \cos \beta)^2(2 + \cos \beta) = (1 - \cos \psi)^2(2 + \cos \psi) - \frac{3v}{\pi R_0^3}. \tag{3.4.18}$$

## 3.5 STATIONARY AXISYMMETRIC FREE SURFACES IN A GRAVITY FIELD

A problem that is relevant for low-gravity science concerns defining the free-surface equilibrium curve for a nonrotating vessel that is in a gravity field. This case is characterized by setting $\omega = 0$ and $b \neq 0$ in equations (3.3.10) and (3.3.11). Under these conditions the governing equations for the free-surface problem, equations (3.3.10) and (3.3.11), reduce to the following:

$$\frac{d^2r}{ds^2} = -\frac{dz}{ds}\left(bz + k - \frac{1}{r}\frac{dz}{ds}\right), \tag{3.5.1}$$

$$\frac{d^2z}{ds^2} = \frac{dr}{ds}\left(bz + k - \frac{1}{r}\frac{dz}{ds}\right). \tag{3.5.2}$$

Equations (3.5.1) and (3.5.2) contain a single parameter, $b$, which may be used for nondimensionalizing the variables. Since $b$ has the dimension $L^{-2}$, then $1/\sqrt{|b|}$

can be used for a length scale. With the aid of $b$, a new set of nondimensional variables can be defined in the following manner:

$$x = r\sqrt{|b|}, \quad y = z\sqrt{|b|}, \quad t = s\sqrt{|b|}, \quad q = k/\sqrt{|b|}.$$

With these dimensionless variables, the equations for the equilibrium free surface take the following form:

$$\frac{d^2x}{dt^2} = -\frac{dy}{dt}\left(\epsilon y + q - \frac{1}{x}\frac{dy}{dt}\right), \tag{3.5.3}$$

$$\frac{d^2y}{dt^2} = \frac{dx}{dt}\left(\epsilon y + q - \frac{1}{x}\frac{dy}{dt}\right). \tag{3.5.4}$$

Note that since $b = \rho g/\sigma$, it can take either positive or negative values depending on the direction of the gravity vector. Thus in the nondimensional equations, $\epsilon = 1$ is used when $b > 0$ and $\epsilon = -1$ is used when $b < 0$. An appropriate set of nondimensional initial conditions can also be defined using the same length scale; these are

$$x(0) = y(0) = \frac{dy}{dt}(0) = 0, \qquad \frac{dx}{dt}(0) = 1. \tag{3.5.5}$$

When equation (3.5.4) is multiplied by $x$ and integrated from the axis of symmetry to the vessel wall, the following relation can be derived, after some algebra:

$$2x \sin \beta = \epsilon x^2 y + q x^2 - (\epsilon/\pi)V_1, \tag{3.5.6}$$

where $V_1 = v_1/|b|$ is the nondimensional volume. Note that the above constraint includes the liquid volume in $V_1$.

The equations for the equilibrium curve, (3.5.3) and (3.5.4), are coupled nonlinear ordinary differential equations that must be solved numerically. A general closed form solution for these equations cannot easily be found. Numerical integration methods are the only alternative for solving this problem for arbitrary values for the physical parameters. Equations (3.5.3) and (3.5.4) together with the initial conditions (3.5.5) form an initial value problem. Solutions for such a system can be determined by using initial value integrators such as the *predictor-corrector method* or the *Runge-Kutta method*.

Initial value integrators produce solutions by starting the integration process with known values for the variables at a specific initial point in $t$, and by proceeding along increasing or decreasing values of $t$. A convenient starting point for the integration of the present problem is at $t = 0$, in which case the initial conditions on all the variables are given by (3.5.5). However, equations (3.5.3) and (3.5.4) are singular at the point $t = 0$ for the initial conditions given in (3.5.5). This singularity does not mean that the solution is singular at $t = 0$, only that the equations are singular at that point. Thus any initial value integrator will fail at the initial point $t = 0$. Reference [17] suggests using the following asymptotic solution to equations (3.5.3) and (3.5.4), which is valid in a small neighborhood of the point $t = 0$:

$$x = t - \frac{1}{24}q^2 t^3 - \frac{1}{160}\left(\epsilon q^2 - \frac{1}{12}q^4\right)t^5 + \cdots,$$

$$y = \frac{1}{4}qt^2 + \frac{1}{64}\left(\epsilon q - \frac{1}{3}q^3\right)t^4 + \cdots. \tag{3.5.7}$$

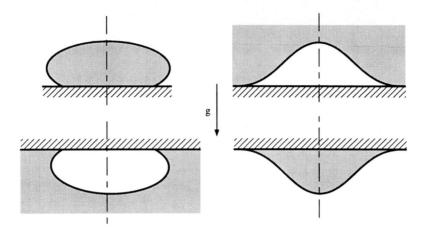

**Figure 3.11**

Thus, whenever numerical means are used to integrate equations (3.5.3) and (3.5.4) from the point $t = 0$, the asymptotic solution given in (3.5.7) should be used to specify the initial conditions at a point very close to this point. The integration process can then proceed from that point forward with the initial value integrator using equations (3.5.3) and (3.5.4).

Equations (3.5.3) and (3.5.4) possess some unique mathematical properties that are quite useful. The system of equations and initial conditions are invariant under the transformations:

$$y \rightarrow -y \quad \text{and} \quad q \rightarrow -q. \tag{3.5.8}$$

From this property it is possible to conclude that as the constant $q$ changes sign, the equilibrium curve for the free surface undergoes a mirror reflection with respect to the $x$-axis. This property allows determination of solutions for $q > 0$ only, since solutions for $q < 0$ are determined from the inverse to the first solution. Reference [17] makes the following general and useful remark concerning this symmetry property: *For every equilibrium state of a liquid in a nonrotating vessel, there exists another corresponding state obtained by a mirror reflection of the entire system with respect to the equipotential plane, if the volume occupied by the liquid is replaced by the gas and vice versa, and the contact angle is replaced by a supplementary angle.* This duality property is illustrated in Figure 3.11.

The solution to equations (3.5.3) and (3.5.4) for the equilibrium curve is presented graphically in Figure 3.12 for two values of the constant $q$. The solution shown is for the positive loading case, i.e., $\epsilon = 1$. We see from Figure 3.12 that the form of the solution curve is strongly dependent on the specific value chosen for $q$. For each value of $q$ the free-surface solution loops around itself several times with increasing $t$. Reference [17] proves that the free-surface solution illustrated in Figure 3.12 is stable only in the continuous section of the curve and unstable in the broken section. This stability criteria means that only the continuous part of the curve represents a physically meaningful solution for the free-surface problem. This is the only portion of the solution that needs to be evaluated in each case. Notice that the stable solution represented by the continuous curve

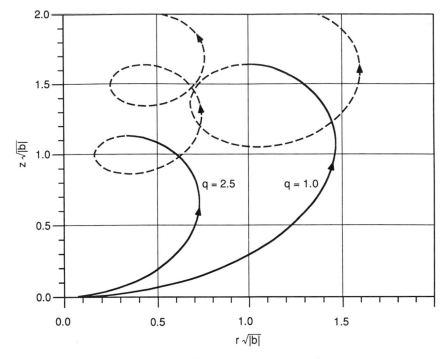

**Figure 3.12**

terminates at the point $(x, y)$ at which $dx/dt = -1$ and $dy/dt = 0$. Figure 3.13 shows several solutions for the free surface obtained from equations (3.5.3) and (3.5.4) for a range of values for the constant $q$ between $q = 0.4$ and $q = 4.0$.

The curves shown in Figure 3.13 constitute only a partial solution to the problem, since the exact shape of the free surface for any problem is a function of the specific value of $q$ appropriate for the physical parameters of the problem. The origin of this undeterminacy lies with the fact that the constant $q$ constitutes another unknown of the problem that must be determined as part of the solution. $q$ depends on the the specific values for the contact angle, $\alpha$, the liquid fill level, $V$, and the vessel wall geometry, $\psi$.

Many techniques have been developed to solve this class of nonlinear problems, the details of which may be found in [17], [16], and [13]. Due to the nonlinearity of the problem, all of these techniques are iterative in nature. Reference [17] discusses one such method, originally developed in reference [20], which is quite straightforward and very easy to implement computationally. This technique is outlined here.

Let us assume that the point of contact $A$ of the equilibrium free surface with the vessel wall $(x_A, y_A)$ is known. Equation (3.5.6) can then be solved for $q$, say $q_A$, in the following manner:

$$q_A = \frac{2}{x_A} \sin(\psi_A - \alpha) - \epsilon y_A + \epsilon \frac{V_{2A} - V}{x_A^2}. \tag{3.5.9}$$

Note that condition (3.3.17) for the physical constraints has been used in deriv-

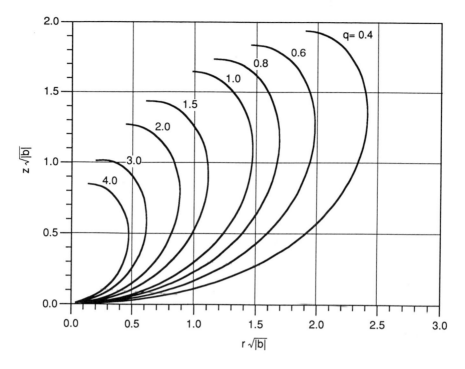

**Figure 3.13**

ing expression (3.5.9). Normally the coordinates of the point of contact must be determined as part of the solution for the full free-surface problem.

   If the assumed value for $(x_A, y_A)$ is the correct one, then integrating equations (3.5.3) and (3.5.4) with that value of $q_A$, from the point $A$ to the axis of symmetry, should produce the exact solution for the free surface. Whenever integrating equations (3.5.3) and (3.5.4) from the vessel wall, the following initial conditions should be used:

$$x = x_A, \quad y = y_A, \quad \frac{dx}{dt} = \cos(\psi_A - \alpha), \quad \frac{dy}{dt} = \sin(\psi_A - \alpha). \qquad (3.5.10)$$

When the coordinates of contact point $A$ are the correct solution for the problem, the conditions at the point $x = 0$ given by (3.5.5) must be recovered. If, on the other hand, the assumed value for $q$ is not the correct one, then the integration of the equations results in a solution for the free surface that eventually turns away from the vessel wall with decreasing $t$. Whenever the assumed contact point lies above or below the solution point, the free-surface curve will turn either upward or downward with increasing $t$, for which $dy/dt = 1$ or $dy/dt = -1$. This behavior of the solution curve is illustrated in Figure 3.14.

   The solution method is initiated by first choosing two widely separated coordinate points for the vessel wall contact point, say $(x_1, y_1)$ and $(x_2, y_2)$, from which $q_1$ and $q_2$ are calculated using equation (3.5.9). Equations (3.5.3) and (3.5.4) are then integrated from the vessel wall. The resulting free-surface solution curves are examined during the integration procedure to determine the direction in which

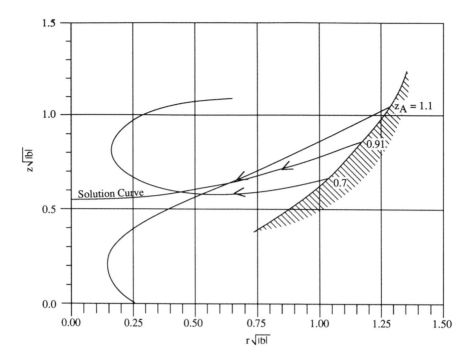

**Figure 3.14**   The general form of the solution curve using the numerical method.

they turn away from the axis of symmetry. If the curves turn in the same direction, then it is concluded that these assumed initial points do not enclose the solution point. Additional points on the vessel surface are chosen, say $(x_3, y_3)$, until two surface curves are found that enclose the solution point as shown in Figure 3.14. New contact points are then chosen that lie inbetween the first two points, and the procedure is repeated with subsequent nested free-surface curves pointing in opposite directions. The procedure is repeated with better estimates for the contact point until the exact contact point is found. The equilibrium curve corresponding to the exact contact point will intersect the $y$-axis with the condition $dy/dt = 0$.

The procedure outlined above can be conveniently automated using the bisection method, or another suitable technique, to converge on the solution curve with the desired accuracy. The refinement outline above is repeated until the actual liquid contact point on the vessel wall is determined to the desired accuracy. Once this point of contact is known, then the actual value of $q$, $q_A$, is calculated using expression (3.5.9), and equations (3.5.3) and (3.5.4) are integrated one final time with the initial conditions (3.5.10) to produce the desired solution for the free surface.

This method is illustrated with the following example from [19]: Consider an elongated ellipsoid of revolution with semiaxes lengths of 100 cm and 50 cm, respectively, partially filled with a liquid whose density is 1 gm/cm$^3$ and surface tension is $\sigma = 49.05$ dyne/cm, having a contact angle $\alpha$ of 30°. The liquid volume comprises 10% of the total vessel volume. The vessel in this example is not rotating, the $z$-axis is directed away from the liquid toward the gas, and the system

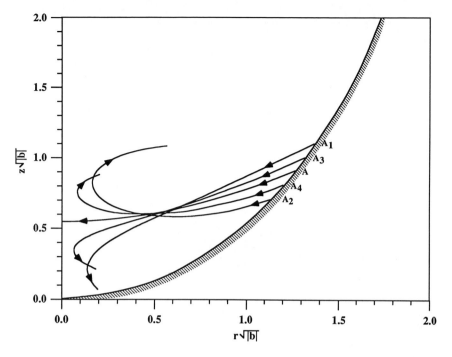

**Figure 3.15**

is subjected to a uniform gravity field in the same direction and is of magnitude $g = -8 \times 10^{-5} \times 981$ cm/sec$^2$.

The free surface can, of course, be determined by integrating equations (3.5.3) and (3.5.4) subject to the physical constraints of the problem. The value for $b$ in this example is $b = \rho g / \sigma = -1.6 \times 10^{-3}$/cm. Using $1/\sqrt{|b|}$ as a length scale results in nondimensional semiaxes values of $W = 2$ and $H = 4$. The nondimensional volume of the ellipse is $V_{el} = 4\pi W^2 H / 3 = 67$. Thus $V$ for the given fill level is $V = 0.1 V_{el} = 6.7$, while $V_2$ and the coordinates of the ellipsoidal vessel wall $\psi(x)$ are

$$\cot \psi = \left( \pm \frac{W}{Hx} \sqrt{W^2 - x_A^2} \right), \qquad V_2 = \frac{2}{3} \pi W H \left( W \mp \sqrt{W^2 - x_A^2} \right). \qquad (3.5.11)$$

In these definitions the upper sign corresponds to the lower half of the ellipsoid and vice versa.

Figure 3.15 shows the solution curves that result from integrating equations (3.5.3) and (3.5.4) for different values of the constant $\{q_i\}$ as determined by expression (3.5.9). It is seen that for values of the contact point $\{A_i\}$, which do not correspond to the correct value, the free-surface curve will turn either upward or downward depending on the specific starting point on the vessel wall. The computations converge to the following coordinates for the liquid contact point: $y_A = 0.9092$, $x_A = 1.2695$. The free surface computed with those coordinates is shown in Figure 3.16. The height of the liquid at the axis of symmetry from the bottom of the vessel, $y_0$, is calculated to be $y_0 = 0.514$ which is dimensionally given by $z_0 = 13.53$ cm.

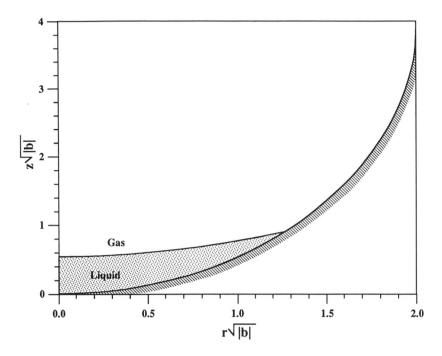

**Figure 3.16**   The solution form of the free surface for the example.

Reference [19] employs a graphical technique for producing the free-surface solution to this example. In this method, a dense family of curves is generated for the functions $\beta$ and $V_1$ in terms of $x$ for the same range of values as for $q$ in Figure 3.13. These curves shown in Figures 3.17 and 3.18 are produced by integrating the following differential equations:

$$\frac{d\beta}{dt} = \epsilon y + q - \frac{1}{x}\frac{dy}{dt}, \tag{3.5.12}$$

$$\frac{dV_1}{dt} = \pi x^2 \frac{dy}{dt}, \tag{3.5.13}$$

subject to the initial conditions

$$\beta(0) = 0, \quad V_1(0) = 0. \tag{3.5.14}$$

With given values for the contact angle $\alpha$, the vessel shape $\psi$, and fill level $V$, two additional curves can be generated in the $(q, x)$ plane, using the nondimensional form of expression (3.3.17) in the following manner:

$$\psi(x) - \alpha = \beta(t) \quad \text{and} \quad V_2(x) - V = V_1(t). \tag{3.5.15}$$

The first of these two curves, $q_1(x)$, is produced from the intersection points of $(\psi(x) - \alpha)$ with the solution curves generated for $\beta(x)$ from Figure 3.17 for different constant values of $q$. The second curve, $q_2(x)$, results from the intersection points of $(V_2(x) - V)$ with the solution curves generated for $V_1(x)|_{q=\text{const.}}$. The

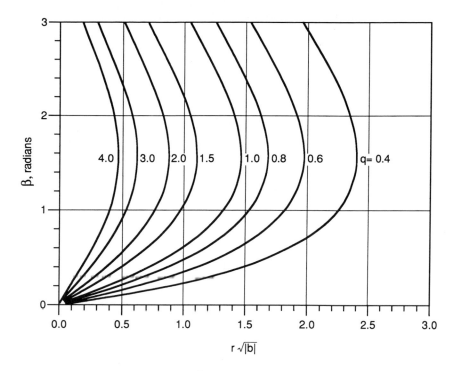

**Figure 3.17**

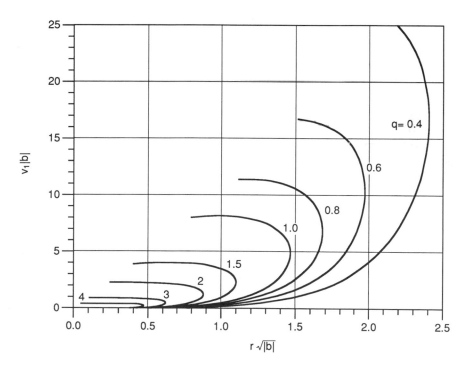

**Figure 3.18**

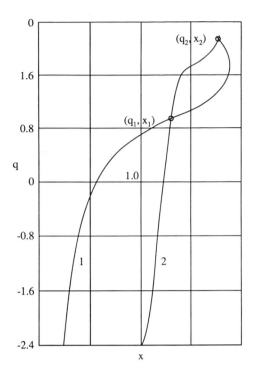

**Figure 3.19** The intersection points in the $(q, x)$ plane using the graphical solution method according to [19].

two curves $q_1(x)$ and $q_2(x)$ for this example are graphed in the $(q, x)$ plane as shown in Figure 3.19. The intersection points of these two curves define all of the possible pairs of values for the constant $q$ and the coordinates of the free-surface contact point, $A$. Once the values of $q$ and $x_A$ are known, it is possible to determine the exact point of contact of the liquid surface with the vessel wall $(x_A, y_A)$ on the specific free-surface curve for the resulting value of $q$. The equations for the free surface, (3.5.3) and (3.5.4), are integrated with the appropriate value for $q$ to yield the required free-surface shape.

   Figure 3.19 shows that there are two points of intersection for this example: $x_1 = 1.27$, $q_1 = 0.94$ and $x_2 = 1.75$, $q_2 = 2.16$. In this example there are two possible solutions for the free-surface contact point that obviously lie in the lower half of the ellipsoid. The first, for which $(x, q) = (1.27, 0.94)$, is the same solution determined above using the computational method. The values of the various parameters for this case are $\psi_A = 58.8°$, $\beta_A = 28.8°$, $V_2 = 7.69$, $V_1 = 0.99$, and $y_A - y_0 = \Delta y = 0.377$. The second intersection point, which was completely missed in the computational method, turns out to be nonphysical. Reference [19] gives the following values for this point: $\psi_A = 74.5°$, $\beta_A = 44.5°$, $V_2 = 17.28$, $V_1 = 10.58$, and $y_A - y_0 = \Delta y = 2.46$. This last value, $\Delta y = 2.46$, does not lie on the vessel wall and thus this solution is not physically meaningful, although it satisfied the governing equations and the imposed constraints.

## *3.6* ROTATING AXISYMMETRIC FREE SURFACES IN ZERO GRAVITY

The problem of defining the equilibrium free surfaces for liquids subjected to rotation is accomplished by allowing the value of the rotation parameter $\omega$ to be nonzero in equations (3.3.10) and (3.3.11). The rotating free-surface problem can be further simplified if the gravity parameter $b$ is also set equal to zero. Such conditions are appropriate for investigating the free surface of a rotating liquid in zero-gravity field. This problem has practical applications in a low-gravity environment. The resulting equations for the free surface with $\omega \neq 0$, $b = 0$, will take the following form:

$$\frac{d^2r}{ds^2} = -\frac{dz}{ds}\left(k - \omega r^2 - \frac{1}{r}\frac{dz}{ds}\right), \tag{3.6.1}$$

$$\frac{d^2z}{ds^2} = \frac{dr}{ds}\left(k - \omega r^2 - \frac{1}{r}\frac{dz}{ds}\right). \tag{3.6.2}$$

Equations (3.6.1) and (3.6.2) possess a natural length scale, namely $\omega^{-1/3}$, which may be used in nondimensionalizing the equations. Upon introducing the nondimensional variables

$$x = r\omega^{1/3}, \quad y = z\omega^{1/3}, \quad t = s\omega^{1/3}, \quad q = q/\omega^{1/3}$$

into equations (3.6.1) and (3.6.2) for the equilibrium free surface, the following set of equations for the coordinates of the free-surface curve result:

$$\frac{d^2x}{dt^2} = -\frac{dy}{dt}\left(-x^2 + q - \frac{1}{x}\frac{dy}{dt}\right), \tag{3.6.3}$$

$$\frac{d^2y}{dt^2} = \frac{dx}{dt}\left(-x^2 + q - \frac{1}{x}\frac{dy}{dt}\right). \tag{3.6.4}$$

For this problem to be consistent, the initial conditions given in (3.3.13) must be nondimensionalized in a similar manner. This procedure results in the following conditions:

$$x(0) = y(0) = \frac{dy}{dt}(0) = 0, \quad \frac{dx}{dt}(0) = 1. \tag{3.6.5}$$

Reference [17] states that the rotating symmetrical problem defined by (3.6.3) and (3.6.4) admits a general symmetry criteria in which the free-surface solution possesses a mirror reflection about the plane $y = const.$; this can be formulated in the following manner: If the solution to equations (3.6.3) and (3.6.4) ever reaches a point in $t$, say $t_c$, for which $|dy/dt| = 1$, then there exists another solution for the equations constituting a mirror reflection of the free-surface curve about the plane $y = y(t_c)$. This property shows that some free-surface solutions may result in a closed surface depicting either a drop or a bubble depending on the direction of the free-surface curvature. The symmetry property also applies for the nonrotating case.

The shape of the free surface for this problem may be determined by integrating the governing equations using initial value integrators, as was done in the previous section. Notice that equations (3.6.3) and (3.6.4) suffer from the same

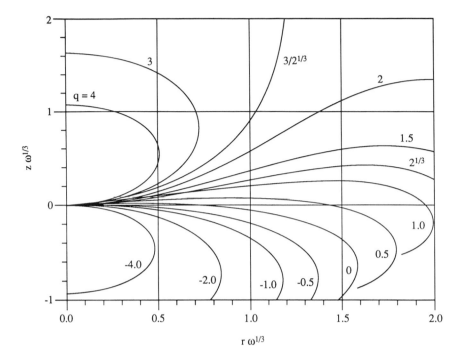

**Figure 3.20**   The shape of the free surface for the case of zero gravity with rotation as a function of the constant $q$.

singular behavior encountered earlier for the nonrotating case. Reference [17] gives the following asymptotic solution which is valid for small values of $t$ in the neighborhood of $t = 0$:

$$x = t - \frac{1}{24}q^2t^3 + \frac{1}{40}\left(q + \frac{1}{48}q^4\right)t^5 + \cdots,$$

$$y = \frac{1}{4}qt^2 - \frac{1}{16}\left(1 + \frac{1}{12}q^3\right)t^4 + \cdots. \tag{3.6.6}$$

Thus equations (3.6.3) and (3.6.4) may be integrated numerically from the point $t = 0$ with the aid of the expansions in (3.6.6).

Figure 3.20 shows the results of the numerical integration of these equations for a range of values as the constant $q$, $-4.0 < q < 4.0$. It should be mentioned again that the specific value of $q$ for any application depends on the physical parameters of the problem including the contact angle $\alpha$, the volume of the liquid $v$, and the shape of the vessel wall $\psi$. Figures 3.21 and 3.22  show the variations of both $\beta$ and $V_1$ for the same range of values of $q$ as in Figure 3.20.

Equations (3.6.3) and (3.6.4) can be combined into a single differential equation for the coordinates of the free surface $(x, y)$, which is independent of the parameter $t$. This equation will take the form

$$\frac{1}{x}\frac{d}{dx}\left(\frac{x\,dy/dx}{\sqrt{1 + (dy/dx)^2}}\right) = -x^2 + q, \tag{3.6.7}$$

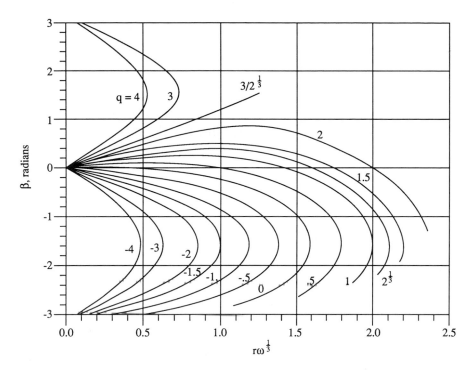

**Figure 3.21**   The variation of $\beta$ with $x$ for a range of values for the constant $q$.

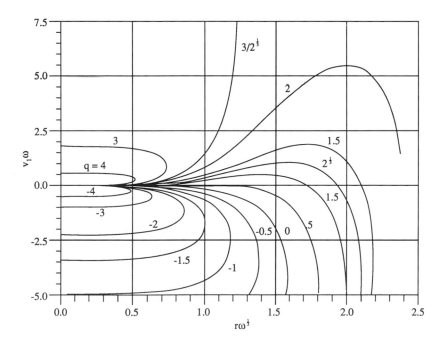

**Figure 3.22**   The variation of $V_1$ with $x$ for a range of values for the constant $q$.

which is subject to the following initial conditions:

$$y(0) = 0, \qquad \frac{dy}{dx}(0) = 0. \tag{3.6.8}$$

For the rotating fluid in zero gravity there exists a closed-form solution for the free-surface equation (3.6.7). The solution is obtained by integrating this equation once with the initial conditions of (3.6.8) resulting in the following equation:

$$\frac{dy/dx}{\sqrt{1 + (dy/dx)^2}} = -\frac{1}{4}x^3 + \frac{1}{2}xq. \tag{3.6.9}$$

Equation (3.6.9) can be integrated once more to yield a closed-form solution in terms of elliptic integrals of the first and second kind. The analytical solution to equation (3.6.9) is quite involved and is limited in its range of applicability. This solution is given in [17] together with its range of validity. The numerical integration of equations (3.6.3) and (3.6.4) is more straightforward and robust.

The solution for the free surface of a rotating liquid can be obtained using either the graphical method or the computational methods outlined in the previous section. For rotating liquid in zero gravity, a special solution for the surface can exist that is unique for this problem. Under special conditions, the equilibrium surface in a rotating vessel may not contact the vessel wall at all. An example of such a surface is a gas bubble whose position with respect to the axis of rotation is determined only by the possibility of its containment inside the vessel. Figure 3.20 shows that such bubbles are possible only for values of the constant $q$ when $q \geq 3/2^{1/3}$. For each such value $q$ there exists a corresponding value for the nondimensional bubble volume $V_g = v_g \omega$, the nondimensional equatorial radius $x_g = r_g \omega^{1/3}$, and the nondimensional bubble height.

Figure 3.20 also shows that a closed equilibrium surface that does not intersect with itself is possible for values of $q \leq q^*$, where $q^* = 2^{1/3}$. This is the value of $q$ for which the equatorial point is at the same height as the pole. These surfaces now contain the liquid inside them, i.e., they represent a rotating drop in equilibrium inside the container.

## 3.7 AXISYMMETRIC FREE SURFACES FOR A ROTATING LIQUID UNDER GRAVITY

In this section the full, general problem is treated for which the value of gravity and the angular velocity are nonzero, i.e., $b \neq 0$ and $\omega \neq 0$. The free surface for this case must be determined by solving the full free-surface equations (3.3.10) and (3.3.11):

$$\frac{d^2r}{ds^2} = -\frac{dz}{ds}\left(bz - \omega r^2 + k - \frac{1}{r}\frac{dz}{ds}\right), \tag{3.7.1}$$

$$\frac{d^2z}{ds^2} = \frac{dr}{ds}\left(bz - \omega r^2 + k - \frac{1}{r}\frac{dz}{ds}\right), \tag{3.7.2}$$

subject to the usual initial conditions (3.3.13), namely

$$r(0) = \frac{dz}{ds}(0) = 0, \qquad z(0) = 0, \qquad \frac{dr}{ds}(0) = 1. \tag{3.7.3}$$

Again, the numerical integration of equations (3.7.1) and (3.7.2) suffers from the same singular behavior encountered in the nonrotating and zero-gravity cases when the integration is started from point $s = 0$. Reference [17] gives an asymptotic solution for $r$ and $z$ that is valid in the neighborhood of $s = 0$ and can be used to initiate the integration process. The asymptotic solution takes the following form:

$$r = s - \frac{1}{24}k^2 s^3 + \frac{1}{40}\left(\omega k - \frac{1}{4}bk^2 + \frac{1}{48}k^4\right)s^5 + \cdots,$$

$$z = \frac{1}{4}ks^2 + \frac{1}{16}\left(-\omega + \frac{1}{16}bk - \frac{1}{48}k^3\right)s^4 + \cdots. \tag{3.7.4}$$

The problem defined by equations (3.7.1) and (3.7.2) possesses three different parameters $b$, $\omega$, and $k$. Any two of these parameters may be used to nondimensionalize the equations. If either of the length scales, $|b|^{-1/2}$ or $\omega^{-1/3}$, is used to nondimensionalize the variables, then the number of free parameters in the problem can be reduced by one. If, for instance, $|b|^{-1/2}$ is used for a length scale, equations (3.7.1) and (3.7.2) take the form

$$\frac{d^2 x}{dt^2} = -\frac{dy}{dt}\left(\epsilon y - \lambda x^2 + q - \frac{1}{x}\frac{dy}{dt}\right), \tag{3.7.5}$$

$$\frac{d^2 y}{dt^2} = \frac{dx}{dt}\left(\epsilon y - \lambda x^2 + q - \frac{1}{x}\frac{dy}{dt}\right), \tag{3.7.6}$$

where $\epsilon = \pm 1$ and $\lambda = \omega|b|^{-2/3}$. Equations (3.7.5) and (3.7.6) in this case only possess two parameters, $\lambda$ and $q$. If, on the other hand, $\omega^{-1/3}$ is used for nondimensionalization, then the following equations result:

$$\frac{d^2 x}{dt^2} = -\frac{dy}{dt}\left(\epsilon \lambda^{-2/3} y - x^2 + q - \frac{1}{x}\frac{dy}{dt}\right), \tag{3.7.7}$$

$$\frac{d^2 y}{dt^2} = \frac{dx}{dt}\left(\epsilon \lambda^{-2/3} y - x^2 + q - \frac{1}{x}\frac{dy}{dt}\right). \tag{3.7.8}$$

Equations (3.7.7) and (3.7.8) emerge from this nondimensionalization with the same two parameters, $\lambda$ and $q$, but in a different functional form than in (3.7.5) and (3.7.6).

The use of the computational method for obtaining the solutions is very convenient and has an advantage over the graphical technique. When using the computational method, there is no added advantage in reducing the number of independent parameters appearing in the equations from three to two. Consequently, the equations may be nondimensionalized using any convenient physical length scale.

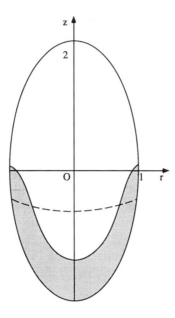

**Figure 3.23**   The shape of the free surface for $\alpha = 30°$ and a fill level of 30%. Solid line: $Ro = 10$; dotted line: $Ro = 0$ (from [21]).

In this case the problem is rendered dimensionless by using a characteristic length scale for the problem, such as the diameter of the vessel. If indeed a physical length scale $L$ is used, then the nondimensional free-surface curve equations will take the following form:

$$\frac{d^2x}{dt^2} = -\frac{dy}{dt}\left(Boy - Rox^2 + q - \frac{1}{x}\frac{dy}{dt}\right),\tag{3.7.9}$$

$$\frac{d^2y}{dt^2} = \frac{dx}{dt}\left(Boy - Rox^2 + q - \frac{1}{x}\frac{dy}{dt}\right).\tag{3.7.10}$$

In equations (3.7.9) and (3.7.10), $Bo = \rho g L^2/\sigma$ is the Bond number, and $Ro = \rho\Omega^2 L^3/2\sigma$ is the Rotation number. This problem can be solved numerically for any given set of values for both $Ro$ and $Bo$.

Note that the graphical method for generating the solution to the general problem given by either equations (3.7.5) and (3.7.6), or (3.7.7) and (3.7.8), becomes very complicated due to the existence of two independent parameters. The free-surface curves in this case must be defined in a three-dimensional space, making the calculations quite intractable.

The full problem is solved by [21] for the example given earlier of an elongated ellipsoid of revolution with semiaxes ratio 1 : 2. Figure 3.23 shows the equilibrium surface for a liquid fill volume of 30% with a contact angle of 30°, for which $Bo = 2$ and $Ro = 10$. The figure also shows the static shape of the free surface under the action of gravity alone. Figure 3.24 shows the equilibrium free surface for a liquid fill volume of 70%, with contact angle $\alpha = 120°$, Bond number $Bo = 1$, and Rotation number $Ro = 5$. Also indicated in the figure is the equilibrium free surface with no rotation (i.e., $Ro = 0$), with the rest of the parameters remaining

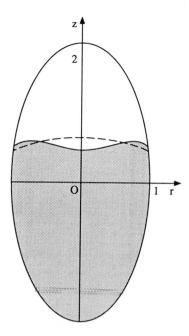

**Figure 3.24**  The shape of the free surface for $\alpha = 120°$ and a fill level of 70%. Solid line: $Ro = 5$; dotted line: $Ro = 0$ (from [21]).

the same. A comparison between the free surfaces with and without rotation helps in drawing a qualitative conclusion regarding the change in shape of the equilibrium surface due to a small increase in the value of the angular velocity, $\Omega$. In this case, a depression is formed at the axis of rotation that gradually increases with angular velocity until the free surface touches the bottom of the vessel. A further increase in the angular velocity makes the surface doubly connected.

Another simple example is to determine the shape of the stable equilibrium surface of a liquid in a cylindrical vessel. The solution for this problem, shown in Figure 3.25, can also be determined using the method outlined above. It can be seen that the shape of the liquid surface for this case is determined by the contact angle $\alpha$ and the dimensionless radius of the cylinder $x_0 = br_0^2 = \rho g r_0^2 / \sigma =$ the Bond number. Figures 3.25 and 3.26 show the equilibrium surface shapes for $Bo = 1000, 100, 10$, and $0$, respectively, for two contact angles $\alpha = 0°$ and $\alpha = 30°$, respectively. The free surfaces shown in Figures 3.25 and 3.26 have been calculated in [14].

## 3.8 PLANAR TWO-DIMENSIONAL FREE SURFACES

In this section, the second class of free surfaces in symmetric geometries is discussed. The two-dimensional plane $(x, z)$ can be considered symmetric about the $y$-axis, making the planar two-dimensional geometry a special case of the general symmetric geometry problem. If the potential function $\Pi$ is independent of the third coordinate (in this case, the $y$-coordinate), then the problem of determining the equilibrium free surface in two dimensions is reduced to the so-called *plane*

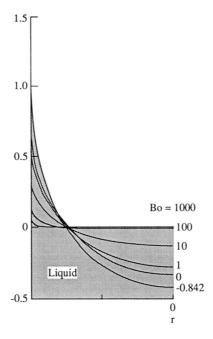

**Figure 3.25** The shape of the free surface for $\alpha = 0°$ in a nonrotating right circular cylinder for $Bo = 1000, 100, 10, 1$, and 0 (from [14]).

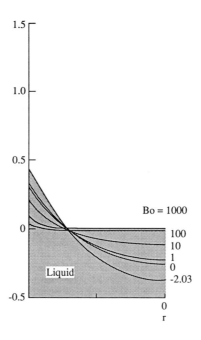

**Figure 3.26** The shape of the free surface for $\alpha = 30°$ in a nonrotating right circular cylinder for $Bo = 1000, 100, 10, 1$, and 0 (from [14]).

*equilibrium problem.* In this case the vessel containing the liquid can be visualized as a channel of infinite span or a finite channel bounded by two planes $y = y_1$ and $y = y_2$.

The sum of the principal radii of curvature for the equilibrium curve can be written in the following form for the two-dimensional case:

$$\frac{1}{R_1} + \frac{1}{R_2} = \pm \frac{(dx/ds)(d^2z/ds^2) - (dz/ds)(d^2x/ds^2)}{\sqrt{[(dx/ds)^2 + (dz/ds)^2]^3}} , \qquad (3.8.1)$$

where $s$ is a parameter representing the length of the free curve $l$. Equation (3.8.1) is the two-dimensional counterpart of equation (3.3.1) for the axisymmetric geometry. The same sign convention is used here as in the previous sections for the axisymmetric problem, where the positive sign should be used when the liquid is to the right of an observer moving in the direction of increasing $s$.

Again, equation (3.8.1) can be separated into two equations for the coordinates $x$ and $z$ of the free-surface curve. Using arguments similar to those used in deriving the free-surface equations for the axisymmetric geometry, the following two differential equations result for the free-surface coordinates:

$$\frac{d^2x}{ds^2} = \mp \frac{dz}{ds}\left[\frac{\rho}{\sigma}\Pi(x,z) + c\right], \qquad (3.8.2)$$

$$\frac{d^2z}{ds^2} = \pm \frac{dx}{ds}\left[\frac{\rho}{\sigma}\Pi(x,z) + c\right]. \qquad (3.8.3)$$

Equations (3.8.2) and (3.8.3) must now be solved subject to the following geometric constraint:

$$\left(\frac{dx}{ds}\right)^2 + \left(\frac{dz}{ds}\right)^2 = 1. \qquad (3.8.4)$$

Since this problem does not have rotational symmetry, the potential function $\Pi$ will consist of only the gravity force function. Substituting $\Pi = gz$ and combining equations (3.8.2) and (3.8.3), the equation for the free-surface curve takes the following form:

$$\frac{d}{dx}\left(\frac{dz/dx}{\sqrt{1 + (dz/dx)^2}}\right) = \pm \left(\frac{\rho g}{\sigma}\right) z + k. \qquad (3.8.5)$$

Similarly equations (3.8.2) and (3.8.3) take the following form with this representation for the potential function:

$$\frac{d^2x}{ds^2} = -\frac{dz}{ds}\left(\pm\frac{\rho g}{\sigma}z + k\right) \qquad (3.8.6)$$

$$\frac{d^2z}{ds^2} = \frac{dx}{ds}\left(\pm\frac{\rho g}{\sigma}z + k\right). \qquad (3.8.7)$$

For the case of *zero gravity* obtained by setting $g = 0$, the free-surface equations will reduce to the following set:

$$\frac{d^2x}{ds^2} = -k\frac{dz}{ds} \qquad (3.8.8)$$

$$\frac{d^2z}{ds^2} = k\frac{dx}{ds}. \qquad (3.8.9)$$

Equations (3.8.8) and (3.8.9) can easily be integrated to yield a solution for the free surface representing a family of all the circles in the $(x, z)$ plane. Thus the equilibrium free-surface shape is circular in zero gravity away from bounding walls. Recall that the same shape resulted for the zero-gravity case in axisymmetric geometry.

In the general case for which $b = \rho g / \sigma \neq 0$, it is possible to use $|b|^{-1/2}$ as a length scale. Thus upon defining the nondimensional variables

$$\xi = x\sqrt{|b|}, \quad \zeta = z\sqrt{|b|}, \quad \tau = s\sqrt{|b|}, \quad \kappa = k/\sqrt{|b|}, \tag{3.8.10}$$

and substituting them into the free-surface equation (3.8.5), the following nondimensional equation for the free-surface curve results:

$$\frac{d}{d\xi}\left(\frac{d\zeta/d\xi}{\sqrt{1 + (d\zeta/d\xi)^2}}\right) = \epsilon\zeta, \tag{3.8.11}$$

where again $\epsilon = \pm 1$. Note that the constant of integration $\kappa$ has been combined with $\zeta$ in equation (3.8.11). This is achieved by using the transformation $\zeta = \zeta^* + \kappa$.

The separated equations can be nondimensionalized in a similar manner to yield

$$\frac{d^2\xi}{d\tau^2} = -\epsilon\frac{d\zeta}{d\tau}\zeta, \tag{3.8.12}$$

$$\frac{d^2\zeta}{d\tau^2} = \epsilon\frac{d\xi}{d\tau}\zeta. \tag{3.8.13}$$

Since in this case the free-surface curvature is given by $\epsilon\zeta$, there exists a point on the equilibrium curve for which $d\zeta/d\xi = 0$. This observation, together with the invariance of equations (3.8.12) and (3.8.13) with respect to the transformations $\tau \to -\tau$ and $\epsilon \to -\epsilon$, leads to the following initial conditions for these equations, which are consistent with the system of equations:

$$\xi(0) = 0, \quad \frac{d\xi}{d\tau}(0) = 1, \quad \zeta(0) = \zeta_0, \quad \frac{d\zeta}{d\tau}(0) = 0. \tag{3.8.14}$$

Equation (3.8.12) can be integrated once to yield the following solution after applying the initial conditions in (3.8.14):

$$\frac{d\xi}{d\tau} = 1 + \frac{\epsilon}{2}(\zeta_0^2 - \zeta^2). \tag{3.8.15}$$

Since the equations for the equilibrium curve are invariant under the transformation $\zeta \to -\zeta$, it is sufficient here to seek solutions only for $\zeta_0 > 0$. The numerical integration of equations (3.8.13) and (3.8.15), subject to initial conditions (3.8.14), yields the solution curves shown in Figure 3.27. The curves in Figure 3.27 are for $\epsilon = -1$ and for $\zeta_0 = 1$, 2, and 2.5, respectively.

Equation (3.8.11) for the free surface can be integrated once to yield

$$\frac{1}{\sqrt{1 + (d\zeta/d\xi)^2}} + \frac{\epsilon}{2}\zeta^2 = C, \tag{3.8.16}$$

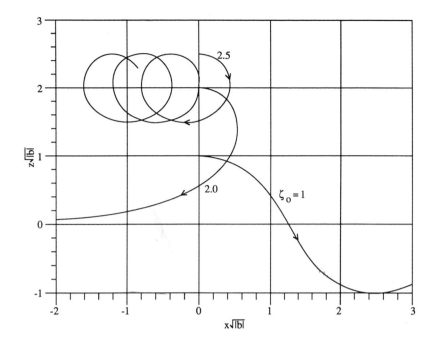

**Figure 3.27** Free-surface solutions in two-dimensional geometry for the zero-gravity case and for values of $\zeta_0 = 1, 2, 2.5$ with $\epsilon = -1.0$.

where $C$ is a constant of integration that can be evaluated using the initial conditions given in (3.8.14). Applying conditions (3.8.14) results in the following expression:

$$C = 1 + \frac{\epsilon}{2}\zeta_0^2. \qquad (3.8.17)$$

Equation (3.8.15) can be integrated once more, resulting in a closed-form solution for the free-surface coordinates in terms of elliptic integrals of the first and second kind. For $\epsilon = -1$, [17] provides the solution

$$\xi = 2E(\lambda, \zeta_0/2) - F(\lambda, \zeta_0/2), \qquad (3.8.18)$$

where

$$\cos \lambda = \zeta/\zeta_0.$$

Meanwhile for $\epsilon = 1$, [17] gives the solution

$$\xi = \pm\left[\lambda(1 + \zeta_0^2/2)F(\nu, \lambda) - \frac{2}{\lambda}E(\nu, \lambda) + \sqrt{2}\sin\delta(1 + \zeta_0^2/2 - \cos\delta)^{-1/2}\right], \quad (3.8.19)$$

where

$$\cos \delta = 1 + (\zeta_0^2 - \zeta^2)/2, \quad \sin \nu = \sqrt{(1 + \zeta_0^2/4)(1 - \cos\delta)/(1 + \zeta_0^2 - \cos\delta)},$$

and

$$\lambda = (1 + \zeta_0^2/4)^{-1/2}.$$

In this solution $F(m, n)$ and $E(m, n)$ are elliptic integrals of the first and second kind, respectively.

In the analysis presented above, the effects of the wall geometry have not been included in evaluating the free-surface shape. In other words, no physical constraints have been considered. When there are walls along the channel, then the treatment of the problem is somewhat different. The analysis for this case follows along lines similar to the axisymmetric problem.

For the two-dimensional problem it is convenient to think of the container as a channel having a cross-sectional form in the $(x, z)$ plane with its axis extending to infinity in the $y$-direction. In addition, the cross section will be assumed symmetric about the $z$-axis. In this manner the equilibrium free surface possesses symmetry with respect to the plane $x = 0$. This condition implies that the equilibrium curve can be determined in a manner similar to the one discussed for the axisymmetric problem. The only difference here is in evaluating the volume of the vessel and the liquid. The volume in two-dimensional geometry will be defined as the cross-sectional area multiplied by a unit length in the $y$-direction:

$$v_1(s, \kappa) = 2 \int_0^s x(s, \kappa) \frac{dz}{ds}(s, \kappa)\, ds. \tag{3.8.20}$$

For the case of liquid contained in a vessel, the equations for the free-surface coordinates must include the constant $\kappa$ in an explicit manner. Thus for the general two-dimensional case, the following equations must be used for evaluating the free-surface coordinates:

$$\frac{d^2\xi}{d\tau^2} = -\frac{d\zeta}{d\tau}(\epsilon\zeta + \kappa), \tag{3.8.21}$$

$$\frac{d^2\zeta}{d\tau^2} = \frac{d\xi}{d\tau}(\epsilon\zeta + \kappa). \tag{3.8.22}$$

Equations (3.8.21) and (3.8.22) must be solved subject to the following conditions:

$$\xi(0) = 0, \quad \frac{d\xi}{d\tau}(0) = 1, \quad \zeta(0) = 0, \quad \frac{d\zeta}{d\tau}(0) = 0. \tag{3.8.23}$$

In conditions (3.8.23) the origin has been translated by an amount of $\zeta_0$ in the positive $\zeta$-direction.

In equations (3.8.21) and (3.8.22) it is assumed that $b = \rho g/\sigma$, where $g$ and $b$ can be of either sign. Figure 3.28 shows the stable free-surface curves resulting from integrating equations (3.8.22) and (3.8.23) for $\epsilon = 1$. These curves are symmetric with respect to the $\zeta$-axis. The curves shown in Figure 3.28 are for different values of $\kappa$. Again, $\kappa$ can be determined from the physical parameters of the problem, including the contact angle $\alpha$, the liquid volume $v$, and the channel cross-sectional geometry. Once the constant $\kappa$ is determined, then the free surface can be constructed by direct integration of the equations.

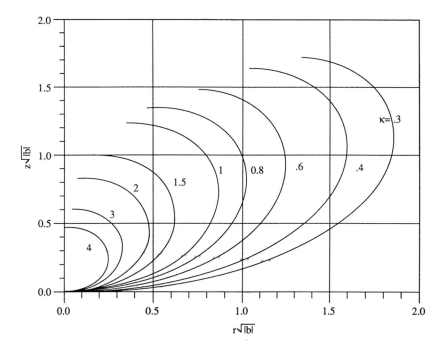

**Figure 3.28** The free-surface solutions for two-dimensional geometry.

## 3.9 *APPROXIMATE SOLUTIONS FOR FREE SURFACES IN TWO DIMENSIONS*

Equations (3.8.22) and (3.8.23) for the equilibrium curve in general two-dimensional geometry are nonlinear, and a general closed-form solution is difficult to obtain with arbitrary constraints. Analytic solutions to this problem can be found under some specific conditions. Two special cases from [17] are presented.

The two-dimensional problem, in which the equilibrium surface adjoins a rigid wall on only one side, is shown in Figure 3.29. As a rule, this situation arises whenever the gravity force is much stronger than the surface tension force, such as menisci in a terrestrial environment. First, it should be observed from Figure 3.29 that the free surface is assumed semi-infinite so that $\zeta \to 0$ as $\xi \to \infty$. Under these conditions the constant of integration $\kappa$ will be zero since it represents the radius of curvature at the axis of symmetry. These criteria simplify considerably the governing equations. Equation (3.8.22) can be integrated once, and upon applying the initial conditions (3.8.21), the following equation for $\xi$ results:

$$\frac{d\xi}{d\tau} = 1 - \frac{1}{2}\zeta^2. \tag{3.9.1}$$

Upon substituting equation (3.9.1) into equation (3.8.23), a single differential equation in terms of $\zeta$ results:

$$\frac{d^2\zeta}{d\tau^2} = \zeta - \frac{1}{2}\zeta^3. \tag{3.9.2}$$

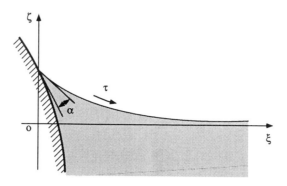

**Figure 3.29**

This equation can be integrated once to yield the following solution:

$$\frac{d\zeta}{d\tau} = -\zeta\sqrt{1 - \frac{1}{4}\zeta^2}. \tag{3.9.3}$$

Equation (3.9.3) can be readily integrated resulting in the following parametric representation for the free-surface coordinates:

$$\xi = \tau - 2\tanh(\tau + c_1) + c_2, \tag{3.9.4}$$

$$\zeta = \pm 2/\cosh(\tau + c_1). \tag{3.9.5}$$

The constants of integration $c_1$ and $c_2$ can be evaluated from the boundary condition at the liquid contact point. If the wall confining the liquid is described by the function $\xi = \psi(\zeta) + const.$, then $c_1$ and $c_2$ can be evaluated with the aid of the algebraic equations

$$2\tanh c_1 - c_2 = 0, \tag{3.9.6}$$

and

$$-(1 - 2/\cosh^2 c_1)d\psi(\zeta_0)/d\zeta \pm 2\sinh c_1/\cosh^2 c_1$$
$$= \cos\alpha\sqrt{1 + (d\psi(\zeta_0)/d\zeta)^2}, \tag{3.9.7}$$

where $\zeta_0 = \pm 2/\cosh c_1$. Equation (3.9.7) is thus transcendental in $c_1$.

For the case of a straight wall for which $d\psi/d\zeta = const.$, equation (3.9.7) becomes quadratic in $\sinh(c_1)$. In that case only one of its roots is physically meaningful. As an example, for a vertical wall for which $d\psi/d\zeta = 0$, equation (3.9.7) takes the form

$$\pm 2\sinh c_1/\cosh^2 c_1 = \cos\alpha, \tag{3.9.8}$$

which has the unique solution

$$\sinh c_1 = (1 + \sin\alpha)/|\cos\alpha| = \cot|\pi/4 - \alpha/2|, \tag{3.9.9}$$

where $\zeta_0 = \sqrt{2}\cos\alpha$.

The second situation in which the equation governing the free-surface shape can easily be integrated is that in which the equations can be linearized. Such condition implies that the free-surface curvature is small. In this case the equilibrium curve equations can be linearized, since $|d\zeta/d\xi|$ is small. For $\epsilon = 1$, the linearized free-surface equation becomes

$$d^2\zeta/d\xi^2 - \zeta = 0, \tag{3.9.10}$$

which has the following solution for the symmetric case:

$$\zeta = c_1 \cosh \xi + c_2. \tag{3.9.11}$$

Again, $c_1$ and $c_2$ are constants of integration.

If the equation describing the half-channel wall geometry is given in the form of $\zeta = \psi(\xi)$, then the boundary conditions assume the following form:

$$\zeta(\lambda) = \psi(\lambda), \quad \frac{1 + (d\zeta/d\xi)(d\psi/d\xi)}{\sqrt{1 + (d\psi/d\xi)^2}} = \cos \alpha,$$

$$\int_0^\lambda (\zeta - \psi)d\xi = \int_0^{\lambda_0} (\zeta_0 - \psi)d\xi = \frac{V}{2}. \tag{3.9.12}$$

Here, $\lambda$ is the location in the $\xi$ direction of the liquid contact point with the vessel wall.

These boundary conditions should also be linearized in order to be mathematically consistent with the governing equation, which takes the form

$$\frac{d^2\psi}{d\xi^2}(\lambda_0)\eta(\lambda_0)\cos^3 \psi_0 - \frac{d\eta}{d\xi}(\lambda_0)\sin \psi_0 = (\alpha - \psi_0)\sin \psi \psi_0, \tag{3.9.13}$$

$$\int_0^{\lambda_0} \eta(\xi)d\xi = 0, \tag{3.9.14}$$

where $\eta = \zeta - \zeta_0$. Defining $c_2 = \zeta_0 + c_3$ we get the following linear system of algebraic equations for $c_1$ and $c_3$:

$$(\cos^3 \psi_0)\frac{d^2\psi}{d\xi^2}(\lambda_0)(c_1 \cosh \lambda_0 + c_3) - (\sin \psi_0)c_1 \sinh \lambda_0 = (\alpha - \psi_0)\sin \psi_0, \tag{3.9.15}$$

$$c_1 \sinh \lambda_0 + c_3 \lambda_0 = 0. \tag{3.9.16}$$

These equations can be solved for $c_1$ and $c_3$, resulting in the following expressions:

$$c_1 = (\alpha - \psi_0)\sin \psi_0/[(\cos^3 \psi_0)(d^2\psi/d\xi^2)(\lambda_0)(\cosh \lambda_0$$
$$- (\sinh \lambda_0)/\lambda_0) - \sin \psi_0 \sinh \lambda_0], \tag{3.9.17}$$

$$c_3 = -c_1(\sinh \lambda_0/\lambda_0). \tag{3.9.18}$$

In the above solution, the contact angle condition was also linearized; a more exact result can be obtained if $\alpha - \psi_0$ is replaced by $(\cos \psi_0 - \cos \alpha)/\sin \psi_0$.

For example, let the channel have a cross section as shown in Figure 3.30. The solution to the linearized problem for this geometry for $\epsilon = 1$ results in the

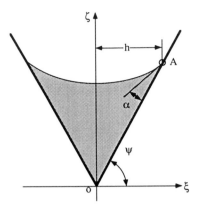

**Figure 3.30**

following transcendental equation for the distance $h$ of the point of contact of the liquid surface with the vessel wall:

$$\frac{\cos\alpha - \cos\psi}{\sin\psi}\left(\coth\lambda - \frac{1}{\lambda}\right) - \frac{1}{2}\lambda\tan\psi + \frac{V}{2\lambda} = 0. \tag{3.9.19}$$

The constants $c_1$ and $c_3$ can easily be expressed in terms of $\lambda$, while for the totally linearized solution the following values are obtained:

$$c_1 = (\psi - \alpha)/\sinh\lambda_0, \qquad c_3 = (\alpha - \psi)/\lambda_0, \tag{3.9.20}$$

where $\lambda_0 = \sqrt{V\cot\psi}$ and $\zeta_0 = \sqrt{V\tan\psi}$. In the limiting case in which $\psi = \pi/2$, describing the equilibrium of a liquid between two vertical sheets separated by distance $2x_0$, we get

$$\zeta = \zeta_0 - \frac{1}{\lambda_0}\cos\alpha + \frac{\cos\alpha}{\sin\lambda_0}\cosh\xi, \tag{3.9.21}$$

where $\lambda_0 = \lambda = \sqrt{b}x_0$. In this case $\cos\alpha$ can still be replaced by $\pi/2 - \alpha$.

The other limiting case in which $\psi = 0$ (i.e., in which a liquid barrier over a horizontal plane is considered) linearization of the problem is not possible. This problem has the solution

$$\zeta = \frac{\tan\alpha}{\sinh\lambda}(\cosh\lambda - \cosh\xi), \tag{3.9.22}$$

where the half-width $\lambda$ of the barrier is found from the equation

$$\lambda\coth\lambda = 1 + \frac{V}{2\tan\alpha}. \tag{3.9.23}$$

## *3.10* SMALL AMPLITUDE SLOSHING

The previous sections were concerned with determining stable equilibrium free surfaces when the liquid is stationary. In this section conditions will be formulated under which undamped waves can arise in the equilibrium free surface that may ultimately destroy the surface. This problem is commonly known as the *sloshing problem* and is of great significance to space fluid management technology.

The sloshing problem can be effectively tackled using hydrodynamic stability methods. These do not solve the problem in its entirety but can provide a great deal of information. It is possible to determine, for instance, the characteristics of the most unstable free-surface waves, or modes, which may be avoided with proper design considerations.

The starting point for the analysis is the equations of motion for an incompressible fluid given by

$$\rho \frac{D\mathbf{v}}{Dt} = -\nabla p + \rho \mathbf{F}_B + \mu \nabla^2 \mathbf{v}. \tag{3.10.1}$$

For an inviscid fluid for which $\mu = 0$, equation (3.10.1) will reduce to

$$\rho \left( \frac{\partial \mathbf{v}}{\partial t} + \mathbf{v} \cdot \nabla \mathbf{v} \right) = -\nabla p - \rho \nabla \Pi, \tag{3.10.2}$$

where $\Pi$ is a potential function for the body force field defined in expression (3.2.3) and $\mathbf{v}$ is the fluid velocity vector.

Furthermore, if the fluid is assumed irrotational,

$$\nabla \times \mathbf{v} = 0, \tag{3.10.3}$$

as then the velocity vector $\mathbf{v}$ may be represented as

$$\mathbf{v}(\mathbf{x}, t) = -\nabla \phi(\mathbf{x}, t), \tag{3.10.4}$$

where $\phi$ is commonly known as the *velocity potential function.*

Upon substituting definition (3.10.4) into equation (3.10.2) and integrating once with the use of the mass conservation equation

$$\nabla \cdot \mathbf{v} = 0, \tag{3.10.5}$$

the following results:

$$-\frac{\partial \phi}{\partial t} + \frac{1}{2} \nabla \phi \cdot \nabla \phi + \frac{p}{\rho} + \Pi(\mathbf{x}) = C(t). \tag{3.10.6}$$

Equation (3.10.6) is commonly known as *Bernoulli's equation* for ideal fluids. $C(t)$ is an arbitrary function of integration.

The velocity field for ideal fluids is normally determined by combining expression (3.10.4) with the mass conservation equation (3.10.5). This procedure results in Laplace's equation for the potential function $\phi$:

$$\nabla^2 \phi = 0. \tag{3.10.7}$$

Normally, equation (3.10.7) is solved for the potential function while the pressure is evaluated from Bernoulli's equation (3.10.6).

The sloshing problem involves movement of the liquid stationary free surface in response to forces applied on the vessel and the liquid. The present formulation is concerned with ideal fluids only, in which the liquid is assumed ideal and the vapor static. Under these conditions the fluid velocity is defined everywhere by the velocity potential function, $\phi$. In order for a unique solution for the velocity potential to exist, equation (3.10.7) must be solved subject to the appropriate boundary conditions.

If the liquid is contained within a vessel whose walls are rigid, then the normal velocity of the liquid must be zero everywhere on the wetted part of the vessel wall, i.e.,

$$v_n = -\nabla\phi \cdot \mathbf{n}_w = -\partial\phi/\partial n = 0. \qquad (3.10.8)$$

In expression (3.10.8), $v_n$ is the fluid velocity component normal to the wall and $\mathbf{n}_w$ is the outward unit normal to the vessel wall. In formulating the sloshing problem, the stationary free surface is assumed to be displaced, in response to the applied forces, by an amount $\eta$ in the direction normal to the surface. In this case the velocity of the free surface in the normal direction can be expressed in terms of the velocity potential in the following manner:

$$\partial\phi/\partial n = -\partial\eta/\partial t. \qquad (3.10.9)$$

Equation (3.10.9) is commonly known as the *kinematic condition* for a free surface. In addition to (3.10.9), the mass of the liquid (the volume for an incompressible liquid) must remain constant throughout the motion. This requirement defines the following additional two conditions on the free surface $S$:

$$\int_S \eta \, dS = 0, \qquad \int_S \frac{\partial\phi}{\partial n} \, dS = 0. \qquad (3.10.10)$$

To fully describe the sloshing problem, the so-called *dynamic condition* (see [17]) must also be imposed on the free surface. The quite complex derivation of this condition is treated rigorously in [17] and is only outlined here.

Let the stationary equilibrium free surface $S$ be displaced, in response to the applied forces, to the positions $S'$ as shown in Figure 3.31. In this manner every point on the stationary surface $S$ is displaced in the normal direction by an amount $\eta$ to corresponding points on the deformed surface $S'$. The coordinates of the deformed surface $S'$ are defined by equation (3.2.8):

$$p_v - p = \sigma\left(\frac{1}{R_1'} + \frac{1}{R_2'}\right), \qquad (3.10.11)$$

where $R_1'$ and $R_2'$ are the principal radii of curvature for the deformed surface, and $p_v$ is the gas pressure. If the motion of the liquid during sloshing is small, as well as the surface deformation $\eta$, then it is possible to neglect the second term in Bernoulli's equation (3.10.6). Under these conditions the equation for the coordinates of the deformed surface are given by the following equation, resulting from combining equations (3.10.11) and (3.10.6):

$$\rho\frac{\partial\phi}{\partial t} + \sigma\left(\frac{1}{R_1'} + \frac{1}{R_2'}\right) - \rho\Pi = C(t). \qquad (3.10.12)$$

The vapor pressure has been absorbed in the arbitrary function of time $C$.

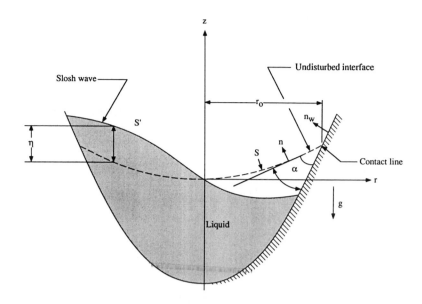

**Figure 3.31**

Following [17], the potential function for the body force $\Pi$ may be decomposed into two parts. The first, $\Pi_0$, corresponds to the potential function for the stationary free surface, the second, $\Pi'$, to the force giving rise to the sloshing motion. In this way the potential function can be written as follows:

$$\Pi = \Pi_0(\mathbf{x}) + \Pi'(\mathbf{x}, t). \tag{3.10.13a}$$

The liquid pressure can also be decomposed in a similar manner:

$$p = p_0(\mathbf{x}) + p_1(\mathbf{x}, t), \tag{3.10.13b}$$

where $p_0$ and $\Pi_0$ are the pressure and force potential function, respectively, appropriate for the stationary surface. The coordinates for this surface are given by the following equation:

$$\sigma \left( \frac{1}{R_1} + \frac{1}{R_2} \right) = \rho \Pi_0 + C. \tag{3.10.14}$$

Upon substituting the decomposition (3.10.13a) into equation (3.10.12), the following equation results for the coordinates of the new surface $S'$:

$$\rho \frac{\partial \phi}{\partial t} + \sigma \left( \frac{1}{R_1'} + \frac{1}{R_2'} \right) - \rho \Pi_0 = \rho \Pi' + C(t). \tag{3.10.15}$$

For small deformations of the free surface, the principle radii of curvature $R_1'$ and $R_2'$ may be represented functionally in the following manner:

$$R_1' = R_1'(\mathbf{x} + \delta\mathbf{x}), \qquad R_2' = R_2'(\mathbf{x} + \delta\mathbf{x}), \tag{3.10.16}$$

where $\delta\mathbf{x}$ is the incremental displacement of the surface $S(\mathbf{x})$ to $S'$. The incremental distance $\delta\mathbf{x}$ can also be decomposed into two components, one in the tangential and another in the normal direction to the surface ($e$ and $n$, respectively). In this manner the variation in the normal direction to the surface $S$ of the new radii of curvature can be expressed as follows:

$$R_1' = R_1'(n + \eta), \qquad R_2' = R_2'(n + \eta). \tag{3.10.17}$$

The stationary component of the body-force potential acting on the deformed surface can also be written as

$$\Pi_0(n + \eta). \tag{3.10.18}$$

Since the deformation of the displaced surface is assumed to be small, all functions of the new position can be expanded in a Taylor series about the stationary surface $S$ in the following way:

$$f(n + \eta) \approx f(n) + \frac{\partial f}{\partial n}\eta. \tag{3.10.19}$$

Upon substituting the above approximations for the various functions into the free-surface equation (3.10.15) and using equation (3.10.14), appropriate for the stationary equilibrium surface, we get

$$\rho\frac{\partial \phi}{\partial t} + \sigma\left[\frac{\partial}{\partial n}\left(\frac{1}{R_1}\right) + \frac{\partial}{\partial n}\left(\frac{1}{R_2}\right)\right]\eta - \rho\frac{\partial \Pi_0}{\partial n}\eta = \rho\Pi' + C(t). \tag{3.10.20}$$

Reference [17] shows that the second and third terms in equation (3.10.20) can be written as

$$\left[\frac{\partial}{\partial n}\left(\frac{1}{R_1}\right) + \frac{\partial}{\partial n}\left(\frac{1}{R_2}\right)\right]\eta = \left[\left(\frac{1}{R_1}\right)^2 + \left(\frac{1}{R_2}\right)^2 + \nabla^2\right]\eta \tag{3.10.21}$$

where $\nabla^2$ is the Laplacian operator. In a Cartesian coordinate system [e.g., $(x, y, z)$], $\nabla^2$ is defined as follows:

$$\nabla^2 = \frac{\partial^2}{\partial x^2} + \frac{\partial^2}{\partial y^2} + \frac{\partial^2}{\partial z^2}.$$

Substituting expression (3.10.21) into (3.10.20), the equation for the dynamical conditions results:

$$\rho\frac{\partial \phi}{\partial t} + \sigma\left[\left(\frac{1}{R_1}\right)^2 + \left(\frac{1}{R_2}\right)^2 + \nabla^2\right]\eta - \rho\frac{\partial \Pi_0}{\partial n}\eta = \rho\Pi' + C(t). \tag{3.10.22}$$

Recapping, the small amplitude sloshing problem is defined by equation (3.10.7) for the liquid velocity, the kinematic and dynamic conditions on the free surface given in equations (3.10.9) and (3.10.22), and the conservation of mass conditions on the free surface by (3.10.10). In addition to these equations, the various boundary conditions on the vessel wall must also be satisfied. These are condition (3.10.8) everywhere on the wetted part of the vessel and the following condition on the contact line $l$

$$\frac{\partial \eta}{\partial e} + \eta \left( \frac{k \cos \alpha - k_w}{\sin \alpha} \right) = 0, \qquad (3.10.23)$$

where $e$ is the tangent to the free surface at the contact line $l$. $k$ and $k_w$ in (3.10.23) are the inverse of the radii of curvature of the free surface and the vessel wall, respectively, at the contact line $l$ in the cross-sectional plane along the normal to the free surface. The derivation of condition (3.10.23) is complex and is given in detail in [17].

Before attempting to solve the linearized sloshing problem, it is convenient to nondimensionalize all the variables in the governing equations and boundary conditions. Once a length scale $L$ is defined, such as the vessel diameter, then the following scales for time, velocity, and force potentials may be determined using $L$, the fluid density $\rho$, and surface tension $\sigma$ in the following manner:

$$t_s = \sqrt{\rho L^2 / \sigma}, \qquad \phi_s = L^3 / t_s, \qquad \Pi_s = \sigma / (L\rho). \qquad (3.10.24)$$

Upon substituting these scales into the governing equations and boundary conditions, the nondimensional system

$$\nabla^2 \phi = 0 \qquad (3.10.25)$$

is obtained for the sloshing problem everywhere in the liquid with

$$\partial \phi / \partial n = 0, \qquad (3.10.26)$$

over the vessel surface wetted by the liquid. Notice that all the variables are now nondimensional.

On the free surface $S$,

$$\partial \phi / \partial t - A_0 \eta + \nabla^2 \eta = \Pi' + c, \qquad \partial \phi / \partial n = -\partial \eta / \partial t, \qquad (3.10.27)$$

where $A_0$ is a function of the stationary state given by

$$A_0 = \frac{\partial \Pi_0}{\partial n} - \left( \frac{1}{R_1} \right)^2 - \left( \frac{1}{R_2} \right)^2. \qquad (3.10.28)$$

Finally,

$$\int_S \eta \, dS = 0, \qquad \frac{\partial \eta}{\partial e} + \eta \left( \frac{k \cos \alpha - k_w}{\sin \alpha} \right) = 0 \qquad (3.10.29)$$

on the contact line $l$.

In the process of determining the solution for the system (3.10.25)–(3.10.29), all the possible sloshing modes can also be determined once the coordinates of the stationary free surface, as well as the external parameters, are known. These parameters include the gravity vector, the rotation rate of the vessel, the vessel geometry, and all of the properties of the fluids including the density, the surface tension, and the contact angle.

The above system is the complete system of equations governing the small amplitude sloshing problem in a general form. The equation for the velocity potential $\phi$, (3.10.25), is a linear partial differential equation, and so are all the boundary conditions. The solution for such a linear system is possible by the method of *separation of variables.* This method allows for writing the functional dependence of time in all of the variables in the following exponential form:

$$\phi(\mathbf{x}, t) = \phi(\mathbf{x})e^{i\omega t}, \quad \eta(x_1, x_2, t) = \eta(x_1, x_2)e^{i\omega t},$$

$$\Pi'(x_1, x_2, t) = \frac{1}{i\omega}\Pi'(x_1, x_2)e^{i\omega t}, \quad C(x_1, x_2, t) = \frac{q}{i\omega}e^{i\omega t}. \tag{3.10.30}$$

In this representation, $\omega$ is the sloshing mode frequency which is usually a complex number, i.e.,

$$\omega = \omega_r + i\omega_i. \tag{3.10.31}$$

Upon substituting these representations into the system (3.10.25)–(3.10.29) and eliminating $e^{i\omega t}$ from the resulting equations, the system

$$\nabla^2\phi = 0 \tag{3.10.32}$$

is obtained everywhere in the liquid with

$$\frac{\partial\phi}{\partial n} = 0 \tag{3.10.33}$$

over the wetted surface of the vessel with

$$A_0\frac{\partial\phi}{\partial n} - \nabla^2\eta\frac{\partial\phi}{\partial n} - q = \omega^2\phi + \Pi', \tag{3.10.34}$$

$$\int_S \frac{\partial\phi}{\partial n}dS = 0, \tag{3.10.35}$$

on the free surface $S$, where $A_0$ is given by expression (3.10.28), and

$$\frac{\partial}{\partial e}\left(\frac{\partial\phi}{\partial n}\right) + \left(\frac{k\cos\alpha - k_w}{\sin\alpha}\right)\frac{\partial\phi}{\partial n} = 0, \tag{3.10.36}$$

on the contact line at the vessel wall.

In the derivation of the above system, the following relationship between the velocity potential function $\phi$ and the surface displacement $\eta$ was used:

$$\eta = \frac{i}{\omega}\frac{\partial\phi}{\partial n}\Big|_S. \tag{3.10.37}$$

The constant $q$ in equation (3.10.34) is the constant of integration in Bernoulli's equation, which must be determined as part of the solution for the full system.

Using the solution form of (3.10.30) reduces the variables in the system of equations (3.10.32)–(3.10.36) to functions of the space coordinates $\mathbf{x}$ only. Also, this system forms what is commonly known as an eigenvalue problem in which $\omega$ is the eigenvalue and $\eta$ is the eigenfunction. Thus a solution, other than the trivial one, exists for this system only for the eigenvalues, which must be determined as part of the solution to the full problem.

This system can be viewed in terms of a stability mechanism in the following way. If the determined eigenvalue $\omega$ possesses a negative imaginary part, then the solution will increase with time. Such a solution is termed as unstable. On the other hand, if the imaginary part of $\omega$ is positive then the perturbation is stable. Sloshing modes for which $\omega_i$ is positive will damp, while those for which $\omega_i$ is negative will amplify and in time will destroy the stationary free surface. The unstable sloshing modes should be avoided in general. The stability of the perturbation sloshing mode waves can thus be analyzed by examining the sign of the frequency (eigenvalue) $\omega$. Once $\omega$ is determined, then the dimensional frequency can be evaluated in the following manner:

$$\omega_{\text{dim}}^2 = \omega^2/(\rho L^3). \tag{3.10.38}$$

## *3.11 SOLUTIONS FOR INFINITESIMAL SLOSHING*

The eigenvalue problem formulated in the previous section for the sloshing problem consists of a linear partial differential system whose solution is straightforward in principle. Ordinarily, the solution to such a system is a function of the geometry of the system only. However, the sloshing problem possesses some difficulties even when the geometry of the system is very simple: one of the boundary conditions, namely (3.10.34), contains a third-order derivative, while the velocity potential equation (3.10.32) is a second-order equation which is of lower order than the boundary condition.

Many solutions exist in the literature for the linearized sloshing problem, all of which are quite involved and very specific to the geometry under investigation. Reference [15] outlines an interesting technique for solving this problem. In that technique, the dynamics of linear sloshing are represented by an equivalent mechanical model composed of a set of pendulums of lengths $L_i$ and masses $M_i$. Each one of these pendulums represents the liquid fraction that participates in each sloshing mode. A rigidly attached mass in this system represents the bulk of the liquid. The advantage of such a system is to recast the problem in terms of structural dynamics where they are many robust finite element numerical codes. In this way, highly sophisticated and validated codes may be used to resolve the sloshing problem.

Reference [17] presents a number of general and useful analytical solutions to the sloshing problem, all with very simple geometries and under very special conditions for the physical boundaries and the stationary state of the system. Such conditions include the vessel geometry, the contact angle, and the initial stationary, equilibrium free surface on which the sloshing modes are superimposed.

The first simplification to be imposed is to assume that the contact angle between the liquid and the vessel wall is $\alpha = 90°$. The next simplification is to assume that the vessel is cylindrical in shape with the coordinate axes $(x, y, z)$ as shown in Figure 3.32. This assumption has practical significance, since the majority of containers and vessels used in space applications are either cylindrical or spherical in shape. It is also assumed that the vessel has a flat horizontal

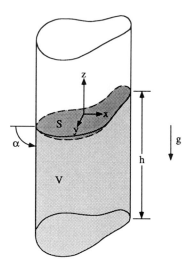

**Figure 3.32**

bottom surface. It will be assumed that the only body force acting on the liquid is the force of gravity, i.e.,

$$\Pi_0 = z\left(\frac{\rho g L^2}{\sigma}\right) = zBo, \qquad (3.11.1)$$

where $Bo$ is the Bond number.

Under these conditions, the stationary equilibrium free surface $S$ will assume a flat horizontal planar shape. To make the analysis easier still, it will be assumed that the stationary surface is located at the position $z = 0$. Placing the origin of the coordinate system at the stationary free surface will locate the vessel bottom surface at $z = -h$, where $h$ is the liquid depth. Under these assumptions the boundary value problem defined by equation (3.10.32) will take the following form:

$$\frac{\partial^2 \phi}{\partial x^2} + \frac{\partial^2 \phi}{\partial y^2} + \frac{\partial^2 \phi}{\partial z^2} = 0 \qquad (3.11.2)$$

everywhere in the liquid and in this case for all $x$, $y$, and $-h \le z \le 0$. Also, the condition on the contact line takes the form

$$\left.\frac{\partial \phi}{\partial z}\right|_{z=-h} = 0 \quad \text{and} \quad \frac{\partial \phi}{\partial e} = 0 \qquad (3.11.3)$$

with these assumptions , while the free-surface conditions become

$$\left[\frac{\partial}{\partial z}\left(Bo\phi - \frac{\partial^2 \phi}{\partial x^2} + \frac{\partial^2 \phi}{\partial y^2}\right) - \omega^2 \phi\right]_{z=0} = 0, \quad \int_S \left.\frac{\partial \phi}{\partial z}\right|_{z=0} dS = 0. \qquad (3.11.4)$$

The constant of integration $q$ has been incorporated with the eigenvalue $\omega$. This is possible because we are interested only in the sign of $\omega$ and not its magnitude.

Due to the cylindrical shape of the vessel, it is possible to split the potential function $\phi$ into horizontal and vertical components. Also, due to the planer shape of the lower vessel surface, a solution for the velocity potential function $\phi$ is possible in the following form:

$$\phi(x,y,z) = F(x,y)\cosh\lambda(z+h). \tag{3.11.5}$$

$F(x,y)$ represents the planform shape of the function $\phi$ and depends on the cross-sectional shape of the cylinder.

Substituting this functional representation for $\phi$ into the governing equation and boundary conditions results in the equation

$$\frac{\partial^2 F}{\partial x^2} + \frac{\partial^2 F}{\partial y^2} + \lambda^2 F = 0 \tag{3.11.6}$$

throughout the liquid with

$$\frac{\partial F}{\partial e} = 0, \text{ on the contact line, and } \int_S F dS = 0. \tag{3.11.7}$$

The first free-surface condition in (3.11.4) gives the following relationship between $\omega$ and $\lambda$ for this problem:

$$\omega^2 = \lambda(\lambda^2 + Bo)\tanh(\lambda h), \tag{3.11.8}$$

with $\lambda$ a positive real number.

The solution for the horizontal shape function $F(x,y)$ can also be determined by the separation of variables method applied to the equation

$$\left( \frac{\partial^2 F}{\partial x^2} + \frac{\partial^2 F}{\partial y^2} \right) = -\lambda^2 F \tag{3.11.9}$$

on the free surface subject to the following boundary condition on the contact line $l$:

$$\frac{\partial F}{\partial e} = 0, \qquad \int_S F \, dS = 0. \tag{3.11.10}$$

The final solution for the velocity potential is then the following function:

$$\phi(x,y,z,t) = F(x,y)\cosh\lambda(z+h)$$
$$\times \exp\{\pm i[\lambda(\lambda^2 + Bo)\tanh(\lambda h)]^{1/2}t\}. \tag{3.11.11}$$

Notice that once $\phi$ is determined, the free-surface displacement $\eta$ can also be evaluated using equation (3.10.37).

The general form of the sloshing wave can then be determined from a sum of all the eigenfunctions $\phi_i$ belonging to the eigenvalues $\lambda_i$. Such a representation of the solution shows that the perturbation function $\phi$ will decay with time to zero when the coefficient to the time exponential function $\omega$ is positive, i.e.,

$$\omega_h = [\lambda(\lambda^2 + Bo)\tanh\lambda\, h]^{1/2} > 0. \tag{3.11.12}$$

If the equilibrium free surface $S$ is finite, then it can be shown that the eigenvalue spectrum is infinite and discrete with a unique limiting point at infinity. Equation (3.11.12) shows that if $Bo \geq 0$, then $\omega_h$ and all other frequencies are real. The Bond number can be positive only if $g$ is positive, i.e., gravity is increasing downward. This analysis shows that *for a vessel with a cylindrical geometry, all infinitesimal sloshing waves on a flat free interface will damp in the presence of terrestrial gravity or even in zero g*. On the other hand, if the Bond number is negative ($Bo \leq 0$), then $\omega_h$ may be a complex number and the sloshing waves could be unstable.

The critical conditions are normally defined as those for which $\omega_h = 0$, representing the boundary between the stable and unstable sloshing modes. The critical value for the Bond number is similarly defined as that value of $Bo$ for which $\omega_h = 0$, which may be calculated from equation (3.11.12). The solution is

$$Bo_c = -\min_i\{\lambda_i^2\}, \tag{3.11.13}$$

where $\lambda_i$ is an eigenvalue of the solution to the shape function $F$.

The simplest cross-sectional geometry, for which this solution method can be applied, is a vessel of circular cross-sectional form. Thus, assuming a cylindrical vessel with circular cross section and radius $r_0$, the equation for the planform function $F(r, \theta)$ [equation (3.11.9)] can be written in terms of the cylindrical coordinates $(r, \theta)$:

$$\frac{1}{r}\frac{\partial}{\partial r}\left(r\frac{\partial F}{\partial r}\right) + \frac{1}{r^2}\frac{\partial^2 F}{\partial\theta^2} = -\lambda^2 F.$$

The planform shape function $F$ can be evaluated using the separation of variables method with the proper boundary conditions. This is accomplished by letting $F$ be represented by

$$F(r, \theta) = R(r)\Theta(\theta). \tag{3.11.14}$$

The solution for the shape function $F$, with the representation of $F$ given by (3.11.14), takes the following form (see [17]):

$$F(r, \theta) = J_0(k_{0n}r)$$

$$+ \sum_{m=1}^{\infty}\sum_{n=1}^{\infty}\{J_m(k_{mn}r)\cos m\theta + J_m(k_{mn}r)\sin m\theta\}.$$

$J_m(k)$ is the Bessel function of order $m$ and $k_{mn}$ is its $n$th positive zero. The eigenvalues are simple for $m = 0$ and double for $m > 0$. Thus the following equality should hold in terms of $\lambda$:

$$\lambda_{mn} = k_{mn}^2; \quad m = 0, 1, 2, \ldots; \quad n = 1, 2, \ldots. \tag{3.11.15}$$

The critical value of the Bond number for this problem can be calculated using equation (3.11.15) in the following form:

$$Bo_c = -k_{11}^2 = -(1.8411)^2 = -3.3896. \tag{3.11.16}$$

For example, let us consider the effect of varying the gravity field on the first (minimum) oscillation frequency of a liquid in a circular cylindrical vessel. The above derivation shows that the square of the dimensional frequency is given by

$$\omega_{11}^2 = \left[ \frac{\sigma}{\rho r_0^3}(k_{11})^3 + \frac{g}{r_0}k_{11} \right] \tanh(k_{11}h). \qquad (3.11.17)$$

Assume that the radius of the cylindrical vessel is $r_0 = 100$ cm and that the vessel is filled with water to a height $H = hr_0 = 100$ cm. In terrestrial conditions (i.e., for $g = 981$ cm/s$^2$), the second term in expression (3.11.17) is much larger than the first, resulting in a value for $\omega$ given by $\omega_{11}^2 \approx 18$/s$^2$. Under microgravity conditions (i.e., for $g = 981 \times 10^{-6}$ cm/s$^2$), the main contribution to the square of the frequency comes from the first term containing the surface tension $\sigma$. In this case $\omega_{11}^2 \approx 0.46 \times 10^{-3}$/s$^2$. This simple example shows that the minimum frequency is considerably lower for this geometry under microgravity conditions.

It is worth noting the role of the oscillations corresponding to the value $m = 1$. In this case a net variable pressure appears on the vessel in the horizontal direction.

Returning to the arbitrary cross-section cylindrical case, it can be seen from equation (3.11.12) that the square of the natural frequency $\omega$, $\omega^2$, is linearly dependent on the Bond number $Bo$. Reference [17] shows that such linear dependence or near linearity holds for different vessel shapes and for different contact angles, $\alpha$.

For liquids of shallow depth (i.e., for cases in which $h$ is small), the following formula for the frequency of oscillation holds:

$$\omega_d^2 \approx \lambda_h(\lambda_h + Bo)h \quad \text{for} \quad \lambda_h h^2 \ll 1. \qquad (3.11.18)$$

For very deep liquid layers, the following holds:

$$\omega_d^2 \approx \sqrt{\lambda_h}(\lambda_h + Bo) \quad \text{for} \quad \lambda_h h^2 \gg 1. \qquad (3.11.19)$$

Expression (3.11.19) shows that the natural frequencies are independent of the liquid layer depth, $h$, and are the same for an infinitely deep vessel. This is physically justified since the oscillations are created in the near-surface layer and dampen exponentially with depth.

For a vessel of infinite depth (i.e., $h \to \infty$), the normal oscillations are derived in a very similar manner with minor adjustments. For this case the dimensional natural frequency of oscillation can be written in terms of the dimensional parameters of the problem in the following manner:

$$\omega_{\text{dim}} = \sqrt{\sigma \omega^2/\rho l^3} = \left[ \lambda_h \left( \frac{\sigma}{\rho L^3}\lambda_h^2 + \frac{g}{L} \right) \tanh \left( \lambda_h \frac{h_{\text{dim}}}{L} \right) \right]^{1/2}. \qquad (3.11.20)$$

This formula shows that $\omega_{\text{dim}}$ increases with $\sigma$, which is expected. For $g > 0$ and $\sigma \to 0$, the characteristic frequency of oscillation in a heavy liquid in a cylindrical vessel results for the limiting case.

Another problem with a simple cross-sectional geometry which can be resolved with this technique is that for a square cross section. If the cylinder cross section is in the form of an infinite channel defined by $|x| \le L$ and $-\infty < y < \infty$, and if $L$ is taken as the length scale, a simple solution for the planform shape function $F(x,y)$ can be formulated (see [17]). Again, using the method of separation of variables in which the function $F$ is represented by

$$F(x,y) = X(x)Y(y), \tag{3.11.21}$$

the solution for the planform shape function takes the following form:

$$F(x,y) = \sum_{n=0}^{\infty} \sum_{m(\text{even})=0}^{\infty} \cos\left(\frac{m\pi x}{2}\right) \cos ny$$

$$+ \sum_{n=0}^{\infty} \sum_{m(\text{odd})=1}^{\infty} \sin\left(\frac{m\pi x}{2}\right) \cos ny. \tag{3.11.22}$$

The corresponding eigenvalues for that solution are given by the following expression:

$$\lambda_{mn} = m^2\pi^2/4 + n^2. \tag{3.11.23}$$

The eigenvalue spectrum in this case is continuous since only one of the two indices denoting the eigenvalue is discrete. For each value of $\lambda > 0$, there is an eigenvalue of finite multiplicity. Using expression (3.11.13), it may be seen that for $Bo < 0$, the equilibrium state is always unstable, i.e., $Bo_c = 0$ for this geometry.

# Chapter 4

# Free Convection

$F$ ree convection may be initiated in fluids by buoyancy forces due to the action of a body force, such as gravity, whenever the material density is not homogeneous. Depending on whether a fluid particle finds itself in heavier or lighter surroundings, it tends to either rise or fall. The conservation of mass requirement will move fresh fluid into the space vacated by the departing fluid, thus bringing about convective motion. The two essential ingredients are the spatial variation of the fluid density and the existence of a body force. If either is absent, buoyancy driven convection cannot take place.

There are many causes for density variance in fluids. In this chapter we will consider inhomogeneities due to (1) temperature variations (temperature gradients), e.g., homogeneous hot and cold fluid, (2) uneven solute distribution (solute concentration gradients), e.g., salt or sugar in water, and (3) both (double diffusive convection).

In terrestrial physical processes, on the surface of the Earth or in its interior, the dominant body force is gravity. Thus, whenever density gradients exist in a terrestrial fluid system, it is always possible to initiate buoyancy-driven convection within the fluid.

Spatial thermal stratification exists in almost all processes consisting of a phase change, due to the latent heats involved. Examples include condensation and evaporation, and solidification and melting. The temperature variation can trigger buoyancy-driven convection. The technological applications involving such processes include materials processing, refrigeration cycles, boiling and combustion. There are numerous other fluid processes that do not involve phase change and yet produce buoyancy-driven convection. Examples include atmospheric circulations, oceanic currents, and convective cooling processes in engineering applications.

Buoyancy-driven convection may be delayed or inhibited when gravity is reduced, or is absent. In this chapter we will identify these effects quantitatively, and develop good estimates both for the onset and the character of free convection in a low-gravity environment.

## 4.1 *THE BOUSSINESQ APPROXIMATION*

The onset and development of convection in fluids can be adequately studied using the complete Navier-Stokes, energy, and related equations. This is normally a very cumbersome and difficult task requiring a major computational fluid dynamics (CFD) effort. Unfortunately, it is also still in the basic research phase and remains beyond most engineering design practices and capabilities. A simpler technique, commonly used for estimating general criteria for the onset of convection, employs hydrodynamic stability analysis. This is more convenient to use than a detailed CFD analysis but, of course, is restricted in the amount of information it provides. However, it is of robust nature and generally supplies sufficient engineering design criteria and estimates to make it quite useful.

Hydrodynamic stability analysis relies on the common observation in fluid dynamics that lighter fluid rises while heavier fluid falls. We also know that a body of homogeneous fluid at rest is in a state of static equilibrium if the weight of a fluid element at every point is exactly balanced by the pressure exerted on it by neighboring elements. This balance continues to hold true even when the element is displaced to another position of rest. Thus, when the density of the fluid $\rho$ varies from one height to another, the pressure $p$ exerted on the fluid element will vary according to the hydrostatic balance equation

$$\nabla p_h = \mathbf{g}\rho_r. \tag{4.1.1}$$

This equation may be integrated for the simplest case to yield

$$p_h = p_0 - g \int_0^z \rho \, dz. \tag{4.1.2}$$

The gravity vector $\mathbf{g}$ is assumed parallel to the vertical coordinate $z$. The fluid element will respond to the imbalance in the pressure expressed by (4.1.2).

Equation (4.1.2) shows that the fluid is in a state of static equilibrium only when the density as well as the pressure are constant in every horizontal plane. This equilibrium stratification state is *stable* when the heavier fluid lies below, since tilting of the density surface will produce a restoring force. When the heavier fluid lies above lighter fluid, the equilibrium state is *unstable* and small displacements of density surfaces from the horizontal will increase and subsequently lead to convective motion, as shown in Figure 4.1. Free convection or buoyancy-driven

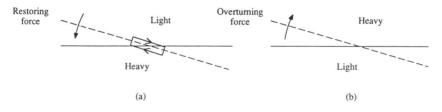

**Figure 4.1**  A sketch of the meaning of thermal instability: (a) stable configuration, (b) unstable configuration.

convection can be established in an environment in which the fluid density strati-fication is unstable. Equation (4.1.2) also reveals that the hydrostatic pressure $p_h$ is zero when the value of the gravity $g$ is also zero.

Buoyancy-driven convection is established in response to the effects of the buoyancy force resulting from minute differences in the fluid density. The origin of the buoyancy force is succinctly demonstrated by [30] to arise from the balance of forces in the equations of motion. This balance of forces, other than the viscous and inertia forces in the equations of motion for fluids, is given by the following two terms:

$$\rho \mathbf{g} - \nabla p.$$

Adding and subtracting $\nabla p_h$, the force balance becomes

$$\rho \mathbf{g} - \nabla p = \rho \mathbf{g} - \nabla p_h - \nabla (p - p_h). \tag{4.1.3}$$

This can be rewritten in the following form after using equation (4.1.1):

$$\rho \mathbf{g} - \nabla p = \mathbf{g}(\rho - \rho_r) - \nabla (p - p_h) = \mathbf{B} - \nabla p_m, \tag{4.1.4}$$

where $\mathbf{B}$ here is the buoyancy force defined by

$$\mathbf{B} = \mathbf{g}(\rho - \rho_r), \tag{4.1.5}$$

$\rho_r$ is a local reference density, and $p_m = (p - p_h)$. Equation (4.1.4) shows that the buoyancy force is due only to differences in the density and not to the absolute value of the density itself.

The fact that buoyancy driven flows are principally due to density differences rather than the absolute value of the density affords another simplification to the equations of motion in the analysis of free convection. Consider the momentum equation for an incompressible fluid in which gravity is the only body force

$$\rho \frac{D\mathbf{v}}{Dt} = -\nabla p + \rho \mathbf{g} + \mu \nabla^2 \mathbf{v}. \tag{4.1.6}$$

Upon expanding $p$ and $\rho$ about their reference hydrostatic equilibrium values, $p_h$ and $\rho_r$,

$$p_m = p - p_h, \qquad \Delta \rho = \rho - \rho_r \tag{4.1.7}$$

and substituting them into the equations of motion, (4.1.6), the following results:

$$\left(1 + \frac{\Delta \rho}{\rho_r}\right) \frac{D\mathbf{v}}{Dt} = -\frac{1}{\rho_r} \nabla p_m + \frac{\Delta \rho}{\rho_r} \mathbf{g} + \frac{\mu}{\rho_r} \nabla^2 \mathbf{v}. \tag{4.1.8}$$

In (4.1.8), the ratio $\Delta \rho / \rho_r$ appears twice, in the inertia term, and in the buoy-ancy term, which is the second term on the right-hand side. When the value of $\Delta \rho / \rho_r$ is small, it produces only a small correction to the inertia term compared to the fluid density $\rho_r$, but it is of primary importance in the buoyancy term. Equation (4.1.8) shows that for small values of $\Delta\rho$, it is reasonable to neglect the ratio $\Delta\rho/\rho_r$ in the inertia term without affecting much the fluid motion. This approximation, which is attributed to J. Boussinesq [26], is commonly known as the *Boussinesq approximation*. Reference [32] gives a detailed account on the first use of this approximation by A. Oberbeck, who derived it in a more formal manner in [39] and used it at a much earlier date than Boussinesq.

In simple terms, the Boussinesq approximation allows for neglecting the variations of the density in so far as it affects the inertia terms, but retains them in the buoyancy terms, where they occur in the body force combination $\mathbf{B} = \mathbf{g}\Delta\rho/\rho_r$. The Boussinesq approximation lies at the foundation of buoyancy-driven convection analysis and has been applied extensively with considerable success.

The advantage in using the above approximation lies in its ability to analyze the effects of variable density on the motion of fluids without resorting to the full set of compressible flow equations. The incompressible flow equations are thus sufficient in this case for resolving free-convection flows when used with this approximation, and lead to great reduction in computational effort. It can also be shown that the fluctuating density variations due to local pressure changes are negligible [47]. This condition implies that fluid flows with density stratification can be effectively treated as incompressible without having to account for sound and shock waves.

The starting point for buoyancy-driven convection analysis is the conservation of mass, momentum, and energy equations for incompressible fluid:

$$\boldsymbol{\nabla} \cdot \mathbf{v} = 0, \tag{4.1.9}$$

$$\rho\left(\frac{\partial \mathbf{v}}{\partial t} + \mathbf{v} \cdot \boldsymbol{\nabla}\mathbf{v}\right) = -\boldsymbol{\nabla}p + \rho\mathbf{g} + \mu\nabla^2\mathbf{v}, \tag{4.1.10}$$

$$\rho c_p\left(\frac{\partial T}{\partial t} + \mathbf{v} \cdot \boldsymbol{\nabla}T\right) = k\nabla^2 T + \Phi + \beta T\frac{Dp}{Dt}. \tag{4.1.11}$$

The effectiveness of the analysis is better demonstrated by assuming only thermally stratified fluid for the time being. In this case the density variation in the fluid can be written as a function of the temperature alone in the following manner:

$$\rho = \rho_r[1 - \beta(T - T_r)], \tag{4.1.12}$$

where $\beta$ is the coefficient of thermal expansion for the fluid defined by

$$\beta = -\frac{1}{\rho}\left(\frac{\partial \rho}{\partial T}\right)_p. \tag{4.1.13}$$

Reference [30] shows that the mechanical energy dissipation term, the fourth term in the energy equation (4.1.11), is also small compared with both the inertia and the buoyancy terms under the Boussinesq approximation. Substituting the Boussinesq approximation, together with the expression for the density variation

(4.1.12), into the momentum equation (4.1.10), and subtracting the hydrostatic balance, the following equations result:

$$\nabla \cdot \mathbf{v} = 0, \tag{4.1.14}$$

$$\frac{\partial \mathbf{v}}{\partial t} + \mathbf{v} \cdot \nabla \mathbf{v} = -\frac{1}{\rho_r}\nabla p_m - \mathbf{g}\beta(T - T_r) + \frac{\mu}{\rho_r}\nabla^2 \mathbf{v}, \tag{4.1.15}$$

$$\rho c_p \left( \frac{\partial T}{\partial t} + \mathbf{v} \cdot \nabla T \right) = k\nabla^2 T. \tag{4.1.16}$$

Equations (4.1.14)–(4.1.16) are called the *Boussinesq equations* for free convection. These are the equations for incompressible flow that account for the density variation in fluids. In this case the expression for the density change, equation (4.1.12), replaces the equation of state for compressible flow.

## 4.2 BUOYANCY-DRIVEN CONVECTION IN HORIZONTAL LAYERS: BENARD CONVECTION

The buoyancy-driven convection equations (4.1.14)–(4.1.16) are nonlinear partial differential equations that are quite difficult to solve. Some simple yet physically meaningful solutions can be obtained for this set under very restricted conditions. However, under general constraints the only possible solution for this set is through numerical approximations. Both approaches for the solution of this set of equations will be discussed in this chapter. We begin with the *stability analysis method.*

One of the simplest examples for which the stability analysis technique can be clearly demonstrated for buoyancy-driven convection is the case in which the fluid consists of a horizontal layer with uniform depth $H$. To initiate convection, the fluid layer must possess a density gradient in the same coordinate direction as the gravity force. It will be assumed that such density variation is due to temperature stratification alone. The gravity body force is assumed to act only in the vertical direction, thus the fluid layer possesses a temperature gradient in the vertical direction $z$. To make the analysis easier without compromising the physics, the fluid layer is assumed to extend to infinity in both horizontal directions, i.e., in both of the $x$- and $y$-directions. With this assumption, the effects of any vertical walls on the interior motion of the fluid are neglected. The governing equations for this flow are the Boussinesq equations written in a Cartesian coordinate system, as shown in Figure 4.2,

$$\frac{\partial u}{\partial x} + \frac{\partial v}{\partial y} + \frac{\partial w}{\partial z} = 0, \tag{4.2.1}$$

$$\frac{\partial u}{\partial t} + u\frac{\partial u}{\partial x} + v\frac{\partial u}{\partial y} + w\frac{\partial u}{\partial z} = -\frac{1}{\rho_r}\frac{\partial p}{\partial x} + \frac{\mu}{\rho_r}\nabla^2 u, \tag{4.2.2}$$

$$\frac{\partial v}{\partial t} + u\frac{\partial v}{\partial x} + v\frac{\partial v}{\partial y} + w\frac{\partial v}{\partial z} = -\frac{1}{\rho_r}\frac{\partial p}{\partial y} + \frac{\mu}{\rho_r}\nabla^2 v, \tag{4.2.3}$$

$$\frac{\partial w}{\partial t} + u\frac{\partial w}{\partial x} + v\frac{\partial w}{\partial y} + w\frac{\partial w}{\partial z} = -\frac{1}{\rho_r}\frac{\partial p}{\partial z} + \frac{\mu}{\rho_r}\nabla^2 w + g\beta(T - T_r), \tag{4.2.4}$$

$$\frac{\partial T}{\partial t} + u\frac{\partial T}{\partial x} + v\frac{\partial T}{\partial y} + w\frac{\partial T}{\partial z} = \frac{k}{\rho_r c_p}\nabla^2 T, \qquad (4.2.5)$$

where

$$\nabla^2 = \left(\frac{\partial^2}{\partial x^2} + \frac{\partial}{\partial y^2} + \frac{\partial^2}{\partial z^2}\right).$$

$u$, $v$, $w$, and $T$ are the Cartesian components of the fluid velocity vector, and the temperature, respectively. $k$, $\mu$, and $c_p$ are the coefficients of thermal conduction, viscosity, and specific heat of the fluid, respectively.

The analysis for buoyancy-driven convection is initiated with a uniformly heated quiescent fluid layer within which the velocity is zero everywhere. When there is no flow in the fluid layer, heat is transported by conduction alone. The governing equations for the layer are obtained by allowing the velocity components $u$, $v$, and $w$ to be zero in equations (4.2.1)–(4.2.5). These considerations result in only the heat conduction equation remaining in the equation set, i.e.,

$$\frac{\partial T}{\partial t} = \frac{k}{\rho_r c_p}\left(\frac{\partial^2 T}{\partial x^2} + \frac{\partial^2 T}{\partial y^2} + \frac{\partial^2 T}{\partial z^2}\right). \qquad (4.2.6)$$

If the temperatures of the bottom and top isothermal surfaces are kept at constant values along the horizontal direction given by $T_1$ and $T_2$, respectively, then equation (4.2.6) admits the following solution for the temperature distribution within the fluid:

$$T(z) = T_1 - (T_1 - T_2)\frac{z}{H}. \qquad (4.2.7)$$

Note that according to the definition of thermal stratification adopted here, the fluid layer is stably stratified when $dT/dz > 0$, and it is unstably stratified when $dT/dz < 0$. Figure 4.2 shows the temperature distribution given by (4.2.7).

The hydrodynamic stability approach for analyzing buoyancy-driven convection in the above layer proceeds in the following fashion: It will be assumed that due to the density stratification in the fluid layer, buoyancy-driven convection would be initiated that leads to fluid motion developing within the quiescent fluid. To analyze the ensuing fluid motion, equations (4.2.1)–(4.2.5) are solved for the flow subsequent to the initiation of convection. Let us define the field variables $u$,

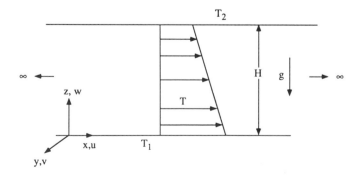

**Figure 4.2**   A sketch for the Benard convection problem.

$v$, $w$, $p$, and $T$ by assuming that they can be decomposed into an averaged value and a fluctuation in the following manner:

$$u = \overline{u}(x, y, z) + u'(x, y, z, t) = 0 + u'(x, y, z, t),$$

$$v = \overline{v}(x, y, z) + v'(x, y, z, t) = 0 + v'(x, y, z, t),$$

$$w = \overline{w}(x, y, z) + w'(x, y, z, t) = 0 + w'(x, y, z, t),$$

$$T = \overline{T}(x, y, z) + T'(x, y, z, t) = \left[ T_1 - (T_1 - T_2)\frac{z}{H} \right] + T'(x, y, z, t),$$

$$p = \overline{p}(x, y, z) + p'(x, y, z, t) = p_h + p'(x, y, z, t).$$

The variables with a bar are the averaged quantities which must satisfy the equations of motion. In this example $\overline{u} = \overline{v} = \overline{w} = 0$ since it is assumed that the fluid layer is initially quiescent, while $\overline{T}$ is given by equation (4.2.7). Substituting these decompositions for the variables into the equations of motion, and subtracting out the averaged values, the following set of equations for the fluctuations result:

$$\frac{\partial u'}{\partial x} + \frac{\partial v'}{\partial y} + \frac{\partial w'}{\partial z} = 0, \tag{4.2.8}$$

$$\frac{\partial u'}{\partial t} + u'\frac{\partial u'}{\partial x} + v'\frac{\partial u'}{\partial y} + w'\frac{\partial u'}{\partial z} = -\frac{1}{\rho_r}\frac{\partial p'}{\partial x} + \frac{\mu}{\rho_r}\nabla^2 u', \tag{4.2.9}$$

$$\frac{\partial v'}{\partial t} + u'\frac{\partial v'}{\partial x} + v'\frac{\partial v'}{\partial y} + w'\frac{\partial v'}{\partial z} = -\frac{1}{\rho_r}\frac{\partial p'}{\partial y} + \frac{\mu}{\rho_r}\nabla^2 v', \tag{4.2.10}$$

$$\frac{\partial w'}{\partial t} + u'\frac{\partial w'}{\partial x} + v'\frac{\partial w'}{\partial y} + w'\frac{\partial w'}{\partial z} = -\frac{1}{\rho_r}\frac{\partial p'}{\partial z} + \frac{\mu}{\rho_r}\nabla^2 w' + g\beta T', \tag{4.2.11}$$

$$\frac{\partial T'}{\partial t} + u'\frac{\partial T'}{\partial x} + v'\frac{\partial T'}{\partial y} + w'\frac{\partial T'}{\partial z} - \left(\frac{T_1 - T_2}{H}\right)w' = \frac{k}{\rho_r c_p}\nabla^2 T'. \tag{4.2.12}$$

Next, assuming that the magnitudes of the fluctuations are small compared to the averaged values of the variables (i.e., $\overline{f} \gg f'$), products of fluctuations are set equal to zero in the above equations. This approximation, usually called *linearization*, results in the following set of linear equations for the fluctuations:

$$\frac{\partial u'}{\partial x} + \frac{\partial v'}{\partial y} + \frac{\partial w'}{\partial z} = 0, \tag{4.2.13}$$

$$\frac{\partial u'}{\partial t} = -\frac{1}{\rho_r}\frac{\partial p'}{\partial x} + \frac{\mu}{\rho_r}\nabla^2 u', \tag{4.2.14}$$

$$\frac{\partial v'}{\partial t} = -\frac{1}{\rho_r}\frac{\partial p'}{\partial y} + \frac{\mu}{\rho_r}\nabla^2 v', \tag{4.2.15}$$

$$\frac{\partial w'}{\partial t} = -\frac{1}{\rho_r}\frac{\partial p'}{\partial z} + \frac{\mu}{\rho_r}\nabla^2 w' + g\beta T', \tag{4.2.16}$$

$$\frac{\partial T'}{\partial t} - \left(\frac{T_1 - T_2}{H}\right)w' = \frac{k}{\rho_r c_p}\nabla^2 T'. \tag{4.2.17}$$

The above equations may be combined with some algebra [27] to result in the following two for the fluctuation vertical velocity $w'$ and temperature $T'$:

$$\left[\frac{\partial}{\partial t} - \nu\nabla^2\right]\nabla^2 w' = g\beta\left(\frac{\partial^2}{\partial x^2} + \frac{\partial^2}{\partial y^2}\right)T', \qquad (4.2.18)$$

$$\frac{\partial T'}{\partial t} - \left(\frac{T_1 - T_2}{H}\right)w' = \frac{k}{\rho_r c_p}\nabla^2 T'. \qquad (4.2.19)$$

$\nu$ is the kinematic viscosity defined by $\nu = \mu/\rho_r$. Equations (4.2.18) and (4.2.19) are coupled linear partial differential equations that can be solved for the fluctuation velocity $w'$ and temperature $T'$.

Any solution to equations (4.2.18) and (4.2.19) must satisfy the appropriate initial and boundary conditions. Due to the fact that the domain in the horizontal direction is infinite in extent, the so-called *periodic boundary conditions* can be imposed on both the velocity and temperature in that direction. Since the depth of the fluid layer is finite, the boundary conditions appropriate to the type of boundaries in the $z$-direction must be applied. Two types of boundaries will be used here. The *no-slip boundary conditions* will be used when the fluid layer is bounded by two rigid walls, while the *stress-free boundary conditions* will be used when any one of the boundaries is free. The stress-free condition requires that the tangential stress on either side of the surface is the same. Since in most applications interest is confined to convection in a liquid layer surrounded by gas, the tangential stress in the surrounding gas is assumed to be zero. This assumption is based on the large difference between the viscosities of the liquid and the gas. Specifically, the following two types of boundary conditions on the fluctuating functions are imposed:

(a) the no-slip, no temperature fluctuations boundary conditions on both the upper and lower rigid surfaces, requiring

$$u' = v' = w' = T' = 0, \qquad z = 0, \; H \qquad (4.2.20)$$

(b) or, if the surfaces are free:

$$w' = \frac{\partial u'}{\partial z} = \frac{\partial v'}{\partial z} = T' = 0. \qquad (4.2.21)$$

Using the mass conservation equation (4.2.13), condition (4.2.21) can be rewritten as follows:

$$w' = \frac{\partial^2 w'}{\partial z^2} = T' = 0. \qquad (4.2.22)$$

Since equations (4.2.18) and (4.2.19) are linear partial differential equations with constant coefficients and with linear boundary conditions, they admit solutions through the method of separation of variables. Due to the fact that the derivatives with respect to time are of first order, the temporal functional dependence of the fluctuations can be represented as $\{e^{\omega^* t}\}$. Thus the velocity and temperature fluctuations may be written as

$$w' = W(z)F(x,y)e^{\omega^* t}, \qquad (4.2.23)$$

$$T' = \theta(z)F(x,y)e^{\omega^* t}, \qquad (4.2.24)$$

where $\omega^*$ is, in general, a complex number, i.e.,

$$\omega^* = \omega_r^* + i\omega_i^*.$$

Substituting these into equations (4.2.18) and (4.2.19) and noting that

$$\frac{\partial}{\partial t} = \omega^* e^{\omega^* t},$$

the following linear ordinary differential equations for the fluctuations amplitude functions, $W$ and $\theta$, result:

$$\left[\omega^* - \nu\left(\frac{d^2}{dz^2} - \alpha^2\right)\right]\left(\frac{d^2}{dz^2} - \alpha^2\right)FW = -g\beta\alpha^2 F\theta, \qquad (4.2.25)$$

$$\left[\omega^* - \frac{k}{\rho_r c_p}\left(\frac{d^2}{dz^2} - \alpha^2\right)\right]F\theta = \left(\frac{T_1 - T_2}{H}\right)FW. \qquad (4.2.26)$$

$F(x,y)$ is the *planform shape function* for the fluctuations, and it results from the separation of variables method, which is a solution to the following equation:

$$\frac{\partial^2 F}{\partial x^2} + \frac{\partial^2 F}{\partial y^2} = -\alpha^2 F, \qquad (4.2.27)$$

where $\alpha$ is an eigenvalue resulting from the separation of variables and also represents a wave number for the planform shape function.

At this point it is convenient to nondimensionalize the variables in equations (4.2.25) and (4.2.26) in the following manner: The fluid layer depth $H$ will be used for a length scale. Time, velocity, and temperature scales can be defined using the coefficient of thermal diffusion $\kappa$, the temperature difference between the boundaries ($T_1 - T_2$), and $H$ in the following manner:

$$t_s = H^2/\kappa, \quad w_s = \kappa/H, \quad T_s = T_1 - T_2, \quad \kappa = k/\rho_r c_p.$$

With these scales, a nondimensional wave number $a$ and a frequency $\omega$ are defined as follows:

$$a = \alpha H, \quad \omega = \omega^* H^2/\kappa.$$

The equations for the fluctuation functions in terms of the nondimensional variables become

$$\left[\frac{\omega}{Pr} - \left(\frac{d^2}{dz^2} - a^2\right)\right]\left(\frac{d^2}{dz^2} - a^2\right)W = -a^2 R\theta, \qquad (4.2.28)$$

$$\left[\omega - \left(\frac{d^2}{dz^2} - a^2\right)\right]\theta = W. \qquad (4.2.29)$$

The temperature and velocity amplitude functions $\theta$ and $W$ in equations (4.2.28) and (4.2.29) are now nondimensional. $R$ and $Pr$ in equations (4.2.28) and (4.2.29), which are the Rayleigh number and the Prandtl number, respectively, are nondimensional numbers resulting from the above process. They are defined in the following manner:

$$R = \frac{gH^3\beta(T_1 - T_2)}{\nu\kappa}, \qquad Pr = \frac{\nu}{\kappa}.$$

The governing equations for the fluctuation amplitudes are linear ordinary differential equations for the perturbation temperature $\theta$ and the perturbation normal velocity $W$. A solution to the above equations can be found once the boundary conditions are also defined in terms of $W$ and $\theta$; these take the following nondimensional forms:

(a)  For the rigid-rigid boundaries case,

$$\theta = W = dW/dz = 0 \quad \text{at} \quad z = 0, 1. \tag{4.2.30}$$

(b)  For the free-free boundaries case,

$$\theta = W = d^2W/dz^2 = 0 \quad \text{at} \quad z = 0, 1. \tag{4.2.31}$$

If one boundary is rigid and the other is free, then the appropriate condition must be used at each boundary. Regardless of whether the bounding walls are rigid or free, either condition is called homogeneous. The theory of differential equations states that for linear homogeneous differential equations with homogenous boundary conditions, only the trivial solution ($W \equiv \theta \equiv 0$) exists, except at specific values for the parameters. Basically, the linear stability problem defined above is an eigenvalue problem in which any one of the parameters is the eigenvalue and the perturbation functions are the eigenfunctions. The parameters in these equations are $\omega$, $Pr$, $a^2$, and $R$, for which the following relationship exists between the eigenvalues:

$$R = f(a^2, \omega, Pr). \tag{4.2.32}$$

The time-dependent representation for the fluctuations in the form of $exp(\omega t)$ implies that the fluctuations will amplify and damp with time depending on the sign of $\omega$. This will occur regardless of the type of solutions for either the amplitude function or the planform shape. This is true because both are solutions to Laplace's equation, and as such are bounded, which means they depend only on the boundary conditions. Due to the fact that periodic boundary conditions were imposed on the planform function $F$, the solution for $F$ can also be represented by periodic functions in the form of a Fourier series. In this case $\alpha$ in equation (4.2.27) can be viewed as a wave number. Thus it is possible to represent the horizontal and temporal dependence of the fluctuations as

$$\theta, W \sim \exp[i(a_x x + a_y y) + \omega t], \quad a_x^2 + a_y^2 = a^2.$$

Such representation defines a wave, with $a$ as the wave number vector propagating in the horizontal direction with speed $\omega$. It is clear that this wave is stationary when $\omega = 0$, and in this case the fluctuations will represent a stationary spatial pattern, such as a cellular pattern. Also, note that because $\omega$ is a complex number, if it is determined that $\omega_r > 0$, the fluctuations will grow exponentially with time, these solutions are called *unstable*. However, if $\omega_r < 0$, then the fluctuations will decay with time and the solutions are called *stable*. Thus the value of $\omega_r = 0$ represents the boundary between stable and unstable fluctuations and provides valuable qualitative information. The locus of the curve in the parameter space for which $\omega_r = 0$ is called the *neutral stability curve*. Such curves provide the *stability maps* that are sought after in most stability analyses.

Unstable solutions for the fluctuations physically represent a departure from the purely conductive stationary state for the present problem. They can be manifested only in the establishment of a convective motion in the fluid. Thus, if the solution for the fluctuation functions indicates instability, it means that the fluid layer is undergoing convective motion. On the other hand, a stable solution will mean that the fluid layer will remain in the stationary conductive state. For the present analysis, it will be assumed here that buoyancy-driven convection will take place whenever the stationary state is unstable.

The analysis for the convection problem described by equations (4.2.28) and (4.2.29) is made easy by setting $\omega \equiv 0$; this is also the method used in constructing the *marginal stability curves*. We obtain

$$\left[ \frac{d^2}{dz^2} - a^2 \right] \theta = -W, \tag{4.2.33}$$

$$\left[ \frac{d^2}{dz^2} - a^2 \right]^2 W = a^2 R \theta. \tag{4.2.34}$$

Note that setting $\omega_r = 0$ does not normally imply that $\omega_i = 0$ as well. However, for the thermal instability problem analyzed here, the *principle of exchange of stability* holds, which leads to the fact that $\omega_i = 0$ whenever $\omega_r = 0$. The proof of that principle is given in [27].

Equations (4.2.33) and (4.2.34) may be combined into a single equation for either one of the two variables $\theta$ or $W$. The equation in terms of $\theta$, for instance, takes the following form:

$$\left[ \frac{d^2}{dz^2} - a^2 \right]^3 \theta = -a^2 R \theta. \tag{4.2.35}$$

At this point the type of boundary conditions imposed must be decided upon in order to determine the solution to equation (4.2.35). The most straightforward solution is for the case when both the upper and lower boundaries are free, the so-called *free-free problem*. These conditions can be written in terms of $\theta$ alone, in the following manner, with the aid of equation (4.2.33):

$$\theta(0) = \theta(1) = \left[ \frac{d^2}{dz^2} - a^2 \right] \theta(0) = \left[ \frac{d^2}{dz^2} - a^2 \right] \theta(1) = 0. \tag{4.2.36}$$

A solution for $\theta$ that also satisfies the above boundary conditions is the following (see [27]):

$$\theta = A_\theta \, \sin(n\pi z) \quad \text{for} \ \ n = 1, 2, 3, \ldots, \tag{4.2.37}$$

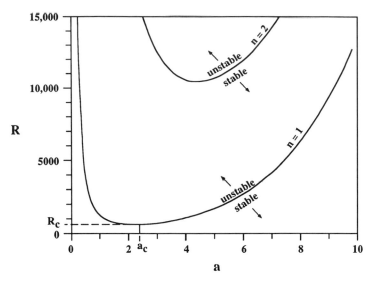

**Figure 4.3**    Marginal stability curves for the free-free case, for $n = 1$ and 2.

where $A_\theta$ is a constant of integration that must be evaluated with the aid of the boundary conditions. Substituting this form of the solution into equation (4.2.35) results in the following algebraic equation:

$$(\pi^2 n^2 + a^2)^3 = a^2 R. \tag{4.2.38}$$

The Rayleigh number $R$ can now be evaluated from (4.2.38) as a function of the wave number $a$:

$$R = \frac{(\pi^2 n^2 + a^2)^3}{a^2}. \tag{4.2.39}$$

Equation (4.2.39) is commonly known as the *dispersion relation*, which is an equation relating the various parameters of the problem to the eigenvalue. It gives the specific functional form for the general relationship given by equation (4.2.32). Note that the Prandtl number does not appear explicitly in equation (4.2.39). It is a distinctive property of this problem that the marginal stability curve is not affected by the Prandtl number for the fluid in question. However, the full convection problem, including the fluctuation growth rate $\omega_r$, is strongly dependent on the Prandtl number.

The curves resulting from the solution to equation (4.2.39) in the $(R, a)$ parameter plane are shown in Figure 4.3 for two values of $n$. These *marginal stability* curves divide the plane into areas where parameters for which the fluctuations are stable are separated from those for which they are unstable. Specifically, the unstable fluctuations in Figure 4.3 lie in the region above the curves, while the stable fluctuations lie below the curves. This means that for Rayleigh number values less than those on the curve, all disturbances with wave number $a$ will be stable. The disturbances will become marginally stable when the Rayleigh number value is on the curve, and when the Rayleigh number exceeds the curve value, the disturbance becomes unstable.

Note that there is a specific value of the wave number $a$ for each curve shown in Figure 4.3, for which $R$ is a minimum. The point in the $(R, a)$ plane at which the minimum value of the Rayleigh number occurs is commonly known as the *critical point*. The minimum value of the Rayleigh number is consequently called the *critical Rayleigh number*, $R_c$. The critical Rayleigh number for the onset of instability is calculated by differentiating equation (4.2.39) with respect to $a$:

$$\frac{\partial R}{\partial a^2} = 3\frac{(\pi^2 + a^2)^2}{a^2} - \frac{(\pi^2 + a^2)^3}{a^4}. \tag{4.2.40}$$

Equating expression (4.2.40) to zero, the following *criticality conditions* result:

$$3a_c^2 = (\pi^2 + a_c^2) \quad \text{or} \quad a_c^2 = \pi^2/2. \tag{4.2.41}$$

The corresponding value of $R_c$ is calculated by substituting the value for $a_c$ into equation (4.2.39) to yield

$$R_c = \frac{(3\pi^2/2)^3}{\pi^2/2} = 27\pi^4/4 = 657.5. \tag{4.2.42}$$

The value for the wave number at which $R_c$ occurs is given by $a_c = \pi/\sqrt{2} = 2.2214$. This value can be translated into a wavelength $\lambda$, characterizing the onset of instability as follows:

$$\lambda_c = \frac{2\pi}{\alpha_c} = \frac{2\pi}{a_c}H = \sqrt{8}H. \tag{4.2.43}$$

Measurements with silicone oil, reported in [31], resulted in values for both $R_c$ and $\lambda_c$ within 10% of the predicted values given in (4.2.41) and (4.2.42).

The fluctuation vertical velocity component $W(z)$ can now be evaluated by substituting the solution for $\theta(z)$ into equation (4.2.34). The solution is given by

$$W(z) = -A_W(\pi^2 + a^2)\sin\pi z. \tag{4.2.44}$$

The higher modes for marginal stability, generated for $n = 2, 3, \ldots$, yield marginal stability curves at larger values of $R$. The general relations for the critical Rayleigh and wave numbers for all $n$ can be obtained similarly to yield

$$R_{c,n} = \frac{27n^4\pi^4}{4} \quad \text{and} \quad a_{c,n} = \frac{n\pi}{\sqrt{2}}. \tag{4.2.45}$$

Figure 4.3 shows the marginal stability curve for $n = 2$.

The case in which the fluid layer is bounded by rigid walls at the top and the bottom surfaces is more interesting. Such a case corresponds to a more realistic model for laboratory experiments, and there are many practical applications for it. Conceptually, the solution to this problem is similar to the free-free case, but the mathematics is slightly more involved. The equations are the same as for the free-free case, but the following boundary conditions must be used:

$$\theta(0) = \theta(1) = W(0) = W(1) = \frac{dW}{dz}(0) = \frac{dW}{dz}(1) = 0. \tag{4.2.46}$$

Because the governing equation for the temperature fluctuations, equation (4.2.35), possesses only even derivatives with respect to $z$ and also because the boundary conditions are symmetric with respect to the centerline between the two boundaries, the solution is symmetric about a plane at the midpoint location in the fluid. It is convenient in this case to translate the origin of the coordinate system to a position midway between the two bounding planes. The governing equations can be written in terms of the vertical velocity fluctuations $W$ as:

$$\left[ \frac{d^2}{dz^2} - a^2 \right]^3 W = -a^2 R W. \tag{4.2.47}$$

This equation must now be solved subject to the no-slip boundary conditions at the solid surfaces located at $z = \pm \frac{1}{2}$

$$W = \frac{dW}{dz} = \left[ \frac{d^2}{dz^2} - a^2 \right] W = 0 \quad \text{at} \quad z = \pm \frac{1}{2}. \tag{4.2.48}$$

Since equation (4.2.47) is even, and the boundary conditions are identical for both locations in $z$, the solution can be written as a sum of even and odd terms. Reference [27] shows that the lowest state will be an even solution with no nodes. The first excited state will be the odd solution with a node at $z = 0$. The general solution to equation (4.2.47) can be written as

$$W(z) = \sum_{j=1}^{6} A_j \exp(\lambda_j z). \tag{4.2.49}$$

Substituting this into equation (4.2.47), the following equation is obtained for $\lambda$:

$$(\lambda^2 - a^2)^3 = -a^2 R. \tag{4.2.50}$$

The six roots of equation (4.2.50) are as follows (see [27]):

$$\lambda_{1,2} = \pm i \lambda_0, \quad \lambda_{3,4} = \pm(\lambda_r + i\lambda_i), \quad \lambda_{5,6} = \pm(\lambda_r - i\lambda_i), \tag{4.2.51}$$

where

$$\lambda_0 = a(\tau - 1)^{\frac{1}{2}},$$

$$\lambda_r = a\left[ \frac{1}{2}\sqrt{(1 + \tau + \tau^2)} + \frac{1}{2}(1 + \tau/2) \right]^{1/2},$$

$$\lambda_i = a\left[ \frac{1}{2}\sqrt{(1 + \tau + \tau^2)} - \frac{1}{2}(1 + \tau/2) \right]^{1/2},$$

$$\tau = (R/a^4)^{1/3}.$$

The constants of integration $A_j$ are evaluated from the boundary conditions. For the even solution, applications of the boundary conditions given in (4.2.48) lead to the following transcendental equation for $a$ in terms of $R$ [27]:

$$-\lambda_0 \tan \frac{1}{2}\lambda_0 = \frac{(\lambda_r + \lambda_i\sqrt{3}) \sinh \lambda_r + (\lambda_r\sqrt{3} - \lambda_i) \sin \lambda_i}{\cosh \lambda_r + \cos \lambda_i}. \tag{4.2.52}$$

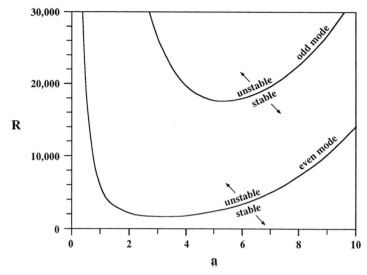

**Figure 4.4** Marginal stability curves for the rigid-rigid case, for the lowest even and odd modes.

Clearly equation (4.2.52) must be solved numerically. Figure 4.4 shows the marginal stability curves for the lowest even mode and the lowest odd mode of this solution. The critical values for the Rayleigh number and wave number can again be determined by differentiating equation (4.2.52) with respect to $a$ and equating the result to zero. This operation yields the critical values corresponding to the lowest even mode:

$$a_c = 3.117, \qquad R_c = 1707.762. \qquad (4.2.53)$$

The odd solution can also be derived in a similar manner to yield the following transcendental equation for the marginal stability curve for the lowest odd mode:

$$-\lambda_0 \cot \frac{1}{2}\lambda_0 = \frac{(\lambda_r + \lambda_i \sqrt{3}) \sinh \lambda_r - (\lambda_r \sqrt{3} - \lambda_i) \sin \lambda_i}{\cosh \lambda_r - \cos \lambda_i}. \qquad (4.2.54)$$

The curve is shown in Figure 4.4. The values for the critical wave number and the Rayleigh number of the lowest odd mode are

$$a_c = 5.365, \qquad R_c = 17610.39. \qquad (4.2.55)$$

The odd solution again has stability modes at increasingly higher values of $R$ as the mode number is increased. These higher modes represent higher wave numbers, which are indicative of shorter wavelengths, and represent disturbances with smaller scales. Also, values of $R_{c,n}$ increase with increasing $n$. Table 4.1 shows the lowest values for both $R_c$, and the corresponding value for $a_c$, for the rigid-rigid, the free-free, and the rigid-free cases.

The stability method for determining the onset of buoyancy-driven convection is extremely valuable in that it provides the regions in the physical parameter space for which the initial conduction state is stable or unstable. However, even

**Table 4.1**   Critical parameters for three types of boundaries.

| Nature of the bounding surfaces | $R_c$ | $a_c$ | $2\pi/a_c$ |
|---|---|---|---|
| Both free | 657.511 | 2.2214 | 2.828 |
| Both rigid | 1707.762 | 3.117 | 2.016 |
| One rigid and one free | 1100.65 | 2.682 | 2.342 |

though instability can be taken to infer the onset of convection, one cannot be absolutely sure that it will take place without experimental validation. Numerous tests have shown that convection does indeed take place when the lowest critical Rayleigh number is attained. Experiments have also confirmed that the wave number at the onset of convection corresponds to the critical wave numbers calculated above.

The stability theory discussed above fails to supply the character or the pattern of the convective motion. Nowhere in the analysis presented here has the actual shape for the planform function $F$ been necessary to calculate either the marginal stability curves or the critical Rayleigh numbers. This is an inherent shortcoming of the linear stability theory, and it follows from the fact that linear stability formulation results in an eigenvalue problem, where the eigenvalue must be determined in addition to the solution. This requirement reduces the number of useful boundary conditions needed for evaluating the solution and thus the solution is calculated up to an arbitrary constant.

In order to determine the form of the convective motion, either a nonlinear stability analysis must be performed or the full problem must be solved via numerical approximations. Both of these analyses are quite complicated and will not be discussed here. There exists a large body of research solely devoted to these types of analyses; see for instance [29] and the references therein.

The first experimental observations on the onset of thermal instability in liquid layers are attributable to H. Benard [24]. The thermal instability problem resulting from vertical temperature gradients is universally known as the *Benard problem*. In fact, Benard's experiments were the principal motivation for the tremendous effort in hydrodynamic stability analyses that followed, and its simplest form was presented above. Many more experimental studies on this type of instability followed with the purpose of validating the analysis discussed above. The original observations of Benard recorded the instability in the form of hexagonal cells distributed evenly throughout the surface of the fluid layer. The fluid was found to rise at the center of the cell and descend at the walls of the cell. Benard performed his experiments with an upper free surface. Later experiments by others using rigid upper and lower boundaries revealed the convection pattern to be in the form of rolls, as shown in Figure 4.5.

It appears at the present time that the form of the convection pattern originally observed by Benard was due to surface tension effects on the free surface. This type of convection, commonly known as the Marangoni instability, is principally due to surface tension gradients and will be discussed in the next section. Reference [33] gives a detailed account of the pattern formation and its development for both the buoyancy-driven and the surface tension-driven convection problems.

**Figure 4.5**   Photograph of a typical convection pattern due to the Benard instability mechanism for the rigid-rigid case [35]. *Courtesy E. L. Koschmieder.*

The easiest and the most straightforward method for determining experimentally the critical Rayleigh and wave numbers is through visual observation. However, when nontransparent walls are used, this technique is not very convenient experimentally.  Much of the quantitative evidence needed for validating the marginal stability limits has been through heat flux measurements. Stability, according to the model analyzed above, implies that the fluid layer is quiescent, while instability implies convection and hence fluid motion. The heat flux, $q$, for the stable situation in which there is no fluid motion is solely due to conduction, i.e.,

$$q = -k\frac{dT}{dz} = k\frac{T_1 - T_2}{H}\,. \tag{4.2.56}$$

On the other hand, when there is motion in the fluid layer, the heat flux is normally calculated through a heat transfer coefficient, $h$:

$$q = h(T_1 - T_2). \tag{4.2.57}$$

When the heat flux in expression (4.2.57) is nondimensionalized with respect to the conduction heat flux, a nondimensional number known as the *Nusselt number* results:

$$Nu = \frac{q}{T_1 - T_2}\frac{H}{k} = \frac{hH}{k}. \tag{4.2.58}$$

It is clear from expression (4.2.58) that when there is no motion in the fluid, $Nu = 1$ and that fluid motion causes the heat flux $q$ to increase from that lower limit.  Measuring $q$ and $T_1 - T_2$ in free convection experiments at increasingly larger values of $R$ should detect the onset of the convective motion in the fluid layer. Thus the Nusselt number is a very accurate indicator of convective transport, as its value increases above its lower limit of 1. This would be expected to correspond, at least approximately, to the marginal stability point.  This is

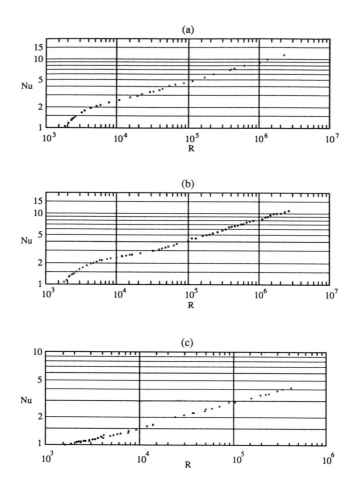

**Figure 4.6**   The variation of the Nusselt number with Rayleigh number from the experiments of [45] on Benard convection: (a) $Pr = 200$, (b) $Pr = 6.8$, (c) $Pr = 0.025$.

another method used by many experimenters to check the stability solutions derived above. Figure 4.6 shows the experimental variation of the Nusselt number as a function of the Rayleigh number for three fluids from [45].

The main difficulty in validating experimentally the analytical results derived above is that they assume an infinite horizontal fluid layer that cannot be achieved experimentally and thus any test validation is only approximate. The more practical way is to examine the influence of side walls on the critical values for both the Rayleigh and the wave numbers; unfortunately this is very complicated. The analysis for the onset of convection when side walls are present is more conveniently tackled through numerical means, as will be seen later in this chapter. This implies that there are no convenient analytical expressions such as equation (4.2.39) for predicting the onset of convection in the side walls problem.

Reference [43] examines experimentally the values of the critical Rayleigh number $R_c$ as a function of the aspect ratio for unstably stratified fluid layers

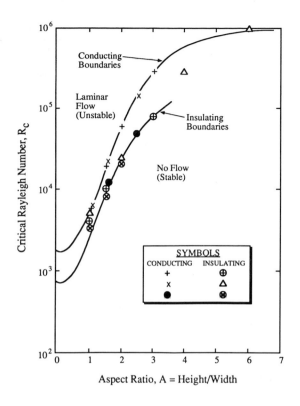

**Figure 4.7** Instability bounds as a function of the aspect ratio, $A$, from [43].

of finite extent. Figure 4.7 shows the variation of the critical Rayleigh number as a function of the cell aspect ratio for both conducting and insulating cell side walls. The *aspect ratio* is defined here as the ratio of the height to the width of the fluid layer. The figure shows that for both cases the minimum critical Rayleigh number indeed occurs for an aspect ratio of zero, and its value increases as the cell width shrinks from its infinite value. These results are comforting in the sense that for practical applications, in which the fluid layer is bounded by side walls, the critical Rayleigh number increases by several orders of magnitude.

The above analysis and discussion shows that convection can be initiated for horizontally infinite layers if they are unstably stratified and the Rayleigh number is above the lowest critical value for that configuration, e.g., if the Rayleigh number is above 657.5 for a fluid layer bounded by two free surfaces and above 1707.8 for a layer bounded by two rigid surfaces. Recall that the Rayleigh number is defined as $R = g\beta H^3 \Delta T / \nu \kappa$, where $\Delta T$ is the temperature difference between the upper and lower surfaces. This definition of $R$ shows that there are several ways of altering its numerical value. One obvious way of changing its value for low-gravity applications is by varying the value for the gravitational acceleration $g$.

For any two identical experiments in which one is performed under terrestrial conditions and the other in low-gravity, it is obvious that each will have a different value for the Rayleigh number since the values of $g$ are different. In a low-gravity environment in which the value of $g/g_0 \approx 10^{-3}$ where $g_0$ is the value of terrestrial gravity (i.e., $g_0 = 980.621$ cm/sec$^2$ = 32.1725 ft/sec$^2$), the value of the Rayleigh number is one one-thousandth its value in the terrestrial environment. This is a very obvious and powerful result that is used extensively in low-gravity design criteria and is the subject of intensive experimental effort. Thus, whenever there are two identical physical processes, one in the laboratory and another in low gravity, it is possible to predict whether convection will occur in the low-gravity environment based on the calculated value for the critical Rayleigh number and the terrestrial experimental results.

## 4.3 *THERMOCAPILLARY CONVECTION—MARANGONI CONVECTION*

Free convection may occur in a different class of processes that do not depend on the buoyancy force for their initiation. Convection can occur in response to variations in the surface tension force distribution on a liquid/gas free surface. Fluid motion occurs in these instances even when there are no density variations in the fluid. Block [25] discovered through experimental observations that convective motion can occur in a thin liquid layer with a free surface due to the action of surface tension gradients on that surface, regardless of the orientation of the surface with respect to the gravity vector. In [44], a linear stability theory was developed in which the value of gravity was set to zero to explain Block's experiments. This type of convective process is extremely relevant to low-gravity applications as gravity is not necessary for its initiation.

The surface tension $\sigma$ for pure compounds depends strongly on the temperature of the fluid. Such dependence can be written in a simple manner in the following form:

$$\sigma = C(T - T_{cr}). \tag{4.3.1}$$

This strong temperature dependence of the surface tension can induce motion in a fluid that has spatial temperature variations. Reference [36] demonstrates, with the following simple example, how convection can be initiated due to surface tension forces alone. In fact, it turns out to be possible to estimate the magnitude of the flow speed as a function of the temperature and the magnitude of the surface tension.

Reference [36] examines a thin horizontal layer of depth $H$ that is bounded laterally by two vertical side walls set at temperatures $T_c$ and $T_h$ with $T_h > T_c$. A two-dimensional configuration as shown in Figure 4.8 is assumed, with the $x$-axis parallel to the temperature gradient while the $z$-axis is normal to the liquid/gas interface located at $z = 0$. Let the distance between the side walls be $L$. Also, for the sake of illustration, the Prandtl number for the fluid is assumed to be sufficiently low so that the induced velocity will not disturb the assumed linear temperature distribution in the layer.

Due to the functional dependence of $\sigma$ on $T$, the gradient of the surface tension at the interface can be calculated in this case as

$$\partial\sigma/\partial x = (\partial\sigma/\partial T)(\partial T/\partial x). \tag{4.3.2}$$

Equation (4.3.1) shows that the value of the surface tension decreases with increasing temperature when evaluated by expression (4.3.2). A surface tension

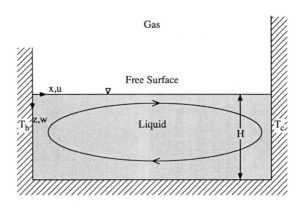

**Figure 4.8**  Sketch of the problem on surface tension effects.

distribution at the surface of the fluid layer that also satisfies equation (4.3.2) can take the following form:

$$\sigma = \sigma(T_c) + \frac{\partial \sigma}{\partial T} \frac{T_h - T_c}{L} x. \tag{4.3.3}$$

Since $T_h - T_c \neq 0$, equation (4.3.3) also represents a nontrivial tangential force acting on the fluid surface. It can be demonstrated easily that such a tangential force on the free surface of the fluid layer can lead to fluid motion in the bulk of the layer.

It is assumed that both the upper and lower surfaces of the fluid layer are perfectly flat with uniform motion only in the $x$-direction. Also, we shall only consider the flow far away from any lateral wall, i.e., the $z$-component of velocity is zero in the region of interest. Under these limitations, the $x$-momentum equation will reduce to the following form:

$$\frac{\partial p}{\partial x} = \mu \left( \frac{\partial^2 u}{\partial x^2} + \frac{\partial^2 u}{\partial z^2} \right), \tag{4.3.4}$$

where $u$ is the $x$-component of the velocity vector. In the region of flow far from either side wall, the variation of the velocity with the horizontal coordinate, $x$, is much less than its variation with the vertical coordinate. With that observation, the $x$-momentum equation will reduce to

$$\frac{\partial p}{\partial x} \approx \mu \left( \frac{\partial^2 u}{\partial z^2} \right). \tag{4.3.5}$$

The $z$-momentum equation, under the same limitations and assuming no hydrostatic force (zero gravity), will reduce to

$$\frac{\partial p}{\partial z} = 0. \tag{4.3.6}$$

Equation (4.3.6) shows that pressure can only be a function of $x$. If it is also assumed that there is no net flow across any plane parallel to the side walls (in this case there is only one large eddy across the layer), then the mass flow rate through any vertical plane must be zero, i.e.,

$$\int_0^H u\,dz = 0. \tag{4.3.7}$$

The no-slip boundary condition for the velocity must be applied at the lower wall, i.e.,

$$u = 0 \quad \text{at} \quad z = -H,$$

while at the upper free surface, a balance between the viscous stress and the surface force must exist at the liquid/gas interface, i.e.,

$$\mu\left(\frac{\partial u}{\partial z}\right)_{z=0} = \left(\frac{\partial \sigma}{\partial T}\right)\left(\frac{\partial T}{\partial x}\right). \tag{4.3.8}$$

Integrating the $x$-momentum equation (4.3.5) twice and applying the boundary conditions results in the following solution for $u$:

$$u = \frac{1}{\mu}\frac{\partial \sigma}{\partial x}(H - z) - \frac{1}{2\mu}\frac{dp}{dx}(H^2 - z^2). \tag{4.3.9}$$

The zero flux condition (4.3.7) may be used to evaluate the pressure gradient in the following manner:

$$\frac{dp}{dx} = \frac{3}{2H}\frac{\partial \sigma}{\partial x}. \tag{4.3.10}$$

Equation (4.3.10) may now be integrated across the two side walls to evaluate the pressure:

$$p = p_0 + \frac{3}{2H}[\sigma(x) - \sigma(0)], \tag{4.3.11}$$

where $p_0$ is an undetermined constant. Finally, the solution for the velocity is determined by substituting expression (4.3.10) into the solution in (4.3.9), resulting into

$$u = \frac{1}{4\mu H}\frac{d\sigma}{dT}(3z^2 - 4zH + H^2)\frac{dT}{dx}. \tag{4.3.12}$$

The liquid velocity at the free surface (i.e., at $z = 0$), is given by

$$u = \frac{H}{4\mu}\frac{d\sigma}{dT}\frac{dT}{dx}. \tag{4.3.13}$$

Expression (4.3.13) shows that the liquid velocity at the interface increases with the thickness of the layer and with the temperature gradient along the layer.

For example, for a water/air interface of a water layer 1.0-mm deep at 20°C, with $\mu = 0.001$ Ns/m$^2$ subjected to a horizontal temperature gradient of 100°K/m and with $d\sigma/dT = -0.5 \times 10^{-3}$ N/(m°K), the interface velocity is estimated at $-12.5 \times 10^{-3}$ m/s $= -12.5$ mm/s.

This rather simple and ideal example shows that a surface tension gradient at an interface can induce a substantial motion in a fluid layer. This motion is commonly known as *thermocapillary* or *Marangoni convection*. It may be difficult to observe under terrestrial conditions as it can be overwhelmed by the strong motion induced through buoyancy-driven convection. However, in a reduced-gravity environment, surface tension-induced convection could be the only convective motion present in a fluid volume. The onset of Marangoni convection has been studied effectively by hydrodynamic stability methods similar to buoyancy-driven convection:

Again, we will consider an infinite horizontal liquid layer with a finite depth $H$ in which the lower surface at $z = 0$ is rigid while the upper surface at $z = H$ is free. The liquid layer will be assumed to support a temperature gradient in the vertical direction $z$ in the form

$$T_2 = T_1 - G_T z, \qquad (4.3.14)$$

where $T_1$ and $T_2$ are the lower and upper surface temperatures, respectively, and the constant $G_T = \partial T / \partial z$ is the vertical temperature gradient, which may be defined through the heat flux $q$ as

$$q = kG_T. \qquad (4.3.15)$$

The stability analysis proceeds by perturbing the stationary conductive basic state described by equation (4.3.14) in a manner similar to the analysis employed in the previous section for buoyancy-driven convection. Assuming the perturbations to be infinitesimally small, the procedure results in a coupled set of linear partial differential equations for both the vertical velocity fluctuation $w'$ and the temperature fluctuation $T'$. These equations take the following form:

$$\left( \frac{\partial}{\partial t} - \nu \nabla^2 \right) \nabla^2 w' = 0, \qquad (4.3.16)$$

$$\left( \frac{\partial}{\partial t} - \kappa \nabla^2 \right) \nabla^2 T' = G_T w'. \qquad (4.3.17)$$

Equations (4.3.16) and (4.3.17) are similar to equations (4.2.18) and (4.2.19), except for one important difference. In the present analysis, the body force is assumed to be zero, which makes the right-hand side of equation (4.3.16) also zero. This implies that the thermocapillary stability analysis performed here is valid for zero-gravity conditions, i.e., $g/g_0 = 0$. In this analysis, the effects of surface tension force alone are being investigated without the added complexity of the buoyancy force.

Solutions to equations (4.3.16) and (4.3.17) are sought subject to the appropriate boundary conditions. If those given in (4.3.3) can be adopted as a reasonable representation for the variation of the surface tension with temperature throughout the fluid layer, then it is possible to describe the surface tension due to small variations in the temperature as

$$\sigma(T_2 + T') \approx \sigma_2(T_2) + T'(\partial \sigma / \partial T)_{T=T_2}, \qquad (4.3.18)$$

which is simply a truncated Taylor expansion for the surface tension as a function of temperature. This representation is only possible for linearized temperature fluctuations. The heat flux at the free surface can be similarly written as

$$q(T_2 + T') \approx q_2(T_2) + T'(\partial q / \partial T)_{T=T_2}, \qquad (4.3.19)$$

where $(\partial q/\partial T)_{T=T_2}$ is the rate of change of the heat loss from the upper surface to the environment. $\sigma_2$ and $q_2$ in (4.3.18) and (4.3.19) are the undisturbed values for the surface tension and the heat flux in the stationary conduction state. $\sigma$ in (4.3.18) is a function of the state of the fluid only, while $q$ is a complicated function of the heat flux between the free surface and the surrounding environment.

The boundary conditions for the velocity fluctuations $w'$ will be the no-slip condition at the rigid wall, given by

$$w' = \frac{\partial w'}{\partial z} = 0 \quad \text{at} \quad z = 0, \tag{4.3.20}$$

while the free surface will be assumed to be flat without curvature, even after the convective motion sets in, i.e.,

$$w' = 0 \quad \text{at} \quad z = H. \tag{4.3.21}$$

This is only an approximation to the true shape of the free surface that results from neglecting the curvature of the surface due to the convective motion. The analysis for the full thermocapillary problem, including the curvature of the free surface, can be found in [46]. Another condition on the free surface can be formulated by seeking a balance between the tangential viscous shear stress and the surface tension force on that surface. The balance of the surface forces can be written, according to reference [46], as

$$\frac{\partial}{\partial x}\left[\mu\left(\frac{\partial u'}{\partial z} + \frac{\partial w'}{\partial x}\right)\right] + \frac{\partial}{\partial y}\left[\mu\left(\frac{\partial v'}{\partial z} + \frac{\partial w'}{\partial y}\right)\right] = \frac{\partial}{\partial x}\left(\frac{\partial \sigma'}{\partial x}\right) + \frac{\partial}{\partial y}\left(\frac{\partial \sigma'}{\partial y}\right) \tag{4.3.22}$$

where

$$\sigma' = \sigma - \sigma_2.$$

Applying condition (4.3.21) and using the conservation of mass equation, equation (4.3.22) reduces to the following form, after applying expression (4.3.18):

$$\mu\frac{\partial^2 w'}{\partial z^2} = \sigma\left(\nabla^2 - \frac{\partial^2}{\partial z^2}\right)T' \quad \text{at} \quad z = H. \tag{4.3.23}$$

The temperature boundary conditions on both the free surface and the wall-bounded surface require great care since the heat flux from the bounded surface is the main driver for the convective motion being investigated. Depending on the thermal properties of the bounding solid surface, whether a high- or a low-conductivity material is used, and on the conductivity of the fluid layer, an almost insulating or a perfectly conducting boundary condition may be imposed. To keep the analysis quite general, the following thermal boundary condition is used at the solid wall surface:

$$T' = \epsilon(\partial T'/\partial z) \quad \text{at} \quad z = 0, \tag{4.3.24}$$

where $\epsilon$ can take the values of either 0 or $\infty$. Condition (4.3.24) is appropriate for the perfectly conducting case when $\epsilon = 0$ and the perfectly insulating condition when $\epsilon \to \infty$.

The temperature boundary condition on the free surface is determined by a balance on the heat flux between the fluid layer and the surrounding environment as

$$-k\frac{\partial T'}{\partial z} = hT' \quad \text{at} \quad z = H, \tag{4.3.25}$$

where $h = (\partial q/\partial T)$ is the heat transfer coefficient of the environment adjacent to the free surface and depends on the state of the environment.

The fluctuation equations can be made dimensionless using the same scales for space, time, velocity, and temperature as those used for the buoyancy-driven convection problem. Since equations (4.3.16) and (4.3.17) are linear with linear boundary conditions, they also admit solutions through the separation of variables method. Substituting

$$T' = \theta(z)F(x,y)e^{\omega t} \tag{4.3.28}$$

$$w' = W(z)F(x,y)e^{\omega t} \tag{4.3.29}$$

into the dimensionless fluctuation equations, the following results for the fluctuations amplitude functions:

$$\left[\omega - Pr\left(\frac{d^2}{dz^2} - a^2\right)\right]\left(\frac{d^2}{dz^2} - a^2\right)W = 0 \tag{4.3.30}$$

$$\left[\omega - \left(\frac{d^2}{dz^2} - a^2\right)\right]\theta = W. \tag{4.3.31}$$

Again, $F(x,y)$ here is the fluctuations planform shape, which is itself a solution to the equation

$$\frac{\partial^2 F}{\partial x^2} + \frac{\partial^2 F}{\partial y^2} + a^2 F = 0 \tag{4.3.32}$$

and $a$ is a constant arising from the separation of variables method.

Similarly, the boundary conditions on both the upper and the lower surfaces take the following nondimensional form:

$$W = \frac{dW}{dz} = 0, \quad \epsilon\frac{d\theta}{dz} = \frac{\theta}{H} \quad \text{at } z = 0, \tag{4.3.33}$$

$$W = 0, \quad \frac{d^2 W}{dz^2} = a^2 M\theta, \quad \frac{d\theta}{dz} = -Bi\theta \quad \text{at } z = 1, \tag{4.3.34}$$

where $M$ and $Bi$ are the *Marangoni* and the *Biot numbers*, respectively, defined in the following manner:

$$M = \frac{(\partial\sigma/\partial T)G_T H^2}{\mu\kappa}, \quad Bi = \frac{hH}{k}. \tag{4.3.35}$$

Equations (4.3.30) and (4.3.31) are solved in a manner similar to the buoyancy-driven convection problem: The onset of thermocapillary convection is determined by examining the sign of the constant $\omega = \omega_r + i\omega_i$. If the real part of $\omega$ is positive, then the fluctuations will grow exponentially and thus destabilize the original conduction solution leading to convection in the fluid. If, on the other hand, the real part of $\omega$ is negative, then the fluctuations will decay with time, regaining the original conductive state and implying a stable configuration. Again, the condition for which $\omega = 0$, representing the marginal stability conditions, defines the boundaries between the stable and unstable solutions. The marginal stability solution is also easy to obtain. Thus, setting $\omega = 0$, the governing equations for the fluctuation amplitudes become

$$\left(\frac{d^2}{dz^2} - a^2\right)\left(\frac{d^2}{dz^2} - a^2\right)W = 0, \tag{4.3.36}$$

$$\left(\frac{d^2}{dz^2} - a^2\right)\theta = -W. \tag{4.3.37}$$

Note that the principle of exchange of stabilities has been invoked in deriving equations (4.3.36) and (4.3.37) without proof. This is commonly done for the thermocapillary convection problem.

Since equation (4.3.36) is in terms of the velocity fluctuations only, it is solved first subject to the velocity boundary conditions. Once $W$ is determined, then equation (4.3.37) is solved for $\theta$ subject to the temperature boundary conditions. The solution to equation (4.3.36) subject to the first of the two conditions in (4.3.33) and the first condition in (4.3.34) is

$$W = A_W[\sinh(az) - (az)\cosh(az) + z(a \coth a - 1)\sinh(az)], \tag{4.3.38}$$

where $A_W$ is a constant of integration. The solution for $\theta$ is complex and can be found in [44]. The equation for the marginal stability curve is obtained by substituting the solution for both $W$ and $\theta$ into the second condition of (4.3.34), relating $W$ to $\theta$. With some algebra, the following equation for the Marangoni number in terms of both the Biot number and the wave number is obtained for the perfectly conducting case, $\epsilon = 0$:

$$M = 8a\left\{\frac{(a \cosh a + Bi \sinh a)(a - \cosh a \sinh a)}{a^3 \cosh a - \sinh^3 a}\right\}. \tag{4.3.39}$$

A similar equation can be derived for the perfectly insulating case (i.e., $\epsilon \to \infty$):

$$M = 8a\left\{\frac{(a \cosh a + Bi \sinh a)(a - \cosh a \sinh a)}{a^3 \sinh a - a^2 \cosh a + 2a \sinh a - \sinh^2 a \cosh a}\right\}. \tag{4.3.40}$$

The marginal stability curves for both a conducting lower boundary, $\epsilon = 0$, and an insulating lower boundary, $\epsilon \to \infty$, are shown in Figures 4.9 and 4.10 for a number of values of the Biot number. The region inside each curve represents values for both the Marangoni number and wave number for which the conduction solution is unstable. The region outside each curve is the stable region.

It can be seen that all of the marginal stability curves converge toward $M = 8a^2$ for large values of $a$. Also, every marginal stability curve has a minimum value of $M = M_c$, which is the critical value of the Marangoni number below which no disturbance can be amplified. Also for each curve there is a corresponding value

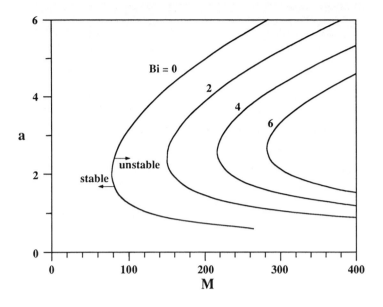

**Figure 4.9**   Marginal stability curves for surface tension-driven convection, for the case of perfectly conducting walls.

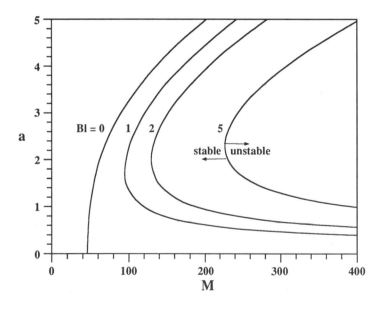

**Figure 4.10**   Marginal stability curves for surface tension-driven convection, for the case of perfectly insulating walls.

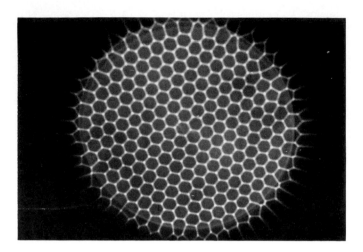

**Figure 4.11**   Photograph of a typical convection pattern with a free upper surface, from [34]. *Courtesy E. L. Koschmieder.*

for the critical wave number at which convection sets in and which is denoted by $a_c$. For the insulating lower boundary, the curve for $Bi = 0$ has a peculiar behavior; the critical Marangoni number occurs at $a = 0$ (i.e., for an infinite wavelength), while $M_c = 48$ for that case. This result shows that the instability at the critical value of the Marangoni number is characterized by a planform pattern corresponding to an infinite wavelength. The general form of the marginal stability curve is similar for both the insulating and the conducting cases. In both the stability region includes larger values for the Marangoni number as the value of $Bi$ is increased.

Again, as in the case of the buoyancy-driven flow, the planform pattern for the fluctuations remains undetermined through the linear stability analysis outlined here. Note that the fluctuation planform shape equation (4.3.30) is identical to its counterpart for the buoyancy-driven convection problem, i.e., to equation (4.2.27). This ambiguity requires the use of other means to determine the exact form of the pattern. Numerous experiments have consistently shown that thermocapillary convection always occurs in the form of uniform hexagonal cells that are evenly distributed across the horizontal free surface. Figure 4.11, taken from the meticulous experiments of [34], shows a typical convection pattern that is commonly observed due to thermocapillary instability.

Experiments on thermocapillary convection are very difficult to perform in a terrestrial laboratory environment due to the complicating buoyancy factor. Any experiment on a horizontal fluid layer, in which there is a vertical temperature gradient, will result in free convection due to both the buoyancy force and the capillary force. The only possibility for separating the effects of these two forces is by reducing the depth of the fluid layer to diminish the effects of the buoyancy force. Since the Marangoni number is proportional to the square of the fluid depth $H$, a full probing of the marginal stability diagram is almost impossible in a terrestrial environment. The ultimate validation of the analysis for thermocapillary convection is through zero-gravity convection experiments. The second-best venue is a low-gravity environment. Thus thermocapillary convection experiments constituted some of the earliest low-gravity flight tests performed. These still remain the most favored low-gravity fluid flow experiments.

The very first set of experiments on thermocapillary convection in low-gravity was conducted on Apollo 14 in 1971 (see [36]). These were exploratory in nature and were not conducted in a controlled laboratory environment. Nevertheless, they revealed that surface tension alone could trigger convection in a fluid layer when the temperature difference reached a certain critical value. Further experiments on thermocapillary convection were conducted on Apollo 17 on convex fluid layers of 2- and 4-mm depths. The critical Marangoni number in these tests was found to be much larger than the critical Marangoni number calculated through linear stability analysis. This value, according to Figure 4.9, is $M_c \approx 80$.

Reference [36] also describes a set of experiments designed to investigate the onset of thermocapillary convection on board a sounding rocket. These experiments, also exploratory in nature, were designed as a precursor for those to be carried out on board NASA's space shuttle. Results from two test sets performed on board a sounding rocket are reported in [36], one for a circular dish and the other for a rectangular dish.

The circular dish had a free surface approximately 75 mm in diameter in which oil was used for a working fluid with depth of approximately 5 mm. A critical Marangoni number of 80 for this fluid will yield a value for the critical vertical temperature difference of $\Delta T_c \approx 1.4\ K$ when the specific fluid thermodynamic properties are used in calculating $M_c$. A cellular pattern, as shown in Figure 4.12, was observed when the imposed $\Delta T$ reached the critical value. Although these cells appear hexagonal in shape, they are not all of a uniform geometry. Due to this nonuniformity of the patterns, a specific value for $a_o$ cannot be determined for this experiment.

The rectangular experimental cell had horizontal dimensions of 40 mm × 60 mm. Silicone oil was used for the working fluid. The fluid layer depth was approximately 5 mm. The critical temperature gradient across the layer for this fluid, according to linear stability theory, was calculated to be $\Delta T_c \approx 0.992$K/mm. A cellular pattern was observed for this geometry when the temperature gradient at the free surface reached a value of 1.24 K/mm. This convection pattern is shown in Figure 4.13. Although it is not regular when compared with ground-based experiments performed in $g/g_0 = 1$, the difference in the critical wave number $a_c$ was no more than $\approx 10\%$. The cell nondimensional wavelengths for $g/g_0 = 1$ were approximately 2.8 × 3.4, as compared with 3.0 × 3.4 for $g/g_0 \approx 10^{-4}$.

## *4.4* COMBINED THERMOCAPILLARY AND BUOYANCY-DRIVEN CONVECTION

The previous section outlined the difficulties associated with differentiating experimentally between the effects due to surface tension force and those due to buoyancy force for free convection in a horizontal fluid layer. The only alternative to the experimental validation is to perform a stability analysis in which both of these effects are included. Reference [38] presents a linear stability analysis

**Figure 4.12**    Photograph of the convection pattern due to surface tension gradients in a circular dish at $10^{-4}$ $g_0$, from [36]. *Courtesy J. C. Legros.*

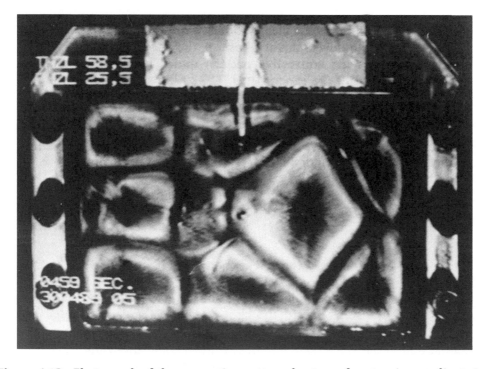

**Figure 4.13**    Photograph of the convection pattern due to surface tension gradients in a rectangular dish at $10^{-4}$ $g_0$, from [36]. *Courtesy J. C. Legros.*

for the combined effects of buoyancy and thermocapillary forces on the onset of thermal convection in horizontal fluid layers. This analysis is briefly outlined here.

The effect of the buoyancy force is to couple the thermal and the velocity fields in the linear equations for the fluctuations. In fact, this coupling already exists in the energy equation. For the combined force analysis, the equation for the vertical velocity fluctuation in this case, equation (4.3.30), should be replaced by the following equation derived for the buoyancy driven problem, equation (4.2.28):

$$\left[\frac{\omega}{Pr} - \left(\frac{d^2}{dz^2} - a^2\right)\right]\left(\frac{d^2}{dz^2} - a^2\right)W = -a^2 R\theta. \tag{4.4.1}$$

The energy equation will remain the same in this case as equation (4.3.31), i.e.,

$$\left[\omega - \left(\frac{d^2}{dz^2} - a^2\right)\right]\theta = W. \tag{4.4.2}$$

The boundary conditions to be imposed for this problem remain the same as those used for the thermocapillary instability problem given by (4.3.33) and (4.3.34), i.e.,

$$W = \frac{dW}{dz} = 0, \qquad \epsilon\frac{d\theta}{dz} = \frac{\theta}{H} \quad \text{at} \quad z = 0, \tag{4.4.3}$$

$$W = 0, \quad \frac{d^2 W}{dz^2} = a^2 M\theta, \quad \frac{d\theta}{dz} = -Bi\theta \quad \text{at} \quad z = 1. \tag{4.4.4}$$

Again, as in the previous sections, only the solution for the marginal stability curve is sought, in which case $\omega$, in equations (4.4.1) and (4.4.2), is set equal to zero. Adopting the terminology of [38], the equations for the marginal stability solution reduce to the following:

$$\left(\frac{d^2}{dz^2} - a^2\right)^2 W_1 = a^2 R_1\theta_1, \tag{4.4.5}$$

$$\left(\frac{d^2}{dz^2} - a^2\right)\theta_1 = W_1, \tag{4.4.6}$$

where $W_1 = W/\pi$, $\theta_1 = \pi\theta$, and $R_1 = R/\pi^4$. Similarly, the boundary conditions will take the following form in the present terminology:

$$W_1 = \frac{dW_1}{dz} = 0, \quad \theta_1 = 0 \text{ at } z = 0, \tag{4.4.7}$$

$$W_1 = 0, \quad \frac{d^2 W_1}{dz^2} = a^2 M_1\theta_1, \quad \frac{d\theta_1}{dz} = -Bi_1\theta_1 \quad \text{at } z = \pi, \tag{4.4.8}$$

where $M_1 = M/\pi^2$ and $Bi_1 = Bi/\pi$. Note that the third condition in (4.4.7) assumes a perfectly conducting lower wall, i.e., $\epsilon = 0$.

The approach used for solving equations (4.4.5) and (4.4.6) is slightly different from the one adopted in the previous section, due to the coupling between $W_1$

and $\theta_1$ in both the equations and the boundary conditions. This means that equation (4.4.5) cannot be solved independently from equation (4.4.6). Reference [38] suggests a solution in terms of a Fourier expansion for both $W_1$ and $\theta_1$ in the following manner:

$$W_1(z) = \sum_{n=1}^{\infty} \left\{ w_n - (2/\pi n^3) \left[ -(-1)^n \frac{d^2 W_1}{dz^2}(\pi) + \frac{d^2 W_1}{dz^2}(0) \right] \right\} \sin nz, \quad (4.4.9)$$

$$\theta_1(z) = \sum_{n=1}^{\infty} \{ \theta_n - (2/\pi n)(-1)^n \theta_1(\pi) \} \sin nz. \quad (4.4.10)$$

Substituting these expansions into equations (4.4.5) and (4.4.6), observing boundary conditions (4.4.7) and (4.4.8), and equating coefficients of $\sin nz$, gives the following two algebraic equations for each value of $n$:

$$(n^2 + a^2)^2 w_n - R_1 a^2 \theta_n = \left( \frac{4a^2}{\pi n} + \frac{2a^4}{\pi n^3} \right) \frac{d^2 W_1}{dz^2}(0)$$
$$+ \left( \frac{2R_1}{\pi n M_1} - \frac{4a^2}{\pi n} - \frac{2a^4}{\pi n^3} \right) (-1)^n \frac{d^2 W_1}{dz^2}(\pi), \quad (4.4.11)$$

$$w_n - (n^2 + a^2)\theta_n = \frac{2}{\pi n^3} \frac{d^2 W_1}{dz^2}(0) + \left( \frac{2}{\pi n M_1} - \frac{2}{\pi n^3} \right)(-1)^n \frac{d^2 W_1}{dz^2}(\pi). \quad (4.4.12)$$

Solving these two equations for $w_n$ and $\theta_n$ and substituting the boundary conditions in both will result in two homogeneous linear equations for $d^2 W_1(0)/dz^2$ and $d^2 W_1(\pi)/dz^2$. In the end, the solutions for these variables result in the following dispersion relation:

$$\sum_{n=1}^{\infty} \frac{n^2(n^2 + a^2)}{\lambda_n} \left[ \sum_{n=1}^{\infty} \left( \frac{n^2 M_1 a^2}{\lambda_n} - \frac{a^2(n^2 + a^2)^2 - R_1 a^2}{\lambda_n} \right) - \frac{1 + \pi B i_1}{2} \right]$$

$$- \sum_{n=1}^{\infty} \frac{(-1)^n n^2}{\lambda_n} \left[ \sum_{n=1}^{\infty} (-1)^n n^2 \left( \frac{M_1 a^2 (n^2 + a^2)}{\lambda_n} + \frac{R_1 a^2}{\lambda_n} \right) \right] = 0, \quad (4.4.13)$$

$$\lambda_n = (n^2 + a^2)^3 - R_1 a^2.$$

Equation (4.4.13) can be solved for the marginal stability curves in terms of the Marangoni number as a function of both the wave number $a$ and the Biot number. Figure 4.14 shows the solution in terms of the normalized critical Marangoni number $M/M_c$ as a function of the normalized critical Rayleigh number $R/R_c$. Figure 4.14 shows the results for both a perfectly insulating and a perfectly conducting lower wall.

## 4.5 *DOUBLE DIFFUSIVE CONVECTION*

It was shown in Section 4.2 that buoyancy-driven convection can occur in a fluid whenever cold fluid lies over hot fluid. In terms of the fluid density, this statement can be reformulated to imply that convection will occur when the heavier fluid is

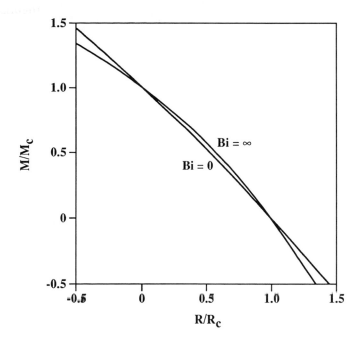

**Figure 4.14** Marginal stability curves for the combined buoyancy-driven and surface tension-driven convection. Values for the parameters are normalized with respect to the critical values (from [38]).

above the lighter fluid. There are other means for altering the density of the fluid in addition to thermal means. One way is to mix two fluids or overlay a denser fluid over a lighter one, such as overlaying salty water over fresh water. These are of common occurrence in the oceans and the atmosphere.

Another situation in which density stratification can occur is in materials processing. Liquids that are a mixture of two components, such as liquid alloys, will freeze at a specific concentration depending on the imposed freezing temperature. If the melt composition is at a different concentration from the value due to thermodynamics, then solute will either be rejected or incorporated into the solid depending on the set freezing temperature and the thermodynamic properties of the material. Thus, it is expected that the melt just ahead of the solid/liquid front will have a different composition than the rest of the melt bulk during the solidification of an alloy. Such solute concentration distribution gives rise to density gradients due to composition in addition to the existing density gradients due to temperature. These density gradients could lead to convection in the melt if the stratification is unstable.

Fluid convection due to both mass and temperature gradients is normally called *double diffusive convection*, since there are two diffusion mechanisms operating to cause convection: one due to thermal gradients, the other to solute gradients. This type of convection is also known in the literature as *thermosolutal*, or *thermohaline*, convection. In this section we will analyze a simple but

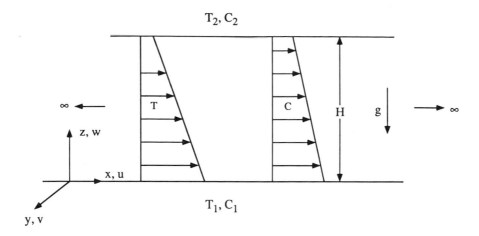

**Figure 4.15**    A sketch of the conditions for a double diffusive instability problem.

typical configuration in which the double diffusive convection mechanism can be established.

Again, as in the previous sections, we will consider a horizontally infinite fluid layer of finite depth $H$. The fluid layer is assumed to support opposing gradients of temperature and solute concentration, as shown in Figure 4.15. The fluid layer is assumed to be bounded from above and below by two horizontal surfaces that may be either free or solid. It is also assumed that the temperature and the solute concentration at each surface are held fixed.

The equations governing the stationary temperature and concentration distributions are the *diffusion equations* for both heat and mass given by

$$\frac{\partial^2 T}{\partial z^2} = 0, \qquad \frac{\partial^2 C}{\partial z^2} = 0,$$

where $T$ and $C$ are the temperature and the solute concentration distributions in the layer, respectively. The solutions to these equations, which are consistent with uniform values for both the temperature and solute concentration at each boundary, are given by

$$T = T_1 - G_T z, \tag{4.5.1}$$

$$C = C_1 - G_C z, \tag{4.5.2}$$

where

$$G_T = \frac{T_1 - T_2}{H}, \qquad G_C = \frac{C_1 - C_2}{H}$$

are the magnitudes of the uniform temperature and concentration gradients, respectively. $G_T$ and $G_C$ in this analysis are positive if the quantities decrease upward. $T_1$ and $C_1$ are the uniform temperature and solute concentration at the lower surface, while $T_2$ and $C_2$ are the respective values at the upper surface. For the present problem, it will be assumed that both $T_1 > T_2$ and $C_1 > C_2$, which means that the thermal stratification is destabilizing while the solutal stratification is stabilizing. The above solution shows that the stationary state temperature and concentration both increase with increasing height $z$. These profiles are shown in the problem sketch of Figure 4.15.

Since the density for the problem here is a function of both the temperature and the solute concentration, and since only buoyancy effects are considered, the Boussinesq approximation, equation (4.1.12), must be modified to the following form:

$$\rho = \rho_r[1 - \beta_t(T - T_r) + \beta_c(C - C_r)], \qquad (4.5.3)$$

where

$$\beta_t = -\frac{1}{\rho}\frac{\partial\rho}{\partial T}\Big|_{C,p}, \qquad \beta_c = \frac{1}{\rho}\frac{\partial\rho}{\partial C}\Big|_{\rho,p}.$$

Equation (4.5.3) is basically a modified Boussinesq approximation that accounts for the density variation due to both temperature and solute concentration variations. It should be stressed that the linear variation of the density with both temperature and concentration described here is only an approximation. There is sufficient evidence that equation (4.5.3) does not universally hold for all fluids. An obvious exception to equation (4.5.3) is water near its freezing temperature.

The full set of equations governing the convective motion for this problem are the conservation of mass for incompressible flow, the momentum equation, the energy equation, and the solute mass conservation equation, namely,

$$\nabla \cdot \mathbf{v} = 0, \qquad (4.5.4)$$

$$\frac{\partial \mathbf{v}}{\partial t} + \mathbf{v} \cdot \nabla\mathbf{v} = -\frac{1}{\rho_r}\nabla p - \mathbf{g}\beta_t(T - T_r) + \mathbf{g}\beta_c(C - C_r) + \frac{\mu}{\rho_r}\nabla^2\mathbf{v}, \qquad (4.5.5)$$

$$\rho c_p\left(\frac{\partial T}{\partial t} + \mathbf{v} \cdot \nabla T\right) = k\nabla^2 T, \qquad (4.5.6)$$

$$\frac{\partial C}{\partial t} + \mathbf{v} \cdot \nabla C = D\nabla^2 C. \qquad (4.5.7)$$

Note that equation (4.5.5) is the Boussinesq equation for the double diffusive problem in which the buoyancy force is due to both temperature and mass stratification. $D$ in equation (4.5.7) is the coefficient of solute mass diffusion, which is assumed to be a constant for this case.

The linear stability method is used here again to investigate the onset of double diffusive convection. Proceeding in a manner similar to the previous sections, the flow variables are decomposed into a fluctuation and a stationary component. The latter is a state for which there is no flow (i.e., $\mathbf{v} = 0$), and for which the temperature and solute distributions are given by expressions (4.5.1) and (4.5.2). When this decomposition is substituted into equations (4.5.4)–(4.5.7) and the fluctuations are linearized, the following linear equations for the fluctuations result:

$$\frac{\partial}{\partial t}\nabla^2 w' = \left(\nabla^2 - \frac{\partial^2}{\partial z^2}\right)(g\beta_t T' - g\beta_c C') + \nabla^2\nabla^2 w', \tag{4.5.8}$$

$$\frac{\partial T'}{\partial t} - G_T w' = \kappa\nabla^2 T', \tag{4.5.9}$$

$$\frac{\partial C'}{\partial t} - G_C w' = D\nabla^2 C', \tag{4.5.10}$$

where $w'$, $T'$, and $C'$ are the fluctuation vertical velocity component, the fluctuation temperature and the fluctuation solute concentration, respectively.

The horizontal boundaries may again be either rigid or free. For a rigid boundary, the no-slip velocity boundary condition is used. This, together with the conservation of mass equation, leads to the following conditions on the normal velocity fluctuation $w'$:

$$w' = \frac{\partial w'}{\partial z} = 0. \tag{4.5.11}$$

If the horizontal boundary is free, then the tangential stress at that boundary will be taken to be zero. These considerations will result in conditions on the normal velocity fluctuation that are identical to the velocity segment of condition (4.2.22). Note that, as explained in Section 4.3, the conditions in (4.2.22) are not strictly correct; they are valid only if the free surface does not deform subject to the ensuing convective motion.

The surface temperature at the boundaries can be kept either fixed, or alternatively, at a fixed heat flux across the boundary, or both. These two conditions can be combined in the following general form, which is commonly known as the *radiation condition*:

$$T' + \lambda\frac{\partial T'}{\partial z} = 0. \tag{4.5.12}$$

Here the sign of $\lambda$ must be chosen to ensure that the fluctuation heat flux is directed out of the fluid layer. A similar condition is imposed on the fluctuation solute concentration:

$$C' + \epsilon\frac{\partial C'}{\partial z} = 0. \tag{4.5.13}$$

The difference between conditions (4.5.12) and (4.5.13) is that $\epsilon$ can take the values of either 0 or $\infty$, while $\lambda$ can take any real value. If the boundary is impermeable, then $\epsilon = 0$ must be used in condition (4.5.13); otherwise, if the concentration is kept constant, then $\epsilon \to \infty$.

Using the same scales for space, time, velocity, and temperature, as used in the Benard problem, gives equations (4.5.8)–(4.5.9) in a nondimensional form. For the present problem, however, the solute concentration fluctuation, $C'$ must be nondimensionalized in the following manner:

$$C' = \frac{C^{*'}}{C_1^* - C_2^*} \tag{4.5.14}$$

where the dimensional variables are denoted with an asterisk. Substituting these variables into the equations for the fluctuations, the following nondimensional set of equations results for the vertical velocity fluctuation $w'$, the temperature fluctuation $T'$, and the concentration fluctuation $C'$:

$$\left(\frac{1}{Pr}\frac{\partial}{\partial t} - \nabla^2\right)\nabla^2 w' = \left(\nabla^2 - \frac{\partial^2}{\partial z^2}\right)(RT' - R_s C'), \tag{4.5.15}$$

$$\left(\frac{\partial}{\partial t} - \nabla^2\right)T' = -w', \tag{4.5.16}$$

$$\left(\frac{\partial}{\partial t} - Le\nabla^2\right)C' = -w'. \tag{4.5.17}$$

Here $R$, $R_s$, $Pr$, and $Le$ are the Rayleigh number, the *solutal Rayleigh number*, the Prandtl number, and the *Lewis number*, respectively. The Rayleigh and Prandtl numbers have already been encountered: the solutal Rayleigh and the Lewis numbers are defined in the following manner:

$$R_s = \frac{\beta_c G_c g H^4}{\nu \kappa}, \qquad Le = \frac{D}{\kappa}.$$

Equations (4.5.15)–(4.5.17) are linear partial differential equations which again can be solved using the separation of variables method. This means that the solution for the fluctuation functions may be represented as

$$w' = W(z)F(x,y)e^{\omega t}, \tag{4.5.18}$$

$$T' = \theta(z)F(x,y)e^{\omega t}, \tag{4.5.19}$$

$$C' = \gamma(z)F(x,y)e^{\omega t}. \tag{4.5.20}$$

Substituting these functional representations into equations (4.5.15)–(4.5.17) results in the following set of ordinary differential equations for the fluctuation amplitude functions:

$$\left(\frac{\omega}{Pr} - \frac{d^2}{dz^2} + a^2\right)\left(\frac{d^2}{dz^2} - a^2\right)W = -a^2(R\theta - R_s\gamma), \tag{4.5.21}$$

$$\left(\omega - \frac{d^2}{dz^2} + a^2\right)\theta = -W, \tag{4.5.22}$$

$$\left[\omega - Le\left(\frac{d^2}{dz^2} - a^2\right)\right]\gamma = -W. \tag{4.5.23}$$

Note that the fluctuations planform shape function $F(x,y)$, which is governed by equation (4.3.30), was also used in deriving equations (4.5.21)–(4.5.23).

The solution for the amplitude function is sought subject to boundary conditions that are appropriate for free upper and lower surfaces with zero temperature and solute concentrations, i.e.,

$$W = \frac{d^2 W}{dz^2} = \theta = \gamma = 0 \quad \text{at} \quad z = 0, 1. \tag{4.5.24}$$

These conditions are unrealistic, and they are used here only to render the solution to equations (4.5.21)-(4.5.23) straightforward. The solutions for the fluctuation amplitude functions can now be represented in the following form (see[22]):

$$(W, \theta, y) = (A_w, A_\theta, A_y) \sin(n\pi z), \qquad n = 1, 2, \ldots \tag{4.2.25}$$

where $A_w$, $A_\theta$, and $A_y$ are constants of integration for the corresponding solutions.

Substituting (4.5.25) into equations (4.5.21)-(4.5.23) and using the boundary conditions (4.5.24) results in the following *dispersion relation*:

$$\omega^3 + \omega^2 \tau^2 (Pr + Le + 1) + \omega[(Pr + 1)(Le + 1)\tau^4 - a^2 Pr(R - R_s)/\tau^2]$$

$$+ \tau^6 Pr Le + a^2 Pr(R_s - LeR) = 0, \tag{4.5.26}$$

where

$$\tau^2 = a^2 + n^2\pi^2.$$

The dispersion relation (4.5.26) is quite complex and its properties have been investigated in great detail in both [22] and [37]. One reason for the complexity of the solution to equation (4.5.26) is that it contains many variables in the form of adjustable parameters. Specifically, the eigenvalue is a function of all of the following variables: the Rayleigh number, the solutal Rayleigh number, the Prandtl number, the Lewis number, the planform wave number, and the vertical wave number.

Another source of complexity for equation (4.5.26) is due to the fact that it is cubic in the eigenvalue $\omega$, which means that there are three roots for the eigenvalue at any value for all the parameters. Reference [22] shows that one of these roots is always real for any value of all the parameters, while the other two may be either real or complex conjugates. For the real roots, the marginal stability curve can be calculated in a manner similar to the previous two sections. For the complex conjugate roots, however, setting $\omega_r = 0$ does not necessarily lead to $\omega_i = 0$. In this case when the growth rate $\omega_r$ is equal to zero, the fluctuations exhibit modulations with time since $\omega_i \neq 0$. These modulations are bounded and the respective fluctuations are oscillatory in nature. The locus of the complex modes for which $\omega_r = 0$ for this case define the *neutral stability curve* in the parameter space.

To render the analysis tractable, the number of adjustable parameters may be reduced by two by fixing both the Prandtl and the Lewis numbers. This is equivalent to solving the double diffusive problem for a specific fluid. The stability boundaries for this problem can then be identified in the $(R, R_s)$ plane by locating the minimum to equation (4.5.26) with respect to the planform wave number $a$. The stability boundaries are delineated by the straight lines $XZ$ and $XW$ in Figure 4.16. The equation for the line $XZ$ is given by

$$R = \frac{R_s}{Le} + \frac{27\pi^4}{4}, \tag{4.5.27}$$

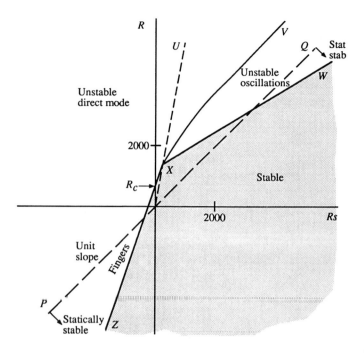

**Figure 4.16** The diagram of the different stability regimes for the double diffusive instability mechanism, from [47].

while the equation for the line $XW$ is given by

$$R = R_s\left(\frac{Pr + Le}{Pr + 1}\right) + (1 + Le)\left(1 + \frac{Le}{Pr}\right)\frac{27\pi^4}{4}. \qquad (4.5.28)$$

These lines are sketched, not to scale, in Figure 4.16.

Note that due to the fact that there are three eigenvalues for each solution of the dispersion relation, the stability map in the $(R, R_s)$ plane is very complex. The anatomy of this plane is roughly sketched here from the detailed analysis of [47]. In the quadrant where $R$ is negative and $R_s$ is positive, both the temperature and the solute are stably stratified and hence all fluctuations are stable. In the upper left quadrant of the $(R, R_s)$ plane, all values of the parameters lead to instability when they fall above the line $XZ$. In the shaded region to the right of the $XZ$ line and below the $XW$ line, all values for the parameters lead to stable fluctuations. The area of that region varies depending on the chosen values for both $Pr$ and $Le$. In the region between the $XW$ line and the $XV$ curve, the instability is manifested in the form of oscillatory modes, i.e., these modes are complex conjugates with $\omega_i \neq 0$. These are also called the *overstable modes*. The modes in all of the region to the left of the $ZXV$ curve are unstable. The modes in this region are real and as such do not have an imaginary part; they are called *direct modes*.

The convective patterns in which double diffusive instability is manifested are also more complex than their counterpart for the buoyancy-driven instability problem. Although the same equation for the planform shape holds in both cases [i.e., equation (4.2.27)], the unstable flow pattern is far richer in the present case. Again, the linear stability analysis performed here is not capable of describing the convection pattern, and reliance must be placed on either more complex analysis or experiments to identify these patterns. In spite of the limitations of linear stability analysis, it appears that it can provide some guidance as to what in fact is observed under highly *supercritical conditions*, defined as those conditions where the Rayleigh number is far greater than the critical value, i.e., $R \gg R_c$.

For all $R$, $R_s$ the most unstable mode is $n = 1$, i.e., the cells extend from top to bottom of the unstable region. On the neutral lines $XZ$ and $XW$, the convection pattern is manifested for $n = 1$ in terms of $a^2 = 1/2$. Throughout the whole top part of the diagram, and the part in which the motion is driven by the temperature difference, the most unstable wave number is near unity. Such a value for the wave number indicates that the convection cells are as wide as they are high. This is true whether the motion is direct or oscillatory. In the first quadrant, where both $R > 0$ and $R_s > 0$, the instability is physically due to the destabilizing thermal gradient, while the solute gradient is stabilizing. In this region the convective instability is manifested in the form of a series of convecting layers forming in succession from the bottom up. This type of convective instability is known as the *diffusive regime*. Figure 4.17 shows a typical example of the layering phenomenon in this regime.

In the lower left quadrant where the instability is manifested in the form of direct modes, the wave number $a$ is relatively large. The cells in this region tend to be tall and thin. Asymptotically, as $|R| \to \infty$ with $Le \ll 1$ the dimensional wavenumber $\alpha$ of maximum growth rate is calculated to be

$$\pi \alpha_{\max} \approx \left( \frac{g \beta_t \Delta T}{H \nu \kappa} \right)^{1/4} .$$

The convective instability in the region in which both $R$ and $R_s$ are negative is manifested in the form of thin columns of high-concentration solute in a background of lower solute concentration. This is in agreement with the observed form of *salt fingers* in their fully developed state. This type of double diffusive instability is commonly known as the *fingering instability*, and occurs frequently in nature. Figure 4.18 shows a typical form of the fingering pattern.

## 4.6 *BUOYANCY-DRIVEN CONVECTION IN INFINITELY LONG VERTICAL CAVITIES*

The results derived in the previous four sections are both powerful and useful for applications in low-gravity environment. However, the assumption of infinite fluid layers in the horizontal direction is quite unrealistic for most engineering and experimental applications. In fact, it is well known that buoyancy-driven convection can be initiated in a fluid-filled cavity even in the absence of vertical temperature gradients within the cavity. In other words, buoyancy-driven convection can take place in non-isothermal enclosures, a case far more realistic for engineering applications than the infinite horizontal layers situation. The fun-

**Figure 4.17** Photograph of the layering mechanism in double diffusive convection from [28]. *Courtesy C. F. Chen.*

damental question that needs to be answered here is "What is the value for the critical Rayleigh number for the onset of convection in enclosures with various temperature gradients?" We provide the answer next.

Clearly, the problem of buoyancy-driven convection in an enclosure will have to be treated in terms of the aspect ratio $A = H/L$: the ratio of the height of the enclosure to its width. A solution to this problem for any finite value of $A$ must reach either limit of $A$ in a smooth manner. This implies that any solution to buoyancy-driven flow in a cavity must produce the Benard problem solution in the limit when $A \to 0$ and must also arrive at the infinite vertical planes solution in the limit when $A \to \infty$. The first case has already been treated in Section 4.2; the second is analyzed in this section.

The problem of establishing free convection flows in a cavity was first studied by [23]. That study gave the solution for the convection between infinite vertical planes as a limiting case for the more general finite dimensional cavity case. This solution is reproduced in here.

**Figure 4.18** Photograph of the fingering mechanism in double diffusive convection. *Courtesy C. F. Chen.*

Consider the flow between two isothermal, infinitely long, vertical walls, each set at different but constant temperatures, say $T_h$ and $T_c$. A sketch of the geometry for this problem is shown in Figure 4.19. The Boussinesq equations in a Cartesian coordinate system for the steady, two-dimensional flow for this case are

$$\frac{\partial u}{\partial x} + \frac{\partial w}{\partial z} = 0, \tag{4.6.1}$$

$$u\frac{\partial u}{\partial x} + w\frac{\partial u}{\partial z} = -\frac{1}{\rho_r}\frac{\partial p}{\partial x} + \frac{\mu}{\rho_r}\left(\frac{\partial^2 u}{\partial x^2} + \frac{\partial^2 u}{\partial z^2}\right), \tag{4.6.2}$$

$$u\frac{\partial w}{\partial x} + w\frac{\partial w}{\partial z} = -\frac{1}{\rho_r}\frac{\partial p}{\partial z} + \frac{\mu}{\rho_r}\left(\frac{\partial^2 w}{\partial x^2} + \frac{\partial^2 w}{\partial z^2}\right) + g\beta(T - T_r), \tag{4.6.3}$$

$$u\frac{\partial T}{\partial x} + w\frac{\partial T}{\partial z} = \frac{k}{\rho_r c_p}\left(\frac{\partial^2 T}{\partial x^2} + \frac{\partial^2 T}{\partial z^2}\right), \tag{4.6.4}$$

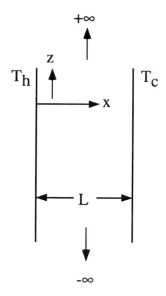

**Figure 4.19** A sketch of the infinitely tall cavity.

where $u$ and $w$ are the $x$- and $z$-components of the velocity vector, respectively. Equations (4.6.1)–(4.6.4) are to be solved subject to the no-slip boundary conditions for velocity and constant temperature along the vertical plates, i.e.,

$$u = w = 0, \quad T = T_h \quad \text{at} \quad x = 0 \tag{4.6.5}$$

$$u = w = 0, \quad T = T_c \quad \text{at} \quad x = L. \tag{4.6.6}$$

Since the vertical cavity is assumed to be infinite in extent, it is reasonable to assume that both velocity components are functions of the horizontal coordinate $x$ only. This assumption can be justified for very tall cavities and especially for far away from the top and bottom end walls. The conservation of mass equation under this assumption reduces to the following form:

$$\frac{\partial u}{\partial x} = 0. \tag{4.6.7}$$

Equation (4.6.7) together with the no-slip boundary condition can be solved for the horizontal velocity component, $u$, resulting in $u = 0$ everywhere.

The $x$-momentum equation (4.6.2) reduces to the following form when the solution for $u$ is used:

$$0 = -\frac{1}{\rho_r} \frac{\partial p}{\partial x}. \tag{4.6.8}$$

The solution to equation (4.6.8) is

$$p = f(z).$$

The $z$-momentum equation with this solution for $u$ and under all of the assumptions imposed above reduces to the following form:

$$0 = -\frac{1}{\rho_r} \frac{\partial p}{\partial z} + \frac{\mu}{\rho_r} \frac{\partial^2 w}{\partial x^2} + g\beta(T - T_r). \tag{4.6.9}$$

Upon differentiating equation (4.6.9) with respect to $x$ and substituting equation (4.6.8), the following coupled equation results for both the temperature and the vertical velocity:

$$\frac{\mu}{\rho_r}\frac{\partial^3 w}{\partial x^3} + g\beta\frac{\partial T}{\partial x} = 0. \tag{4.6.10}$$

Since the vertical plates on each side of the cavity are kept at a constant temperature along their entire lengths, it is reasonable to expect the temperature inside the cavity to be a function of the horizontal coordinate $x$ only. With this assumption on the temperature together with those already made on the velocity field, $u$ and $w$, the energy equation (4.6.4) reduces to the form

$$\frac{\partial^2 T}{\partial x^2} = 0. \tag{4.6.11}$$

The solution to equations (4.6.10) and (4.6.11) together with the boundary conditions (4.6.5) and (4.6.6) will give both the temperature and the velocity distributions within the cavity. It is convenient at this point to nondimensionalize the variables in equations (4.6.10) and (4.6.11). The dimensional velocity $w$, the coordinate distance $x$, and the temperature $T$ may be written in the following manner:

$$W = \frac{wL}{\kappa}, \qquad X = \frac{x}{L}, \qquad \theta = \frac{T - T_c}{T_h - T_c}$$

where $L$ is the horizontal distance between the vertical plates. With this nondimensionalization, equations (4.6.10) and (4.6.11) can be rewritten as

$$\frac{\partial^3 W}{\partial X^3} = -R\frac{\partial \theta}{\partial X}, \tag{4.6.12}$$

$$\frac{\partial^2 \theta}{\partial X^2} = 0, \tag{4.6.13}$$

where $R$ is the Rayleigh number defined here as

$$R = \frac{g\beta(T_h - T_c)L^3}{\nu\kappa}.$$

This definition for the Rayleigh number is basically identical to the definition used for the solutions obtained earlier, except it is given in terms of these horizontal temperature difference instead of the vertical. In engineering applications, these horizontal temperature gradients are normally written in terms of the *Grashof number*, $Gr$, instead of the Rayleigh number. It is defined as follows:

$$Gr = \frac{g\beta(T_h - T_c)L^3}{\nu^2} = RPr.$$

As seen from the reduced energy equation (4.6.13), heat is transported across the cavity by conduction alone. The temperature distribution across the cavity is linear and independent of the flow field. Since the flow is driven by buoyancy, the velocity distribution depends on the temperature variation across the layer. Also, because the two vertical surfaces are assumed to be infinite with no end plates, the condition of zero net vertical mass flow at any given cross section is used instead. Due to the nature of the equations and boundary conditions, the velocity distribution must be antisymmetric about the midpoint plane. This condition implies that $W = 0$ at that point. With this condition it is sufficient to resolve the velocity field across only half of the cavity width with the following boundary conditions replacing (4.6.5) and (4.6.6):

$$W(0) = W(1/2) = 0, \qquad \theta(0) = 1, \qquad \theta(1) = 0. \qquad (4.6.14)$$

The solution to the energy equation (4.6.13), with the boundary conditions (4.6.14), is given by

$$\theta = 1 - X. \qquad (4.6.15)$$

Substituting this solution for the temperature in equation (4.6.12) gives for the vertical velocity $W$ the following solution, which is consistent with the imposed boundary conditions:

$$W = \frac{R}{12}X(1 - X)(1 - 2X). \qquad (4.6.16)$$

The distributions of the vertical velocity $W$ and the temperature $\theta$ as a function of the horizontal distance $X$ are shown in Figure 4.20. This idealized and simple solution due to buoyancy effects shows that the fluid will rise along the hot wall and descend along the cold. It also shows that effectively there is no horizontal mixing of the fluid inside the cavity. This implies that heat is transported across the cavity by conduction alone.

For this purely conductive transport case, the heat flux $q$ across the fluid layer is obtained as $k(T_h - T_c)/L$. This gives the heat transfer coefficient $h$, based on the temperature difference $T_h - T_c$, as $k/L$. The Nusselt number for this flow is defined by

$$Nu = hL/k, \qquad (4.6.17)$$

which in this case takes the value of 1.0.

Note that the solution for the velocity field $W$ was found to be a function of the Rayleigh number for all values of $R$, i.e., the critical Rayleigh number is zero. This solution is different from the buoyancy-driven solution for a horizontal layer in which there was a nonzero value of $R$ below which there is no convection. The simple solution (4.6.16) shows that as long as $R \neq 0$, there is always convective motion in such a configuration. This solution also indicates that *no matter how small the value of g, there is always convective motion as long as there is a horizontal temperature gradient.* This result is radically different from that for the Benard problem for a horizontally infinite fluid layer.

Both solutions can be summarized in the following statement: Any temperature gradient in a fluid in the normal direction to the gravity vector will always lead to convection in the fluid, while temperature gradients in the direction of gravity may lead to convection.

The pressure gradient in the vertical direction for this case can be evaluated from the vertical momentum equation (4.6.9). It should be remembered that these solutions apply only in the region far from the ends of the vertical cavity. There

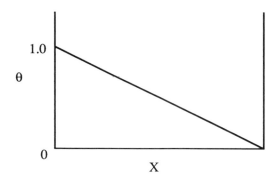

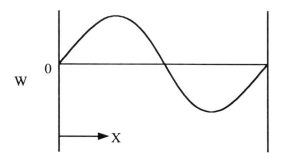

**Figure 4.20**   Velocity and temperature distributions in an infinitely long vertical cavity.

have been many solutions for convection in a vertical slot, including walls having temperature distributions that vary with the vertical distance $z$. The stability of these flows have also been investigated together with the effects of the end walls and transient effects (see [42]).

## 4.7 *CONVECTION IN CAVITIES OF FINITE GEOMETRIES*

The above analysis for buoyancy-driven flow in a vertical fluid layer revealed an essentially different mechanism for the onset of free convection than that for infinite horizontal layers. However, both configurations are basically idealizations for buoyancy-driven convection in finite-dimensional cavities of real practical configurations. Buoyancy-driven convection in enclosures is encountered in many engineering applications including crystal growth, solar collectors, and fires in enclosures, among others. These flows are inherently three-dimensional in nature, usually with complex boundary conditions. This section outlines an analytical approach for investigating buoyancy-driven convection in two-dimensional rectangular cavities. The results of this approach can be extended to fully three-dimensional practical configurations without seriously impacting the physics of the problem.

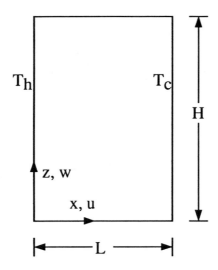

**Figure 4.21**

Consider a vertical cavity bounded by vertical and horizontal surfaces that are held at different temperatures. The horizontal walls will be taken either as insulated or with temperatures varying linearly between the two vertical surfaces. Figure 4.21 shows a sketch of the rectangular cavity assumed here. It should be kept in mind that the interest here is in the nature of the free convective flow that arises, as well as the rate of heat transport between the confining walls. The governing equations are the Boussinesq equations for a steady, two-dimensional flow in a rectangular Cartesian coordinate system. These equations are given by (4.6.1)–(4.6.4) and they must be solved in this case subject to the following boundary conditions on all the walls surrounding the cavity:

$$u^* = w^* = 0, \quad T^* = T_h \ \text{at} \ x^* = 0 \ \text{ for } \ 0 \le z^* \le H, \tag{4.7.1}$$

$$u^* = w^* = 0, \quad T^* = T_c \ \text{at} \ x^* = L \ \text{ for } \ 0 \le z^* \le H, \tag{4.7.2}$$

$$u^* = w^* = 0 \ \text{at} \ z^* = 0, H \ \text{ for } \ 0 \le x^* \le L, \tag{4.7.3}$$

and either

$$T^* = T_h - (T_h - T_c)\frac{x}{L}, \quad \text{or} \quad \frac{\partial T^*}{\partial z^*} = 0 \ \text{at} \ z^* = 0, H, \ \ 0 < x^* < L. \tag{4.7.4}$$

$u^*$ and $w^*$ are the horizontal and vertical velocity components, respectively. Conditions (4.7.4) for the temperature on the horizontal end walls, $z^* = 0, H$, imply either a linearly varying surface temperature or an insulated horizontal wall.

Since the geometry for this problem is two-dimensional, the solution can be formulated in terms of the *stream function* $\psi^*$ defined in the following manner:

$$u^* = \frac{\partial \psi^*}{\partial z^*} \qquad w^* = -\frac{\partial \psi^*}{\partial x^*}. \tag{4.7.5}$$

The governing equations (4.6.1)–(4.6.4) are nondimensionalized using the cavity height $H$ for a length scale. Also, as in the previous section, the thermal diffusivity

$\kappa$ will be used to scale the velocity field inside the cavity. Thus, the variables are nondimensionalized in the following manner:

$$x = \frac{x^*}{L}, \quad z = \frac{z^*}{L}, \quad u = \frac{u^*L}{\kappa}, \quad w = \frac{w^*L}{\kappa},$$

$$\theta = \frac{T^* - T_c}{T_h - T_c}, \quad \psi = \frac{\psi^*}{\kappa}. \tag{4.7.6}$$

Introducing the nondimensional variables into equations (4.6.1)–(4.6.4) and using the definition for the stream function $\psi$, the vertical and horizontal momentum equations can be combined into the following single equation in terms of the stream function and the temperature:

$$\nabla_1^2(\nabla_1^2\psi) = \frac{1}{Pr}\left[\frac{\partial\psi}{\partial z}\frac{\partial}{\partial x} - \frac{\partial\psi}{\partial x}\frac{\partial}{\partial z}\right]\nabla_1^2\psi - R\frac{\partial\theta}{\partial x}. \tag{4.7.7}$$

Similarly, the nondimensional energy equation takes the following form in terms of the stream function $\psi$:

$$\nabla_1^2\theta = \frac{\partial\psi}{\partial z}\frac{\partial\theta}{\partial x} - \frac{\partial\psi}{\partial x}\frac{\partial\theta}{\partial z}, \tag{4.7.8}$$

where

$$\nabla_1^2 = \frac{\partial^2}{\partial x^2} + \frac{\partial^2}{\partial z^2}.$$

$R$ is the Rayleigh number, which has already been defined in the previous sections.

The boundary conditions in terms of the nondimensionalized variables become

$$x = 0 \quad \text{for} \quad 0 \leq z \leq A: \quad \psi = \frac{\partial\psi}{\partial x} = 0, \quad \theta = 1, \tag{4.7.9}$$

$$x = 1 \quad \text{for} \quad 0 \leq z \leq A: \quad \psi = \frac{\partial\psi}{\partial x} = 0, \quad \theta = 0, \tag{4.7.10}$$

$$z = 0, A \quad \text{for} \quad 0 \leq x \leq 1: \quad \psi = \frac{\partial\psi}{\partial x} = 0, \quad \theta = 1 - x, \quad \text{or} \quad \frac{\partial\theta}{\partial z} = 0. \tag{4.7.11}$$

Here $A = H/L$ is the aspect ratio.

The above equations possess the following parameters: the Rayleigh number $R$, the Prandtl number $Pr$, and the aspect ratio $A$. For insulated horizontal surfaces, the thermal energy is transported only through the vertical surfaces. If the horizontal surfaces are kept at adiabatic conditions, the Nusselt number for the net nondimensional heat flux $q$ between the vertical walls is defined as

$$Nu = \frac{hL}{k} \quad \text{where} \quad h = \frac{q}{T_h - T_c}.$$

$h$ is the heat transfer coefficient, which is defined above in terms of the heat flux.

Equations (4.7.7) and (4.7.8) with boundary conditions (4.7.9)–(4.7.11) form a coupled partial differential boundary value problem. The general solution for this problem for arbitrary values of the parameters can only be obtained through numerical means. However, approximate analytical solutions can be reached for large ($\to \infty$) or small ($\to 0$) values of the parameters. Before the advent of high-speed computing machines, the latter approach was the only practical method for obtaining solutions for such a problem. In the previous sections, solutions were developed, for instance, for the two limiting values of the aspect ratio $A$, i.e., for $A \to 0$ and $A \to \infty$.

Reference [23] outlines the following solution valid for small values of the Rayleigh number $R$. Under this assumption the stream function and the temperature may be expanded in power series in terms of $R$:

$$\psi(x,z) = R\psi_1(x,z) + R^2\psi_2(x,z) + \cdots, \tag{4.7.12}$$

$$\theta(x,z) = z + R\theta_1(x,z) + R^2\theta_2(x,z) + \cdots. \tag{4.7.13}$$

The first term in the stream function expansion was found in [23] to be

$$\psi_1 = \frac{2}{3}\frac{(A-z)^2}{1+A^4}z^2x^2(1-x)^2, \tag{4.7.14}$$

while the maximum value for $\theta_1$ was found to be

$$\theta_{1\{max\}} = \left(\frac{1}{2880}\right)\frac{A^3}{1+A^4}\left[1 - \frac{32}{\pi^3}\sum_{n=0}^{\infty}\frac{(-1)^n}{(2n+1)^3\cosh(2n+1)\pi A/4}\right]. \tag{4.7.15}$$

This maximum value for the temperature was found in [23] to occur at $x = 1/2$ and $z = A/4$. For a value of $A = 2$, the maximum value for the nondimensional temperature was found to be $0.97 \times 10^{-4}$, while for $A = 4$ it was found to be $0.79 \times 10^{-4}$. The perturbation method of [23] shows that for small values of the Rayleigh number there is little increase in the heat flux over its value for the conduction case only for $R < 1000$ when $A$ is large. The same analysis shows that conduction was the only mode of heat transfer as $A \to \infty$ for all values of the Rayleigh number. This was also the result of the previous section.

With the widespread availability of computers and general solver routines, solving the two-dimensional problem defined by (4.7.7)-(4.7.11) has become a straightforward task. These numerical solutions can be easily determined as long as the values for both $R$ and $A$ are moderate. They become increasingly difficult to obtain when the values for these two parameters become large. The numerical solution for the three-dimensional case possesses difficulties even for moderate values for the Rayleigh number. At the present time there are many commercially available numerical codes for solving the free convection problem for most reasonable geometries.

In an early review of the free convection problem in an enclosure, reference [42] formulated the nondimensional heat transfer coefficient (i.e., the Nusselt number), in terms of the Rayleigh number and the aspect ratio, using results from several investigations. In [42], a single equation was formulated for the Nusselt number that combines the results of different investigations in the following manner:

$$Nu = a\,R^b\,A^c, \tag{4.7.16}$$

where the constants $a$, $b$, and $c$ are given in Table 4.2. The results of equation (4.7.16) are shown in Figure 4.22 for different values of the aspect ratio $A$.

Many more solutions for the free convection problem, including the evaluation of the heat transfer coefficient in finite cavities under various conditions, can be found in [30].

**Table 4.2** The Nusselt number as a function of the Rayleigh number (see [42]).

| Symbol | $a$ | $b$ | $c$ | $H/L$ |
|---|---|---|---|---|
| ———— | 0.0782 | 0.3594 | — | 1 |
| - - - - - - | 0.199 | 0.3 | $-0.1$ | $2 \leq H/L \leq 20$ |
| . . . . . . . | 0.18 | 0.25 | $-0.111$ | $2 \leq H/L \leq 20$ |

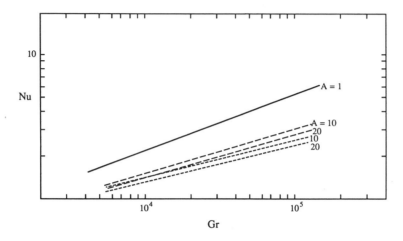

**Figure 4.22** The variation of the Nusselt number as a function of the Rayleigh number and aspect ratio $A$ from different calculations (from [42]).

## *4.8 SCALE ANALYSIS IN BUOYANCY-DRIVEN CONVECTION*

As was demonstrated in the previous sections, the heat flux $q$ is normally defined in terms of the parameters of the problem through a relationship between the Nusselt number and the remaining parameters in the following manner:

$$Nu = f(R, Pr, A). \qquad (4.8.1)$$

Equation (4.8.1) is only a functional statement whose exact form depends on the specific problem under consideration. For an infinite horizontal layer, for instance, the equation to use is either (4.2.39), (4.2.52), or (4.2.54). For an infinite vertical layer it is given by (4.6.17). For flow conditions that are more complex than these two idealized geometries, such a relationship can only result from extensive analysis, numerical or otherwise. Equation (4.7.16) is one example of such a complicated expression.

Order of magnitude estimates may be very useful in place of exact relationships when these are not known or exact results are not needed. Such estimates can be determined from the conservation equations by balancing the various terms in the governing equations. With this method only a handful of data for any specific process, including low-gravity processes, can provide reasonable estimates for the heat flux without resorting to complicated numerical solutions or an extensive experimental program. The order-of-magnitude estimate analysis developed by [40] is reviewed in this section. Such analysis can serve as a very useful engineering tool. This technique is especially useful for applications in low-gravity processes since no data base exists or, at most, the existing data is very sparse.

The starting point for the analysis is again the conservation equations, which include the conservation of mass, momentum, energy, and species equations. The steady-state forms of these equations are used here to simplify the analysis:

$$\frac{\partial v_i^*}{\partial x_i^*} = 0, \tag{4.8.2}$$

$$\rho v_j^* \frac{\partial v_i^*}{\partial x_j^*} = -\frac{\partial p^*}{\partial x_i^*} - \beta_t \rho g_i (T^* - T_r^*) + \beta_c \rho g_i (C^* - C_r^*) + \mu \frac{\partial}{\partial x_j^*} \frac{\partial v_i^*}{\partial x_j^*}, \tag{4.8.3}$$

$$\rho c_p v_j^* \frac{\partial T^*}{\partial x_j^*} = k \frac{\partial^2 T^*}{\partial x_j^* \partial x_j^*}, \tag{4.8.4}$$

$$v_j^* \frac{\partial C^*}{\partial x_j^*} = D \frac{\partial^2 C^*}{\partial x_j^* \partial x_j^*}, \tag{4.8.5}$$

where $v_j^*$ is the dimensional velocity component in the $x_j^*$ direction and $C^*$ is the dimensional species concentration. Equation (4.8.4) is the modified Boussinesq equation derived in Section 4.5, in which the density is assumed to be a linear function of both the temperature and the species concentration, in the following manner:

$$\rho = \rho_r [1 - \beta_t (T^* - T_r^*) + \beta_c (C^* - C_r^*)]. \tag{4.8.6}$$

An essential step for the order-of-magnitude estimate analysis is to write all of the variables in the conservation equations in nondimensional form. Thus all of the variables in equations (4.8.2)–(4.8.5) are made dimensionless in the following manner:

$$u_i = v_i^*/U, \quad x_i = x_i^*/L, \quad \theta = (T^* - T_r^*)/(T_w^* - T_r^*),$$

$$\phi = (C^* - C_r^*)/(C_w^* - C_r^*), \quad p = p^*/\rho U^2.$$

$U$ in the above definitions denotes an as yet undetermined reference velocity which serves as a velocity scale in the analysis. $L$ is a characteristic length scale, while the subscripts $w$ and $r$ refer to two different reference points. Substituting the nondimensional variables into equations (4.8.2)–(4.8.5), the following dimensionless equations result:

$$\frac{\partial u_i}{\partial x_i} = 0, \tag{4.8.7}$$

$$u_j \frac{\partial u_i}{\partial x_j} = -\frac{\partial p}{\partial x_i} + \frac{\beta_t L g_i \Delta T^*}{U^2} (\theta + N\phi) + \frac{\nu}{UL} \frac{\partial}{\partial x_j} \frac{\partial u_i}{\partial x_j}, \tag{4.8.8}$$

$$u_j \frac{\partial \theta}{\partial x_j} = \frac{\kappa}{UL} \frac{\partial^2 \theta}{\partial x_j \partial x_j}, \tag{4.8.9}$$

$$u_j \frac{\partial \phi}{\partial x_j} = \frac{D}{UL} \frac{\partial^2 \phi}{\partial x_j \partial x_j}. \tag{4.8.10}$$

$\Delta T^* = T_w^* - T_r^*$ is an imposed temperature difference, and $N$ is the ratio of the buoyancy force due to concentration stratification to the thermal buoyancy defined as

$$N = \beta_c \Delta C^* / \beta_t \Delta T^*, \qquad (4.8.11)$$

where $\Delta C^* = C_w^* - C_r^*$.

The value of the parameter $N$ could vary over a wide range depending on the specific application. Positive values for $N$ imply that the combined driving forces augment each other, while negative values for $N$ imply that they oppose each other. The rest of the parameters appearing in the equations are the Reynolds number, the *Peclet number*, the *Schmidt number*, and the Grashof number, defined in the following manner:

$$Re = UL/\nu, \quad Pe = \kappa/(UL) = Pr\,Re$$

$$Sc = D\nu/(U^2 L^2), \quad Sc\,Re = D/(UL)$$

$$Gr = g\beta L^3 \Delta T^*/\nu^2, \quad R/(Pr\,Re^2) = Gr/Re^2 = g\beta L \Delta T^*/U^2.$$

$R$ in the above definitions is the Rayleigh number.

Reference [40] shows that a value for the reference velocity, $U$, can be determined by proper balance considerations of the various terms in the conservation equations. This is true for any process that can be described by the above equations when the values of the parameters are known. If it is assumed, for instance, that the flow is primarily due to thermal effects (i.e., $N \ll 1$), then $U$ may be determined in the following manner.

Since the buoyancy force is the driving mechanism in the above flow, the buoyancy term must always be included in the balance analysis. When the flow is driven by the buoyancy force alone, equation (4.8.8) shows that there must be a balance between the buoyancy and viscous forces. Hence the coefficients of the second and third terms in equation (4.8.8) must be assigned the same order of magnitude, i.e.,

$$\frac{\beta_t g L \Delta T^*}{U^2} \approx \frac{\nu}{UL}. \qquad (4.8.12)$$

Equation (4.8.12) may be solved for the velocity scale $U$ to yield

$$U \approx \beta g \Delta T^* L^2/\nu = Gr(\nu/L). \qquad (4.8.13)$$

If, on the other hand, a thermal boundary layer exists, then the thickness of the layer should be considered as the fundamental length scale, since the buoyancy force acts over that length. This requires stretching the coordinate in the direction normal to the boundary layer to allow for the buoyancy and the viscous terms, in equation (4.8.8), to be of the same order of magnitude. This analysis was developed in [41] resulting in the following values for $U$:

$$U \approx \sqrt{\beta g \Delta T^* L} = (\nu/L)\sqrt{Gr} \quad \text{for } Pr \ll 1, \qquad (4.8.14)$$

$$U \approx \sqrt{\beta g \Delta T^* L/Pr} = (\nu/L)\sqrt{Gr/Pr} \quad \text{for } Pr \gg 1. \qquad (4.8.15)$$

When there is only a velocity boundary layer, a case in which the flow is driven by an external force, a balance between the inertia and buoyancy forces must exist. This requirement leads to a reference velocity similar to that given by expression (4.8.14).

In situations where the buoyancy, the inertia, and the viscous terms are all of the same order, then either equation (4.8.14) or (4.8.15) can be used. When the reference velocity obtained in this manner is substituted into the definitions for the Reynolds and the Peclet numbers, criteria in terms of buoyancy-driven flow parameters for the use of these velocities are obtained. Reference [40] shows that such considerations lead to the following estimates for the reference velocity scale $U$:

$$U \approx Gr(\nu/L) \quad \text{for} \quad Gr \le 1 \quad \text{and} \quad R \le 1,$$

$$U \approx (\nu/L)\sqrt{Gr} \quad \text{for} \quad \sqrt{Gr} > 1 \quad \text{and} \quad Pr < 1,$$

$$U \approx (\nu/L)\sqrt{Gr/Pr} \quad \text{for} \quad \sqrt{Gr} > 1 \quad \text{and} \quad Pr > 1.$$

For situations in which the flow is primarily due to concentration differences, $N \gg 1$, $\beta_c \Delta C^*$ should replace $\beta_t \Delta T^*$ in the definition for the Grashof number. When both thermal and concentration effects are of the same order (i.e., for $N \approx 1$), then $\Delta \rho / \rho$ should replace $\beta_t \Delta T^*$.

# Chapter 5

# Materials Processing

**M**aterials processing in space is an important example of the commercial utilization of the space environment. The tremendous reduction in gravity presents a decisive advantage for producing high-quality crystalline materials. It has also led to the manufacturing of true spherical shapes for much larger solid volumes than was possible in the terrestrial environment. In addition to the obvious advantages of low gravity, there are many other reasons that make manufacturing of materials in space very attractive.

The production of higher quality crystals in space processing invariably implies larger crystals without defects and alloys with compositional uniformity. The planarity of the solid/melt interface during solidification leads to crystals without, or with minimum, facets. The homogeneity of an alloy crystal, on the other hand, is achieved primarily by suppressing fluid convection in the melt during solidification.

The solidification of materials naturally involves temperature gradients across the solid/melt interface in pure materials, and solute concentration gradients in alloys. Such solute and temperature gradients could induce free convection in the melt. On the other hand, the accompanying generation of mass and heat at the interface could in turn strengthen or weaken these gradients. There exists a complex coupling between the shape and the propagation speed of the interface and these gradients. In order to achieve the desired results in space processing, strict controls of the thermal and solute environments in the melt, crystal, and ampoule are necessary. Such controls mandate a thorough understanding of the thermal properties in the solidification process.

This chapter develops the essentials of solidification processes in terms of heat transfer and fluid dynamics. First we treat the *Stefan problem*, in which the effects of the thermal gradients alone on the speed and shape of the crystal/melt interface are described. Next, the combined effects of both the solute gradients and the temperature gradients are analyzed. The morphological instability due to small departures from the theoretical limits of the energy and mass fluxes is examined. This instability could lead to the formation of uneven interfaces and, consequently, defective crystals. Finally, the role played by CFD in space processing is explored.

## 5.1 *SOLIDIFICATION OF PURE MATERIALS*

The simplest solidification process is that of a pure substance from its melt, such as the freezing of ice in a sample of pure water. The reverse process of melting is equally interesting, and all processes described in this section have reverse analogues. For clarity, however, it is easier to think primarily in terms of solidification. For the case of a pure substance, the solidification process is governed entirely by heat fluxes. The rate of solidification at any point along the solid/liquid interface is governed by the speed with which the latent heat, generated or absorbed at that point, is conducted into the bulk of the sample or removed at the boundaries. Quantitatively, this statement can be formulated in the following manner.

If $T_m$ is the freezing (melting) temperature of a substance, then the temperature of both the solid and the liquid must be $T_m$ at the solid/liquid interface, which is designated here by $S(t)$, i.e.,

$$T_s = T_l = T_m \quad \text{at} \quad x = S(t). \tag{5.1.1}$$

The subscripts $s$ and $l$ stand for solid and liquid, respectively.

Another thermal criterion that must be satisfied at the interface deals with the absorption or liberation of the latent heat at the interface. Let the region $x > S(t)$ be designated as containing liquid at a temperature $T_l(x, t)$, and the region $x < S(t)$ as containing solid at a temperature $T_s(x, t)$ and let $h_L$ cal/gm be the latent heat of fusion of the substance. As the solid/liquid interface moves a distance $dS$, a quantity of heat equivalent to $(\rho h_L dS)$ per unit area is liberated. This heat must be removed by conduction alone when there is no motion in the liquid. The energy balance across a one-dimensional solid/liquid interface requires the following condition to be satisfied (see [50]):

$$k_s \frac{\partial T_s}{\partial x} - k_l \frac{\partial T_l}{\partial x} = \rho h_L \frac{dS}{dt}, \tag{5.1.2}$$

where $k_s$ and $k_l$ are the coefficients of thermal conduction for solid and liquid, respectively.

For a pure material, conditions (5.1.1) and (5.1.2) must be satisfied simultaneously at the solid/liquid interface. For a general three-dimensional geometry, the interface condition (5.1.2) can be written in the following alternative form:

$$k_s |\nabla T_s| - k_l |\nabla T_l| = \pm h_L \rho \frac{\partial T_s / \partial t}{|\nabla T_s|} = \pm h_L \rho \frac{\partial T_l / \partial t}{|\nabla T_l|}. \tag{5.1.3}$$

The applications of these conditions can be demonstrated by considering the following simple example from [52]. Consider a semi-infinite liquid region initially at the equilibrium freezing temperature $T_m$, as shown in Figure 5.1. At time zero,

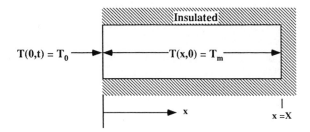

**Figure 5.1** Sketch for the Stefan problem.

one end of the region is suddenly cooled to a lower temperature $T_0$ corresponding to the temperature in the solid. At that time, freezing commences through heat conduction in the solid. At a later time there will be two regions that are separated by the interface; one region is solid next to the boundary and the other is liquid. Assuming all of the thermophysical properties to be independent of temperature, the temperature in both the solid, $T_s$, and the liquid, $T_l$, are governed by the conservation of energy equations. These equations in one-dimensional geometry and in the absence of motion in both regions take the form

$$\rho_s c_{ps} \frac{\partial T_s}{\partial t} = k_s \frac{\partial^2 T_s}{\partial x^2} \quad \text{for } 0 \le x \le S(t) \tag{5.1.4}$$

$$\rho_l c_{pl} \frac{\partial T_l}{\partial t} = k_l \frac{\partial^2 T_l}{\partial x^2} \quad \text{for } x > S(t) \tag{5.1.5}$$

where $c_{ps}$ and $c_{pl}$ are the specific heats for the solid and liquid, respectively.

At the solid/liquid interface both conditions (5.1.1) and (5.1.2) must hold. Assuming that the liquid ahead of the interface remains at the constant temperature $T_m$, equation (5.1.5) need not be solved. With this assumption, the interface energy conservation condition (5.1.2) is reduced to the form

$$k_s \frac{\partial T_s}{\partial x} = \rho h_L \frac{dS}{dt} \quad \text{at } x = S(t). \tag{5.1.6}$$

Condition (5.1.6) shows that the position of the interface is a function of the temperature distribution in the solid. A major source of difficulty in solidification problems is the coupling between the field equations and the interface condition, as shown by condition (5.1.6). For the example considered here, the following additional boundary conditions must be imposed:

$$T = T_0 \quad \text{at } x = 0, \tag{5.1.7}$$

and

$$\frac{\partial T}{\partial x} = 0 \quad \text{at } x = X. \tag{5.1.8}$$

An initial condition on the position of the interface is needed to complete the specification of the problem. For the present example, the following initial condition is used

$$S = 0 \quad \text{at } t = 0. \tag{5.1.9}$$

Equation (5.1.4) is commonly known as the *diffusion equation*, which is a parabolic partial differential equation. Also, since equation (5.1.4) is linear, a solution for this equation is attempted through the *similarity method*. A similarity variable $\eta$ is introduced for this purpose, defined by [50] as

$$\eta = \frac{x}{2\sqrt{\kappa_s t}}, \tag{5.1.10}$$

where $\kappa_s = k_s/(\rho_s c_{ps})$. Since many combinations of values for $x$ and $t$ give the same $\eta$, the answer depends only on $\eta$. Such a combination for the variables also means that there are many points in $x$ and $t$ that have a "similar" answer. To transform equation (5.1.4) in terms of $\eta$, the following derivatives are also needed:

$$\frac{\partial}{\partial t} = \frac{\partial \eta}{\partial t}\frac{d}{d\eta} = -\frac{\eta}{2t}\frac{d}{d\eta},$$

$$\frac{\partial}{\partial x} = \frac{\partial \eta}{\partial x}\frac{d}{d\eta} = \frac{1}{2\sqrt{\kappa_s t}}\frac{d}{d\eta},$$

$$\frac{\partial^2}{\partial x^2} = \frac{1}{4\kappa_s t}\frac{d^2}{d\eta^2}.$$

Substituting these relations into equation (5.1.4) results in the following ordinary differential equation for $T_s$ as a function of $\eta$ alone:

$$\frac{d^2 T_s}{d\eta^2} + 2\eta\frac{dT_s}{d\eta} = 0. \tag{5.1.11}$$

Equation (5.1.11) can be integrated once to yield

$$\frac{dT_s}{d\eta} = A_1 \exp(-\eta^2). \tag{5.1.12}$$

Upon integrating (5.1.12) once more, the following solution for $T_s$ results:

$$T_s = A_1 \int_0^\eta \exp(-\xi^2)d\xi + A_2. \tag{5.1.13}$$

The integral on the right-hand side of equation (5.1.13) is known as the *error function*, defined by

$$\text{erf}(\eta) = \frac{2}{\sqrt{\pi}} \int_0^\eta \exp(-\xi^2)\,d\xi, \tag{5.1.14}$$

where $\text{erf}(0) = 0$ and $\text{erf}(\infty) \to 1$. The boundary condition given by (5.1.7) can be used to evaluate the constant $A_2$ in equation (5.1.13). This will result in the following solution for the temperature in the solid:

$$T_s = A_1 \left(\frac{\sqrt{\pi}}{2}\right) \text{erf}\left(\frac{x}{2\sqrt{\kappa_s t}}\right) + T_0. \tag{5.1.15}$$

The temperature distribution in the solid must also satisfy both conditions (5.1.1) and (5.1.2) at the liquid/solid interface $S(t)$. Upon substituting expression

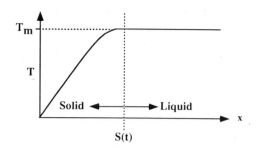

**Figure 5.2** Temperature distribution in the solid for the Stefan problem.

(5.1.15) into these conditions, the following two equations result:

$$T_m = A_1 \left( \frac{\sqrt{\pi}}{2} \right) \text{erf} \left( \frac{S}{2\sqrt{\kappa_s t}} \right) + T_0, \tag{5.1.16}$$

$$A \rho_s c_{ps} \exp \left( -\frac{S^2}{4\kappa_s t} \right) = \rho h_L \left( \frac{S}{2\sqrt{\kappa_s t}} \right). \tag{5.1.17}$$

Inspection of equation (5.1.16) shows that the argument for the error function, $S/(2\sqrt{\kappa_s t})$, must be a constant as both $T_m$ and $T_0$ are not functions of time. This observation leads to the definition of another constant, $\lambda$, in the following manner:

$$\lambda = \frac{S}{2\sqrt{\kappa_s t}}. \tag{5.1.18}$$

Equations (5.1.16) and (5.1.17) can now be solved for the two constants $\lambda$ and $A_1$. $A_1$ is evaluated from equation (5.1.16). Upon substituting the value for $A_1$ into equation (5.1.17), the following transcendental equation for the constant $\lambda$ results:

$$\lambda \exp(\lambda^2) \text{erf}(\lambda) = \frac{\rho_s c_{ps}}{\rho h_L \sqrt{\pi}} (T_m - T_0). \tag{5.1.19}$$

Finally, when the value for the constant $A_1$ is substituted into equation (5.1.16), the following solution for the temperature distribution in the solid results:

$$T_s(x, t) = T_0 + \frac{T_0 - T_m}{\text{erf}(\lambda)} \text{erf} \left( \frac{x}{2\sqrt{\kappa_s t}} \right). \tag{5.1.20}$$

Equation (5.1.19) shows that the constant $\lambda$ is only a function of the thermodynamic properties of the material and the boundary conditions. Once $\lambda$ is evaluated for the specific material, then the interface position $S(t)$ can be determined with the aid of equation (5.1.18) in the following manner:

$$S(t) = 2\lambda\sqrt{\kappa_s t}, \quad \text{and} \quad \frac{dS}{dt} = \frac{\lambda\sqrt{\kappa_s}}{\sqrt{t}}. \tag{5.1.21}$$

Note that the solution for the interface propagation rate $dS/dt$ given in (5.1.21) contains an artificial singularity as $t \to 0$. This singularity is caused by the instantaneous change in the temperature at the boundary $x = 0$. Figure 5.2 shows a schematic of the temperature distribution given by (5.1.20).

It is important to note that the thickness of the solidified layer as measured from the beginning of the solidification process is proportional to $\sqrt{t}$. This relationship makes the corresponding interface velocity inversely proportional to the square root of time. For example, consider the solidification of pure aluminum for which $T_m = 660°C$ against a perfectly conducting wall set at 25°C, where $\kappa_s = 0.981$ cm$^2$/sec. $\lambda$ in this case is found to be $\lambda = 0.693$, which leads to a value for the interface velocity of $dS/dt = 0.686$ cm/sec$^{1/2}/\sqrt{t}$.

The solution given in (5.1.20) is one of the simplest solutions for the solidification problems encountered in materials processing, namely, the solidification of a pure material in one space dimension. Reference [50] states that this problem was first formulated by J. Stefan in a study of the thickness of polar ice. Problems concerned with freezing and melting have come to be known subsequently as *Stefan problems.*

The complexity of the solution, derived in (5.1.20), is obvious even for this simple model. Although the field equations [the diffusion equations given in (5.1.4) and (5.1.5)] are linear, the conditions to be satisfied at the solid/liquid interface [conditions (5.1.1) and (5.1.2)] make the problem highly nonlinear. The nonlinearity of the problem stems from the fact that the interface position $S(t)$ in this case is another unknown that must be determined as part of the solution itself. This difficulty, which is associated with the nonlinearity of the problem, remains even for one-dimensional geometry and when fluid convection in the liquid is neglected. Adding one more dimension and liquid motion will complicate the problem even further, and only numerical solutions are possible for such a problem.

Much of the work on solidification has correlated the crystal microstructure with the interface velocity and the temperature gradients during freezing. However, even this simple model is too complex to be used for an actual solidification model, due to the fact that the interface speed and the temperature gradients vary continuously during the solidification process. Clearly, the difficulty associated with the Stefan problem can be reduced considerably by assuming the interface propagation rate $dS/dt$ to be an externally imposed parameter. Under these conditions, the problem becomes manageable and may be applied easily to materials processing. Bridgman developed a solidification furnace in which the interface velocity is controlled; this furnace will be discussed in a later section. However, the furnace pull rate $U$ must obey certain criteria in order for the solidification front to remain stable. These criteria relate the pull speed to the internally determined interface speed $dS/dt$, which as demonstrated is a function of the imposed temperature gradients across the interface. Such criteria are determined through the morphological instability analysis, which will also be studied in a later section of this chapter.

## 5.2 *SOLIDIFICATION OF BINARY ALLOYS*

The difficulties encountered in the previous section, when solidifying a pure material, are further compounded for materials that are composed of two component alloys. Such materials are commonly known as *binary alloys.* The difficulty in this case is mainly due to the inclusion of the mass diffusion equations into the system as well as to imposing mass conservation across the solid/liquid interface. To understand the problems connected with freezing of binary alloys, the process is assumed to be achieved by imposing an externally controlled solidification rate. This constrained growth rate corresponds to the case in which a very long sample is made to freeze unidirectionally at a constant rate. This can be accomplished, for instance, by a traveling furnace. This is a very important model, because it approximates many practical crystal growth processes.

Following [60], the simplest model for alloy solidification is examined here. This is the basic problem of steady-state freezing of a dilute binary alloy at a constant velocity $U$. When dealing with binary mixtures, the solute transport across the solid/liquid interface must be accounted for in addition to the heat transport treated in the previous section. To achieve this, the boundary conditions for the conservation of solute mass across the interface must be developed. The solute mass flux at the interface must be balanced by the solute mass deficit from one phase to another; in other words, in a one-dimensional geometry, the following condition must be applied:

$$D_s \frac{\partial C_s}{\partial x} - D_l \frac{\partial C_l}{\partial x} = (C_l - C_s)\frac{dS}{dt}, \qquad (5.2.1)$$

where $C$ is the solute concentration and $D$ is the coefficient of solute mass diffusion. For most materials, the coefficient of mass diffusion in the solid is several orders of magnitude smaller than in either the liquid phase or the gas phase, i.e., $D_s \ll D_l$. If solute diffusion in the solid is ignored, then the mass flux balance at the solid/liquid interface reduces to the following condition for a general three-dimensional geometry:

$$D\nabla C_l = (C_s - C_l)\frac{dS}{dt}. \qquad (5.2.2)$$

Also, for any mixture, a special thermodynamic and kinetic balance at the interface must hold as given by the *phase diagram* for that mixture. Assuming that local equilibrium prevails at the solid/liquid boundary, then such a balance translates into the following interface condition on the solute concentration:

$$C_s = KC_l, \qquad (5.2.3)$$

where $K$ is known as the *distribution* or *segregation* coefficient. This coefficient is a constant and can be evaluated from the phase diagram for the specific alloy under consideration.

The phase diagram can be extremely complicated for most alloys; Figure 5.3 shows a typical example. For dilute alloys, in which interest is confined to either of the extreme ends of the phase diagram, it is possible to assume the solidus and liquidus curves to be straight lines, as shown in Figure 5.4. Case (a) corresponds to the rejection of solute upon freezing for which $K < 1$, while case (b) corresponds to enhanced incorporation of solute upon freezing and thus to $K > 1$.

In addition to the segregation coefficient, the phase diagram defines another parameter $m$, the slope of the liquidus line, where the sign is given by $m(1 - K)$

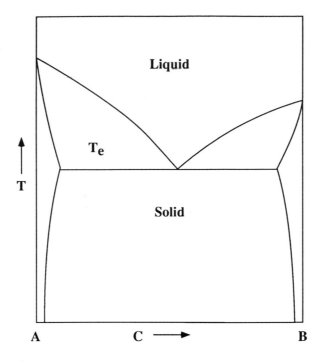

**Figure 5.3**    Sketch of a typical phase diagram.

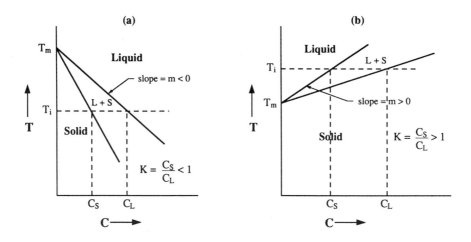

**Figure 5.4**    Linear approximation for the phase diagram appropriate for a dilute binary alloy: (a) $K < 1$ and (b) $K > 1$.

< 0. Thus, the solidification temperature for a binary alloy at the solid/liquid interface can be determined from the phase diagram by using

$$T_i = T_m + mC_l \quad \text{at} \quad x = S(t),  \tag{5.2.4}$$

where $T_m$ is the solidification temperature for the pure substance and $T_i$ is the temperature at the solid/liquid interface. Condition (5.2.4) is the counterpart of condition (5.1.1) for a binary mixture, derived by assuming a planar interface. For an arbitrarily curved solid/liquid interface, condition (5.2.4) must be modified to the general form

$$T_i = T_m + mC_l - \frac{T_m \sigma}{\rho h_L}\left(\frac{1}{R_1} + \frac{1}{R_2}\right) \quad \text{at} \quad x = S(t),  \tag{5.2.5}$$

where $\sigma$, $R_1$, and $R_2$ are the surface tension and the principal radii of curvature for the interface, respectively.

Since it is assumed that the solidification process, and hence the solid/liquid interface, is moving with a constant speed $U$, it is convenient to introduce a new coordinate system that moves with the interface speed. To this end, a new space variable $z$ is defined as follows:

$$z = x - Ut.  \tag{5.2.6}$$

The solute and energy conservation equations for the one-dimensional problem can be formulated in terms of the new coordinate, $z$, in the following manner:

$$-U\frac{\partial C_l}{\partial z} = D_l \frac{\partial^2 C_l}{\partial z^2},  \tag{5.2.7}$$

$$-U\frac{\partial T_l}{\partial z} = \kappa_l \frac{\partial^2 T_l}{\partial z^2},  \tag{5.2.8}$$

$$-U\frac{\partial T_s}{\partial z} = \kappa_s \frac{\partial^2 T_s}{\partial z^2},  \tag{5.2.9}$$

$$-U\frac{\partial C_s}{\partial z} = D_s \frac{\partial^2 C_s}{\partial z^2}.  \tag{5.2.10}$$

In its original form, the liquid solute concentration equation (5.2.7) is identical to the energy conservation equation, either (5.1.4) or (5.1.5), with $C$ replacing $T$ and $D$ replacing $\kappa$.

For a typical solidification process relevant to crystal growth, $D_l/U \approx 50\ \mu\text{m}$ whereas $\kappa_s/U \approx \kappa_l/U \approx 50$ cm. In most realistic crystal growth processes, interest is confined to changes in $T_s$, $T_l$, and $C_l$ taking place in the immediate vicinity of the solid/liquid interface making variations over distances of $\approx 50\ \mu\text{m}$ more appropriate. Based on this observation, a length scale on the order of $D_l/U$ can be defined for the purpose of nondimensionalizing the spatial coordinate in the following manner:

$$z^* = zU/D_l.$$

Using this, equations (5.2.7)-(5.2.10) can now be written in terms of $z^*$:

$$-\frac{\partial C_l}{\partial z^*} = \frac{\partial^2 C_l}{\partial z^{*2}}, \tag{5.2.11}$$

$$-\frac{D_l}{\kappa_l}\frac{\partial T_l}{\partial z^*} = \frac{\partial^2 T_l}{\partial z^{*2}}, \tag{5.2.12}$$

$$-\frac{D_l}{\kappa_s}\frac{\partial T_s}{\partial z^*} = \frac{\partial^2 T_s}{\partial z^{*2}}, \tag{5.2.13}$$

$$-\frac{D_l}{D_s}\frac{\partial C_s}{\partial z^*} = \frac{\partial^2 C_s}{\partial z^{*2}}. \tag{5.2.14}$$

For the specific case in which it is assumed that $\kappa_{l,s} \gg D_l$ and $D_l \gg D_s$, equations (5.2.11)-(5.2.14) can be reduced to the following nondimensional form:

$$-\frac{\partial C_l}{\partial z^*} = \frac{\partial^2 C_l}{\partial z^{*2}}, \tag{5.2.15}$$

$$\frac{\partial^2 T_l}{\partial z^{*2}} = 0, \tag{5.2.16}$$

$$\frac{\partial^2 T_s}{\partial z^{*2}} = 0. \tag{5.2.17}$$

Equations (5.2.15)-(5.2.17) are the appropriate equations for use in the immediate vicinity of the interface.

For consistency, the interface and other boundary conditions must also be written in the new coordinate system. The location for the origin of the coordinates in the moving system must be placed at the solid/liquid interface, i.e., $S(z^* = 0) = 0$. The interface conditions now take the form

$$T_l = T_s = T_0 = T_m + mC_0, \quad KC_0 = C_\infty \text{ at } z^* = 0, \tag{5.2.18}$$

$$\frac{dT_l}{dz^*} = G_l \text{ at } z^* = 0, \tag{5.2.19}$$

where $G_l$ is assumed given. Note that with the transformation to the moving coordinate system, the subscript $i$ has been replaced with 0. To complete the specification of the problem, a condition is needed on the solute concentration in the liquid far from the interface. We write following [60]:

$$C \to C_\infty \text{ as } z^* \to \infty. \tag{5.2.20}$$

Equation (5.2.15) can now be integrated twice to yield the solution

$$C = A_1 - A_2 \exp(-z^*),$$

where $A_1$ and $A_2$ are constants of integration. Using conditions (5.2.18) and (5.2.20) to evaluate the constants $A_1$ and $A_2$, we obtain for the solute distribution in the liquid:

$$C = C_\infty + C_\infty \left(\frac{1-K}{K}\right) \exp(-Uz/D). \tag{5.2.21}$$

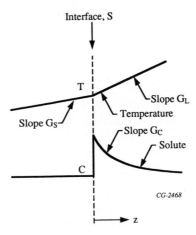

**Figure 5.5** Temperature and solute concentration distributions for the binary alloy solidification problem.

Equations (5.2.16) and (5.2.17) can be integrated easily for the temperature distributions in both the solid and the melt. The solutions for $T_l$ and $T_s$ take the following form after applying the interface conditions:

$$T_l = T_0 + G_l z, \tag{5.2.22}$$

$$T_s = T_0 + G_s z, \tag{5.2.23}$$

where $G_l$ and $G_s$ are the constant temperature gradients in the liquid and the solid, respectively.

In the above solutions, the subscript $l$ is dropped from the solute concentration since no solute diffusion in the crystal is assumed. Note that the temperature gradient in the solid, $G_s$, is not an independent parameter but is determined from the energy balance at the interface. $G_s$ can be evaluated by applying the interface energy condition (5.1.6), which may be written for the present case in the form

$$U = \frac{1}{\rho h_L}(k_s G_s - k_l G_l), \tag{5.2.24}$$

where $U$ and $G_l$ have specified values. Similarly, the solute concentration gradient at the interface is given by

$$G_C = \frac{dC}{dz}\bigg|_{z=0} = -C_\infty \left(\frac{U}{D}\right)\frac{1-K}{K}. \tag{5.2.25}$$

A sketch of these ideal solutions is shown in Figure 5.5 for the case $K < 1$. Also note that the solute rich boundary layer in the melt next to the interface drops off significantly in a distance the length of which is proportional to $\approx D/U$.

The solutions shown in Figure 5.5 for the ideal solidification case clearly show the existence of strong temperature and solute concentration gradients in the melt. These gradients are unavoidable during any solidification process. Depending on the physical orientation of the melt in relation to the gravity vector, free

**Figure 5.6**    A typical example of the striation phenomenon in alloy solidification from [55].

convection can be established in the melt in response to these gradients. Two types of convection processes are possible in the melt in the neighborhood of the crystal/melt interface. If the melt is that of a pure material and the gravity vector is such that the interface is above the melt, buoyancy-driven convection may take place due to the temperature gradients alone. For a binary alloy melt, depending on whether the segregation coefficient $K$ is such that $K < 1$ or $K > 1$, double diffusive convection may occur regardless of the position of the melt with respect to the interface. The different kinds of free convective processes that can possibly occur in a solidification configuration were discussed in Chapter 4. The obvious advantage of space processing is in the possibility of suppressing or diminishing free convection in the melt during solidification.

One of the most severe problems associated with processing electronic materials is the occurrence of compositional striations in the solid. The striations are manifested in the form of very narrow bands of different material compositions, as shown in Figure 5.6. Such striations are responsible for altering the compositional homogeneity of the crystallized alloy material and thereby pose a serious problem for the semiconductor industry. Several hypotheses have been advocated concerning the origin of these striations, the most acceptable of which traces it to the onset of oscillatory convection modes in double diffusive convection (see [49] for instance). These modes can occur in fluids for which the values for the Rayleigh numbers are such that $R > 0$ when $R_s > 0$ according to Figure 4.15. It

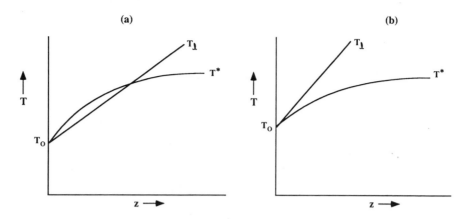

**Figure 5.7** Definition of constitutional supercooling from [60]: (a) for $K < 1$ and (b) for $K > 1$.

is obvious that the elimination of compositional defects is of great value, thus performing research on the problem of striations has become one of the major thrusts in space processing by the semiconductor industry.

The solutions derived in this section apply for ideal solidification processes. Such processes are difficult to achieve under realistic crystallization conditions. To investigate how the solution differs for a process under realistic conditions, it is appropriate to examine how small departures from the ideal geometry would affect these solutions. Following reference [60], let us assume that the solutions obtained in (5.2.21)–(5.2.25) are valid even when the interface is perturbed slightly from the imposed planar geometry. The thermodynamic freezing temperature in the liquid may be calculated by substituting the solution from (5.2.21) into condition (5.2.4) for the solidification temperature. This will result in the freezing temperature

$$T^* = T_m + m \left[ C_\infty + C_\infty \left( \frac{1-K}{K} \right) \exp(-Uz/D) \right],$$

or

$$T^* = T_0 - mC_\infty \left( \frac{1-K}{K} \right) [1 - \exp(-Uz/D)], \qquad (5.2.26)$$

where $T^*$ is the solidification temperature at a very small distance from the original interface, caused by small perturbations of the interface. $T_0$ is given by equation (5.2.18).

It is instructive to compare the temperature in (5.2.26) with the actual melting temperature given by (5.2.22). These two temperature distributions are shown graphically in Figure 5.7 for both cases (a) $K < 1$ and (b) $K > 1$. The sign convention adopted here is as before: $m(1 - K) < 0$. For case (a) there is a zone in which $T^*$ lies above $T_l$, indicating that the freezing temperature is below the actual liquid temperature. When such a condition exists, the liquid in this zone is said to be *supercooled*. The thermodynamic process, in which the liquid is actually being supercooled due to the presence of a solute concentration, is commonly known as *constitutional supercooling*. For case (b), on the other hand, there is no zone of constitutional supercooling.

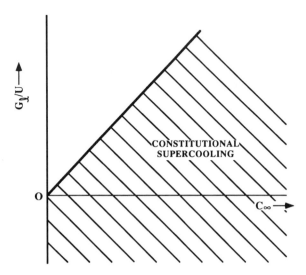

**Figure 5.8**　Constitutional supercooling stability boundaries as given by equation (5.2.30), from [60].

The initial gradients for the liquid and the solidification temperature shown in Figure 5.7 suggest, based on equation (5.2.26), that constitutional supercooling occurs whenever

$$\left.\frac{dT^*}{dz}\right|_{z=0} > G_l, \tag{5.2.27}$$

or

$$-mC_\infty \left(\frac{1-K}{K}\right)\left(\frac{U}{D}\right) > G_l. \tag{5.2.28}$$

Inequality (5.2.28) may be rewritten in the following manner with the aid of equation (5.2.25):

$$mG_C > G_l. \tag{5.2.29}$$

The boundaries separating the conditions for which there is constitutional supercooling can be delineated by replacing the inequality sign in (5.2.28) with an equality sign. The equation for the line representing this boundary is given by

$$\frac{G_l}{U} = -m\left(\frac{1-K}{DK}\right)C_\infty. \tag{5.2.30}$$

This line is shown graphically in Figure 5.8 in a plane where the coordinates are given by $G_l/U$ and $C_\infty$.

## 5.3 MORPHOLOGICAL INSTABILITY

The solutions obtained for the simple solidification models discussed in the previous sections are possible only when the ideal conditions assumed in these models are maintained. Situations such as uniform heat fluxes over large boundaries are

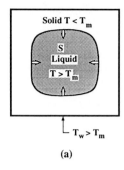
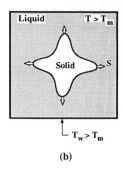

(a)                              (b)

**Figure 5.9**  Example of processes leading to morphological instability from [57].

very difficult to keep during a realistic solidification process. Also, variations in thermal and physical conditions are always present even for closely controlled processes. Indeed, imperfections are very difficult to eliminate in certain processes, that results in the ever present dendritic growths.

The problems associated with nonuniformities in solidification are very well illustrated by the following example from [57]: Consider the two simple experimental configurations shown schematically in Figure 5.9. These experiments show a pure liquid contained in a vessel with walls held at some temperature $T_w$, which is less than the melting temperature $T_m$. In case (a), the liquid is initially at a temperature $T \geq T_m$, and the solidification front $S$ propagates from the wall inward. It moves smoothly and uniformly toward the center at a rate proportional to the rate of heat conduction through the surrounding solid. Such a configuration is completely stable. In the second case, the liquid is initially undercooled to a temperature $T < T_m$, and the solidification process is initiated at a seed crystal located at the center of the vessel. The latent heat in this case must be conducted through the liquid. Even in the absence of convection in the melt for this case, the interface $S$ breaks up into dendrites that grow relatively rapidly out from the central seed. This growth is intrinsically unstable.

The essential difference between the two processes is that in the unstable case, the solid front advances into a metastable phase, that is, into an undercooled liquid. This type of instability can lead to dendritic growths resulting in perturbances that are typically on the order 10–100 $\mu m$. These growth features have no direct connection to smoothness or roughness on an atomic scale, but could lead to facets in crystals. The instability resulting in such growths is known as *morphological instability* and is caused by perturbations in the solidification process. Such perturbations may be due to actual crystalline imperfections, thermal fluctuations, inhomogeneities in composition, or other causes. At various stages during the growth process, these imperfections may cause small fluctuations in the shape of the solid/liquid interface, such as depicted by the curves at the left in Figure 5.10. Under stable conditions, the perturbations will decay with time and the interface will return to a smooth shape. For unstable conditions, the perturbations will be amplified, causing greater disturbance, lateral spreading, and eventually a distorted and often corrugated interface.

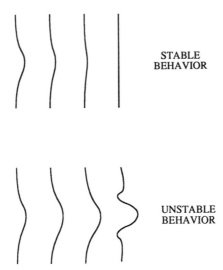

STABLE
BEHAVIOR

UNSTABLE
BEHAVIOR

**Figure 5.10**   Graphical representation of interface perturbations growth and decay.

This example shows that morphological instability may take place whenever there is any degree of supercooling in the melt. Supercooling can easily occur in alloy solidification due to the lowering of the solidification temperature by the solute concentration as demonstrated in Section 5.2. Reference [60] shows that it is possible to expect morphological instability whenever constitutional supercooling is present.

Morphological instability can be semi-quantitatively understood by examining equations (5.2.26)–(5.2.29) and Figure 5.8. It can be seen that a large positive value for $G_l$ provides a stabilizing influence that may be less effective with higher pull velocity $U$. Impurity concentration $C_\infty$ is destabilizing. Also note that the above considerations will always predict instability if $G_l < 0$; for $C_\infty = 0$, this would be supercooling in the ordinary sense.

Next, a formal derivation for the morphological instability criteria is presented using the hydrodynamic instability mechanism. This derivation proceeds along the same lines as adopted in Chapter 4 on thermal instability in fluids. The starting point for the derivation is the two-dimensional counterpart of the energy and species conservation equations in Section 5.2. Under the same assumptions as in Section 5.2, namely for a constant solidification rate $U$ and in the absence of fluid convection in the melt, these equations take the following form:

$$\frac{\partial C^*}{\partial t^*} - U\frac{\partial C^*}{\partial z^*} = D\left(\frac{\partial^2 C^*}{\partial z^{*2}} + \frac{\partial^2 C^*}{\partial y^{*2}}\right), \qquad (5.3.1)$$

$$\rho_j c_{pj}\left(\frac{\partial T_j^*}{\partial t^*} - U\frac{\partial T_j^*}{\partial z^*}\right) = k_j\left(\frac{\partial^2 T_j^*}{\partial z^{*2}} + \frac{\partial^2 T_j^*}{\partial y^{*2}}\right), \quad j = l, s. \qquad (5.3.2)$$

It is convenient to nondimensionalize the variables in equations (5.3.1) and (5.3.2) in the following manner:

$$(z, y) = (z^*, y^*)U/D, \quad t = t^* U^2/D, \quad C = C^* U/G_C D, \quad T_j = T_j^* U/GD,$$

where $U$ is the uniform solidification rate, $D$ is the solute diffusion coefficient in the liquid, $G_C$ is the solute concentration gradient at the interface defined in Section 5.2, and $G$ is the thermal conductivity weighted temperature gradient given by

$$G = \frac{k_s G_s + k_l G_l}{k_s + k_l}, \tag{5.3.3}$$

where $k_{s,l}$ and $G_{s,l}$ are the thermal conductivities and the stationary temperature gradients in the solid and liquid, respectively. The variables with an asterisk are dimensional and the subscript $l$ for the solute concentration equation has been dropped since, as in Section 5.2, no variation of the solute concentration in the solid is considered. Substituting the nondimensionalized variables into equations (5.3.1) and (5.3.2), we get:

$$\frac{\partial C}{\partial t} - \frac{\partial C}{\partial z} = \left( \frac{\partial^2 C}{\partial z^2} + \frac{\partial^2 C}{\partial y^2} \right), \tag{5.3.4}$$

$$Le_j \left( \frac{\partial T_j}{\partial t} - \frac{\partial T_j}{\partial z} \right) = \left( \frac{\partial^2 T_j}{\partial z^2} + \frac{\partial^2 T_j}{\partial y^2} \right), \quad j = l, s. \tag{5.3.5}$$

$Le_{s,l}$ are the *Lewis numbers* for the solid and liquid phases, respectively, defined by $Le = D/\kappa$.

Next, the temperature and the concentration functions are resolved in terms of an average stationary component and a fluctuation, in the following manner:

$$T_{s,l}(y, z, t) = \overline{T}_{s,l}(z) + T'_{s,l}(y, z, t), \tag{5.3.6a}$$

$$C(y, z, t) = \overline{C}(z) + C'(y, z, t). \tag{5.3.6b}$$

Upon substituting these expansions and subtracting the averaged values, equations for the temperature and concentration fluctuations result:

$$\frac{\partial C'}{\partial t} - \frac{\partial C'}{\partial z} = \left( \frac{\partial^2 C'}{\partial z^2} + \frac{\partial^2 C'}{\partial y^2} \right), \tag{5.3.7}$$

$$Le_j \left( \frac{\partial T'_j}{\partial t} - \frac{\partial T'_j}{\partial z} \right) = \left( \frac{\partial^2 T'_j}{\partial z^2} + \frac{\partial^2 T'_j}{\partial y^2} \right), \quad j = l, s. \tag{5.3.8}$$

Notice that the stationary temperature and concentration distributions, $\overline{T}_j$ and $\overline{C}$ in this case, are given by equations (5.2.21)-(5.2.23).

Equations (5.3.7) and (5.3.8) are linear and if, in addition, it is assumed that the boundary conditions on the fluctuations are linear as well as homogenous, then the equations admit solutions through the separation of variables method. This method allows the solution for the fluctuations to be represented in the following manner (see Chapter 4):

$$T'_{s,l}(y, z, t) = \theta_{s,l}(z) \exp(iay + \omega t), \tag{5.3.9}$$

$$C'(y, z, t) = \gamma(z) \exp(iay + \omega t), \tag{5.3.10}$$

where $a$ is the fluctuation wave number and $\omega$ is the fluctuation frequency. The dimensional wave number $\alpha$ and the dimensional frequency $\omega^*$ are given by

$$a = \alpha D / U \quad \text{and} \quad \omega = D\omega^* / U^2,$$

respectively.

Substituting these solution forms into the fluctuation equations (5.3.7) and (5.3.8), the following equations result for the fluctuation amplitudes:

$$\left[ Le_s \left( \frac{d}{dz} - \omega \right) + \left( \frac{d^2}{dz^2} - a^2 \right) \right] \theta_s = 0, \tag{5.3.11}$$

$$\left[ Le_l \left( \frac{d}{dz} - \omega \right) + \left( \frac{d^2}{dz^2} - a^2 \right) \right] \theta_l = 0, \tag{5.3.12}$$

$$\left[ \left( \frac{d}{dz} - \omega \right) + \left( \frac{d^2}{dz^2} - a^2 \right) \right] \gamma = 0. \tag{5.3.13}$$

Morphological instability is mainly concerned with the occurrence of perturbations in the crystal/melt interface. For that reason, the variation of the interface shape from its stationary initial form at $z = 0$ must also be included in the analysis. Thus, let us assume that the interface is perturbed in the following manner:

$$z' = \eta \, \exp(iay + \omega t), \tag{5.3.14}$$

where $\eta$ is the amplitude of the interface perturbations.

At this point the interface conditions must be formulated in a manner consistent with the linear perturbation analysis for the fluctuation equations. At the solid/liquid interface, the fluctuations must satisfy the conditions for the conservation of both energy and mass. Conditions (5.1.2), (5.2.2), (5.2.3), and (5.2.5) must be appropriately derived for the fluctuations. It should be noted that the interface position in this case is located at the perturbed position, i.e., $z = z_0 + z'$ where $z_0 = 0$. By assuming that $z'$ is small, all functions of $z$ may then be expanded using Taylor's series:

$$T_j(z_0 + z') \approx T_j(z_0) + z' \frac{\partial T_j}{\partial z} \bigg|_{z=z_0}, \quad C(z_0 + z') \approx C(z_0) + z' \frac{\partial C}{\partial z} \bigg|_{z=z_0}. \tag{5.3.15}$$

The interface velocity may also be written in terms of a stationary and a fluctuating component:

$$v \approx U + \frac{\partial z'}{\partial t}. \tag{5.3.16}$$

With these provisions, and using the expansions in (5.3.6) together with the definitions for the fluctuations in (5.3.9) and (5.3.10), the conservation of energy condition across the interface takes the following form (see [54]):

$$k_s \left( \frac{d\theta_s}{dz} - Le_s \frac{G_s}{G} \eta \right) - k_l \left( \frac{d\theta_l}{dz} - Le_l \frac{G_l}{G} \eta \right) = \frac{h_L U}{G} \omega \eta. \tag{5.3.17}$$

Similarly, the condition for the conservation of mass across the interface takes the form

$$\frac{dy}{dz} + (1 - K)y - (K + \omega)\eta = 0. \tag{5.3.18}$$

In addition to conditions (5.3.17) and (5.3.18), the following for the equilibrium temperature at the interface must hold:

$$\theta_l - Sy - \left( S - \frac{G_l}{G} - \mathcal{A}a^2 \right) \eta = 0, \tag{5.3.19}$$

where

$$S = \frac{mG_C}{G}, \qquad \mathcal{A} = \frac{T_m \sigma}{\rho h_L} \left( \frac{U^2}{GD^2} \right).$$

Conditions (5.3.17), (5.3.18), and (5.3.19) are the fluctuation counterparts for conditions (5.1.2), (5.2.2), and (5.2.5), respectively. $\mathcal{A}$ is a capillary nondimensional parameter. $S$ is a nondimensional number that represents the ratio of stationary state solute concentration gradient to the temperature gradient at the interface. Reference [54] identifies this ratio as the *Sekerka number* in recognition of the work of R. F. Sekerka, who first recognized this ratio together with W. Mullins in [58] to be an important parameter for the morphological instability mechanism.

Finally, the far field boundary conditions must also be specified before attempting to solve the fluctuation amplitude equations. The following conditions will be used here:

$$\theta_l, \to 0 \quad \text{and} \quad y \to 0 \quad \text{as} \quad z \to \infty, \tag{5.3.20}$$

$$\theta_s \to 0 \quad \text{as} \quad z \to -\infty. \tag{5.3.21}$$

The fluctuations amplitude equations (5.3.11)–(5.3.13) are linear, homogeneous ordinary differential equations, with homogeneous boundary conditions. Such a problem defines an eigenvalue problem similar to those encountered in Chapter 4 on free convection. In such a formulation the stability criteria can be determined by examining the sign of $\omega_r$, where $\omega = \omega_r + i\omega_i$. The system will be called unstable if $\omega_r > 0$ and stable if $\omega_r < 0$.

The eigenvalue $\omega$ is normally determined by solving the appropriate dispersion equation for $\omega$ as a function of all the parameters involved in the problem. This dispersion relation is determined by first identifying the solutions to the amplitude functions from the linear equation set (5.3.11)–(5.3.13). When these solutions are substituted into the interface and boundary conditions, conditions (5.3.17)–(5.3.21), the dispersion equation for this problem will result. Reference [61] shows that the principle of exchange of stability holds for this problem, so that $\omega_r = 0$ implies $\omega_i = 0$. This property makes it convenient to determine the conditions of marginal stability for the problem; they are obtained by setting $\omega = 0$ in equations (5.3.11)–(5.3.13), resulting in the following set of equations for the fluctuation amplitudes:

$$\left[ Le_s \frac{d}{dz} + \left( \frac{d^2}{dz^2} - a^2 \right) \right] \theta_s = 0, \tag{5.3.22}$$

$$\left[ Le_l \frac{d}{dz} + \left( \frac{d^2}{dz^2} - a^2 \right) \right] \theta_l = 0, \tag{5.3.23}$$

$$\left[ \frac{d}{dz} + \left( \frac{d^2}{dz^2} - a^2 \right) \right] \gamma = 0. \tag{5.3.24}$$

The solution to equations (5.3.22)–(5.3.24), which is also compatible with the imposed boundary conditions, may be expressed in the following manner:

$$\theta_j = A_{1j} e^{\lambda_j z} + A_{2j} e^{-\lambda_j z}, \qquad j = l, s, \tag{5.3.25}$$

$$\gamma = A_{1c} e^{\lambda_c z} + A_{2c} e^{-\lambda_c z}. \tag{5.3.26}$$

The different $\lambda$'s in (5.3.25) and (5.3.26) are the roots to the following quadratic equations:

$$\lambda_j^2 + Le_j \lambda_j - a^2 = 0, \qquad j = l, s, \tag{5.3.27}$$

$$\lambda_c^2 + \lambda_c - a^2 = 0. \tag{5.3.28}$$

Substituting the solution forms of (5.3.25) and (5.3.26) into the equations for the boundary conditions, a set of algebraic equations for the constants of integration result. The dispersion relation is determined from the solution of the algebraic equations for the integration constants. The resulting dispersion relation is very complicated due to the large number of parameters involved in this problem. Any reduction in the number of these parameters can lead to a more manageable solution for the stability criteria.

If the solidification process is a thermally steady state (i.e., $Le_s = Le_l = 0$), then the solution for the perturbation equations with the imposed boundary conditions yields the following dispersion relation (see [54]):

$$S = \frac{2K - 1 + \sqrt{1 + 4a^2}}{-1 + \sqrt{1 + 4a^2}} [1 + \mathcal{A}a^2]. \tag{5.3.29}$$

In equation (5.3.29) the terms have already been rearranged to give $S$ in terms of the remaining parameters of the problem.

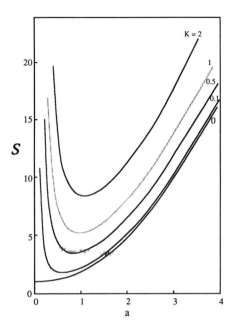

**Figure 5.11**  Marginal stability curves for the morphological stability problem, from [54].

The full dispersion relation for the general problem in which $\omega \neq 0$, and under the same assumptions, takes the following form, as shown in [54]:

$$S = \frac{2K - 1 + \sqrt{1 + 4(a^2 + \omega)}}{-1 + \sqrt{1 + 4(a^2 + \omega)} - 2\omega} \left[ 1 + \mathcal{A}a^2 + \frac{\omega h_L U}{aG(k_s + k_l)} \right]. \qquad (5.3.30)$$

Figure 5.11 shows a plot of $S$ as a function of the perturbation wave number, according to expression (5.3.29), with the capillary parameter set at $\mathcal{A} = 1$ and for several values for the segregation coefficient $K$. The critical value for $S$ is obtained by setting $dS/da = 0$ to yield the following equation for the critical wave number $a_c$ as a function of the rest of the parameters:

$$\mathcal{A} \left( \sqrt{1 + 4a_c^2} - 1 \right)^2 \left( \sqrt{1 + 4a_c^2} + K \right) = 4K. \qquad (5.3.31)$$

The critical wave number is the real root to equation (5.3.31). The minimum value for the Sekerka number $S_c$ is evaluated by substituting that value for $a_c$ into the dispersion relation (5.3.29).

There are several ways to physically interpret the instability criteria derived above. First, the marginal stability curves shown in Figure 5.11 indicate that in a solidification processes for which the values of $S$ are above the proper curve for the specific value of $K$, any departure from uniformity in the solid/liquid interface will be amplified leading to perturbances in the interface itself. This is

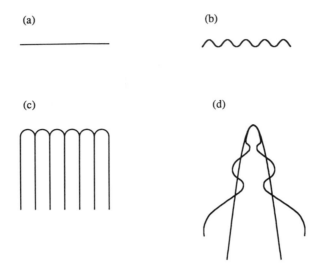

**Figure 5.12**    Graphical interpretation of the role of morphological instability in causing dendritic structures (see [53]).

one mechanism for the formation of dendritic structures in solids. The parameter

$$S = \frac{mUC_\infty}{DK} \frac{(1-K)(k_s + k_l)}{k_s G_s + k_l G_l} \tag{5.3.32}$$

can be varied in several ways. For the same solidification process, it is seen that $S$ increases with an increase in the growth velocity $U$. Thus, to avoid the formation of dendrites, the growth velocity must be slow enough to ensure that the value of $S$ is below the critical value.

Another method, in which the interface undulations can be regulated is by controlling at the interface the solute and the temperature gradients: $G_s$, $G_l$, and $G_C$, respectively. Instability can arise, for instance, for the constitutional supercooling case, which is composed of a destabilizing solute gradient in the melt generated by the crystallization process and a stabilizing temperature gradient that is externally superimposed. The redistribution of the heat flux through the solid/liquid system following interfacial deformation exerts an influence that is stabilizing or destabilizing depending on whether the thermal conductivity of the solid is less than or greater than the melt. Morphological instability can also be driven by the capillarity of the interface through the value of $\mathcal{A}$, which influences the value of $S$, as can be seen from the dispersion relation. In the limiting case $\mathcal{A} \to 0$, the critical Sekerka number $S_c \to 1$, which is known as the *modified constitutional supercooling* condition. On the other extreme, when the capillarity parameter $\mathcal{A}$ becomes very large, $S_c \to K\mathcal{A}$. This limit is referred to as the condition of *absolute stability.*

The morphological instability theory presented above shows that in practical solidification processes, the interface will remain planar until the value for the translation speed $U$ reaches the critical value $U_c$. In the vicinity of $U_c$, nearly two-dimensional steady cells will appear as shown in Figure 5.12. Subsequently, these cells will deepen as $U$ is increased beyond the critical value. As $U$ is increased

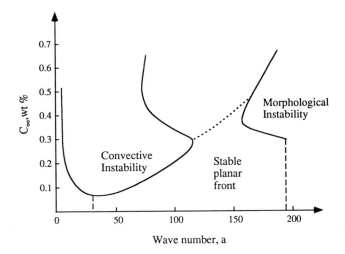

**Figure 5.13**   Comparison between the convective instability and morphological instability boundaries for the solidification of Pb-Sn alloy, from [51].

further, there will be a dendritic transition in which the deep cells develop side branches according to the stability criteria of Figure 5.11.

The above stability calculations are appropriate for a purely diffusion dominated solidification process. In the case of convection in the fluid, the analysis becomes more complicated, but similar stability criteria can still be derived. In a realistic materials processing situation, convection in the fluid can give rise to, or amplify, the morphological instability mechanism, but very little feedback exists between the convective instability mechanism and the morphological instability. This lack of coupling between the two instability mechanisms is due to the large separation in values of the critical wave numbers for each mechanism. In the morphological instability, the critical wave lengths are in the 10–100 $\mu$m range, while in the convective instability, the critical wave length is proportional to the length scale of the container, which could be 1–100 mm. Figure 5.13 clearly illustrates the difference between the values of the critical wave numbers for each of the two instability modes. These marginal stability curves are from the stability calculations performed for the *Pb-Sn* alloy in [51].

Notice that in the derivation of the criteria for morphological instability, gravity did not play any role. This implies that a solidification process that is morphologically unstable or stable remains so regardless of the value of gravity in the environment in which the process is taking place. Thus, the stability criteria developed in this section should apply equally well for space processing. The derived morphological instability limits must be included in the design criteria for any space processing apparatus.

## 5.4 *SPACE PROCESSING*

The discussion in the previous sections was concerned with diffusion-controlled processes in which fluid dynamics in the melt were neglected. However, fluid dynamics plays an important role in most material processes due to the thermal

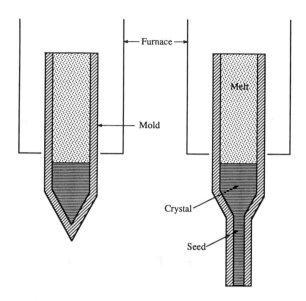

**Figure 5.14**   Schematic of the Bridgman furnace.

and concentration gradients present in the melt. Such gradients can ultimately give rise to complicated circulation patterns in the melt that may significantly alter the solidification process. A very lucid discussion on the fluid dynamic effects during solidification processes may be found in the review article [49]. For materials processing, space has been advocated as the ultimate environment for suppressing fluid flows by reducing the effects due to buoyancy.

It has been clearly demonstrated that when processing of materials in terrestrial environment, the unavoidable and normally unwanted phenomena of buoyancy driven convection and morphological instability are usually present. These instabilities are mainly due to the presence of temperature and solute concentration gradients in the fluid. Processing of materials in low-gravity environment may help in stabilizing the former, while the latter is unaffected by gravity. However, morphological instability can be controlled by modifying the interface speed, a condition that is achieved in two solidification configurations with relative ease. These are the *Bridgman-Stockbarger* furnace and the *float zone* method, which have been strongly advocated as the most suitable methods for space processing.

In the Bridgman process the melt containing ampoule, with its long axis in the vertical direction, is slowly lowered through the furnace. This motion will allow solidification to begin either at a point at the bottom, or on a seed that is on a downward extension of the mold, as shown in Figure 5.14. The Bridgman furnace is mainly used for solidification from melt. In the currently used designs for the vertical Bridgman-Stockbarger furnace, the axial temperature gradient necessary to induce solidification is established by separating the hot and cold zones with an adiabatic zone. The radial heat flux from the ampoule to the furnace is suppressed in this adiabatic zone. Figure 5.15 shows the basic concepts for such a design.

In the float-zone process, a heating element is gradually driven over a polycrystalline solid, melting it and leaving a single crystal after the hot zone has passed, as shown in Figure 5.16. This can be done in either a boat or a crucible, but nearly containerless growth can be achieved in terrestrial environment by using a variant process called *float-zone melting*. In that process, the hot zone is passed over a rod supported only at its end. The crystal and nutrient rods are sometimes differentially rotated about their axes in order to improve the crystal uniformity. This is achieved by providing a viscous shear layer that tends to isolate the growth interface from the fluid convection deeper in the melt. In terrestrial applications of float-zone melting, the gravitational force limits the float zone size which can be increased in low-gravity environment.

The analysis for the solidification process in all of these methods can become very complex when convection in the melt is included. Buoyancy-driven convection in the melt is unavoidable in both of the furnace designs described above due to the solute and thermal gradients that are always present. Also, realistic limitations on the furnace design for all solidification processes give rise to horizontal temperature gradients in the melt in addition to the axial gradients. Due to the complicated thermal conditions in such furnaces, convection may always be present, even in low-gravity environment.

If convection in the melt is included, the simple analytical techniques for describing the solidification process outlined in the previous sections cannot be used. The only available tool for including all of the effects taking place in any realistic furnace configuration is with the use of CFD. CFD techniques have become widely used in many applications including space processing. Due to the low pull speed normally needed for materials of interest in space processing, CFD methods are easier to implement in such applications. The slow speed eliminates the added complexity of turbulent flows in the fluid phase that may exist in moderately fast flows. Also, use of the Boussinesq approximation allows the fluid to be considered incompressible, resulting in a considerable reduction in effort.

To fully simulate solidification processes in the above furnaces, the full set of conservation equations must be solved with the proper geometry and boundary conditions. The only variations in simulating the different solidification processes lie with the boundary conditions. The field equations are the same in all of these processes. The governing equations are the conservation of mass, momentum, energy, and species in the melt, and the conservation of energy in the crystal, which are

$$\nabla \cdot \mathbf{v} = 0, \tag{5.4.1}$$

$$\rho \left( \frac{\partial \mathbf{v}}{\partial t} + \mathbf{v} \cdot \nabla \mathbf{v} \right) = -\nabla p + \rho g [1 - \beta_t (T - T_0) + \beta_c (C - C_0)] \mathbf{e}_z$$
$$+ \rho \mathbf{F}_b (\mathbf{v}, \mathbf{x}) + \mu \nabla^2 \mathbf{v}, \tag{5.4.2}$$

$$\left( \frac{\partial C}{\partial t} + \mathbf{v} \cdot \nabla C \right) = D \nabla^2 C, \tag{5.4.3}$$

$$\left( \frac{\partial T_l}{\partial t} + \mathbf{v} \cdot \nabla T_l \right) = \kappa_l \nabla^2 T_l, \tag{5.4.4}$$

$$\left( \frac{\partial T_s}{\partial t} + \mathbf{U} \cdot \nabla T_s \right) = \kappa_s \nabla^2 T_s, \tag{5.4.5}$$

where $\mathbf{v}$ is the fluid velocity in the melt and $\mathbf{U}$ is the sample pull rate.

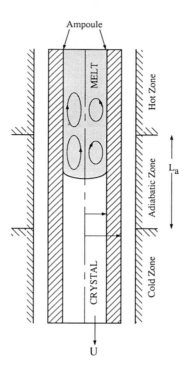

**Figure 5.15** Sketch for the MIT Bridgman furnace, from [48].

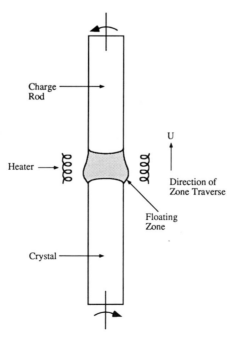

**Figure 5.16** Schematic for the float-zone process.

The vector function $\mathbf{F}_b$ in equation (5.4.2) represents a body force that may exist in the melt in addition to the force of gravity. The body force caused by an imposed magnetic field $\mathbf{B}(\mathbf{x}, t)$, for instance, is the Lorentz force given by

$$\mathbf{F}_b \propto (\mathbf{v} \times \mathbf{v} \times \mathbf{B}).$$

The coriolis and centrifugal forces in a rotating ampoule can also be represented by such a body force, in this case given by

$$\mathbf{F}_b \propto 2\Omega \times \mathbf{v} + \Omega \times \Omega \times \mathbf{r},$$

where $\Omega$ is the angular velocity of the sample. The species diffusion coefficient in the solid is normally assumed to be much smaller than that in the melt (i.e., $D_s \ll D_l$), thus making $D_s/D_l \approx 0$. This assumption will lead to the conclusion that the solute concentration in the solid, $C_s$, remains constant with its value everywhere in the solid being the same as at the solid/liquid interface.

In all of these processes, the growth rate $\mathbf{U}$ is assumed be constant and equal to the crystal pull rate, which is imposed externally. Under these conditions, the temperatures, melt velocity, and solute concentration at the interface must satisfy the following constraints:

$$k_s(\mathbf{n} \cdot \nabla T_s) - k_l(\mathbf{n} \cdot \nabla T_l) = \rho h_L(\mathbf{U} \cdot \mathbf{n}), \tag{5.4.6}$$

for the conservation of energy and

$$D(\mathbf{n} \cdot \nabla C) = C(\mathbf{U} \cdot \mathbf{n})(1 - K), \tag{5.4.7}$$

for the conservation of solute mass. Also, for slow growth rates, for which local thermodynamic equilibrium is assumed, the shape of the solid/liquid interface is given by

$$T_s = T_l = T_m + mC - \frac{T_m \sigma}{\rho h_L}\left(\frac{1}{R_1} + \frac{1}{R_2}\right), \tag{5.4.8}$$

where $T_m$ is the melting temperature for a pure material.

In addition to conditions (5.4.6)–(5.4.8), the following constraints on the fluid velocity at the interface must be satisfied:

$$(\mathbf{n} \cdot \mathbf{v}) = (\mathbf{U} \cdot \mathbf{n}) \quad \text{and} \quad (\mathbf{t} \cdot \mathbf{v}) = (\mathbf{U} \cdot \mathbf{t}) \tag{5.4.9}$$

where $\mathbf{n}$ and $\mathbf{t}$ are the unit normal and tangent vectors to the interface, respectively. To realistically model the processing furnace, specific boundary conditions on the ampoule for the Bridgman process and conditions for the free surfaces in the float-zone process must also be specified.

The above equations and boundary conditions represent a very complex system whose solutions can only be obtained through numerical approximations. CFD has advanced to the point where accurate solutions for the above problem can be produced fairly routinely.

The general procedure, normally adopted for simulating a solidification process in low-gravity environment, can be described as follows: Initially the problem is solved for gravity values $g/g_0 = 1$, which may be validated through ground-based experiments. Once a sufficient level of confidence is gained in the numerical approximations, numerical solutions are produced for different values of the

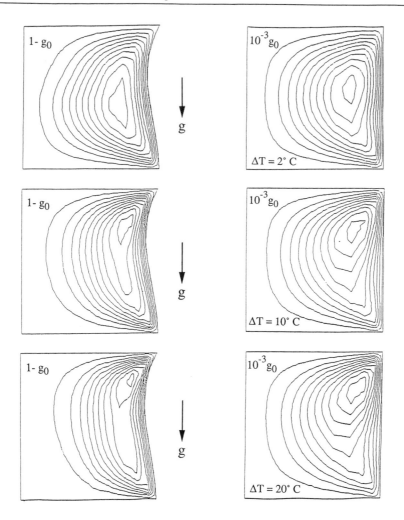

**Figure 5.17** Streamline distributions in the melt for a float-zone process from a model solution of [59] for $g/g_0 = 1$ and $10^{-3}$.

gravity. Solutions applicable for low-gravity environment are obtained by using the appropriate value for $g$ in equations (5.4.1)–(5.4.5).

Figure 5.17 shows the streamline pattern for the flow field in the melt as well as the free surface shape for a float-zone model at both $g/g_0 = 1$ and $g/g_0 \approx 10^{-3}$. This figure is from [59], in which the numerical simulation method described above was used. The results shown in Figure 5.17 were computed with a commercially available, finite element computer code.

Figure 5.18 shows the streamlines and isotherms as well as the free surface and solid/melt interface shape for a zone melting model at $g/g_0 = 1$ and $g/g_0 \approx 0$ through numerical simulation for a $Si$ crystal. These results are from [56].

Figure 5.19 shows the streamline patterns in the melt and the interface shape for a Bridgman furnace simulation in which the value for the gravity varies between $g/g_0 = 1$ and $g/g_0 \approx 10^{-3}$. The solutions in Figure 5.19 are from [48] for a $GaGe$ binary alloy, for values of the Rayleigh number $R$ ranging from $10^5$ to $5 \times 10^7$.

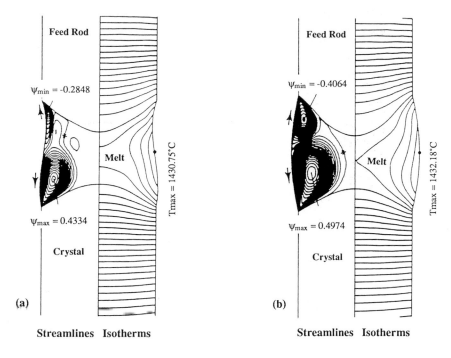

**Figure 5.18** Isotherm and streamline distributions in the melt for a float-zone process from [56]: (a) for $g/g_0 = 1$ and (b) for $g/g_0 = 0$.

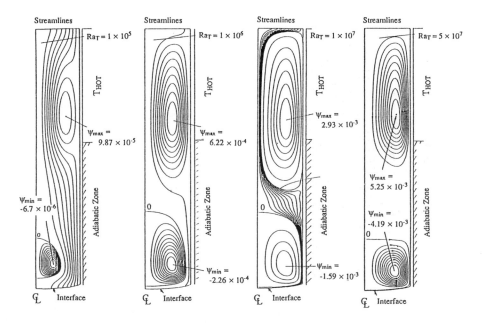

**Figure 5.19** Isotherms and streamlines in the melt for the Bridgman furnace from [48] for a range of values of the Rayleigh number.

# Chapter 6

# Drops and Bubbles

**W**ithout the hydrostatic pressure of gravity, many fluid systems can exist in space with only surface tension affecting the liquid surface. Such capillary phenomena are manifested in terrestrial environment in liquid films, foams, drops, bubbles, and other configurations of fluid surfaces that separate bulk phases.

The behavior of drops and bubbles, which may be present already in a system or may be generated through some physical or chemical means, varies depending on the gravitational environment. In the terrestrial case, the density difference and gravity force combine to yield a net hydrostatic force on the drops and bubbles. This causes them to move to the top or bottom of the fluid column, and a considerable effort must be expended to keep them from settling. In contrast, the background gravity levels inside orbiting vehicles are small, and it can be expected that any suspended object in a fluid experiences only a negligible hydrostatic force and therefore remains suspended.

The possibility of suspending drops and bubbles freely in a low-gravity environment can be used for both research and technological purposes. This environment can be exploited for some fundamentally important experiments, including the development of new measurement techniques for liquid viscosity and surface tension. This is one of the reasons why low-gravity studies of drop dynamics and coalescence have flourished.

In materials processing, the avoidance of physical contact between the sample and the container wall improves contamination control and thus reduces the probability for heterogeneous nucleation in an undercooled melt. This further implies an enhanced capability for attaining deep undercooling. Scientific interests in the deeply undercooled liquid state include, but are not restricted to, the determination of thermophysical properties, the partial control of cooling rate during solidification, the selection of possible metastable solid structures upon crystallization of alloys, and the study of the subtle effects influencing nucleation of the solid phase, including small amplitude perturbations and flows.

In this chapter, we will first examine the dynamic states of simple drops under the action of both gravity and surface tension. The effect of external fields such

171

as acoustic and electric fields, will also be investigated. Next, the dynamic states of rotating drops will be examined. The thermohydraulic effects on drops and bubbles within a fluid host will also be treated. Finally, the thermocapillary effects on bubble dynamics will be discussed.

## 6.1 *DYNAMICS OF SIMPLE DROPS*

The dynamics of liquid drops, held together by surface tension or by an external field, such as an electrical charge, has been of theoretical and experimental interest for the past century. Rotating liquid drops held in a neutrally buoyant liquid were used in early experiments as models for the shape of rotating heavenly bodies. It was assumed that the attractive forces of surface tension would mimic the gravitational attraction of solar-sized bodies. Femtometer-sized liquid drops, electrically charged and rigidly rotating, have also been postulated as models for heavy atomic nuclei and later used in model calculations for nuclear fusion.

There exists a large body of theoretical analyses on the equilibrium shapes and dynamics of such drops. However, little experimental confirmation of these analyses has been undertaken due to the difficulties of levitating reasonably sized drops on Earth. Some data is available for drops in free fall for periods on the order of few seconds, or in liquid/liquid systems where a drop is made to suspend neutrally buoyant in an outer surrounding liquid. Although long durations are feasible with the second system, the results are usually influenced by the more complex fluid dynamics due to the interaction between the two fluids.

Normally, the equilibrium shape of a free liquid sample, under the sole influence of surface tension, is spherical, as was demonstrated in Chapter 3. Secondary stresses acting on a drop give rise to some interesting variations in its shape. These stresses are usually due to perturbations induced by either natural sources such as gas flows, interparticle collisions, and electrical charges, or non-natural effects such as forces used for levitation and positioning. There has been, for example, considerable experimental and theoretical research on the shape distortion of nonrotating drops caused by a distributed electric or acoustic field.

The shape oscillations of freely supported drops and bubbles that are subjected to a stress field have been studied extensively. Rayleigh formulated the earliest analysis on the behavior of an oscillating liquid drop about its spherical equilibrium shape. It was confined to small-amplitude, axisymmetric oscillations in which the liquid inside the drop was assumed to be inviscid. Also, the analysis was performed under the assumption of zero gravity. Rayleigh's analysis is repeated here.

Following [66], consider a spherically symmetric fluid volume whose radius is $a(t)$, expanding radially at the rate $da/dt$. If the fluid inside and outside the volume is assumed to be inviscid and irrotational, the velocity field $\mathbf{v}$ both inside and outside the volume can be described by a scalar velocity potential $\phi$ in the following manner:

$$\mathbf{v} = -\nabla\phi. \tag{6.1.1}$$

Substituting this definition for the velocity field into the conservation of mass equation

$$\nabla \cdot \mathbf{v} = 0,$$

we get

$$\nabla^2 \phi = 0. \tag{6.1.2}$$

The fluid pressure under these assumptions can be evaluated using Bernoulli's equation

$$\frac{p}{\rho} - \frac{\partial \phi}{\partial t} + \frac{1}{2}(\nabla \phi)^2 = C(t), \tag{6.1.3}$$

where $C$ is a constant of integration that is a function of time only. The solution to equation (6.1.2) for a spherically symmetric fluid volume is given in [62] as

$$\phi = \frac{a^2}{r}\frac{da}{dt}, \tag{6.1.4}$$

with $r$ the radial coordinate.

Now let the original spherical shape of the fluid volume be slightly perturbed to a new shape whose surface is defined by the new radius $r_s$ as

$$r_s = a(t) + \eta(t)Y_n^m(\theta, \varphi), \tag{6.1.5}$$

where $Y_n^m$ is a spherical harmonic of degree $n$ while $\theta$, $\varphi$, and $r$ are spherical coordinates and $\eta$ is the perturbation amplitude. In the derivation followed here, the magnitude of $\eta$ is assumed to be small compared to the radius, i.e.,

$$|\eta(t)| \ll a(t). \tag{6.1.6}$$

For infinitesimal values of the perturbation $\eta$, the velocity potentials inside and outside of the deformed surface, $\phi_i$ and $\phi_o$, are given to a first-order estimate in $\eta$ by [66] as follows:

$$\phi_i = \frac{a^2}{r}\frac{da}{dt} - \frac{r^n}{na^{n-1}}Y_n^m\left[\frac{d\eta}{dt} + 2\frac{\eta}{a}\frac{da}{dt}\right], \quad r < a \tag{6.1.7a}$$

$$\phi_o = \frac{a^2}{r}\frac{da}{dt} + \frac{a^{n+2}}{(n+1)r^{n+1}}Y_n^m\left[\frac{d\eta}{dt} + 2\frac{\eta}{a}\frac{da}{dt}\right], \quad r > a. \tag{6.1.7b}$$

The subscripts $i$ and $o$ designate properties inside and outside the volume, repectively.

The shape of any fluid surface is normally defined by the balance of forces acting on the surface. In the absence of gravity, the interfacial surface tension force $\sigma$ is balanced by the pressure difference across the surface, according to Laplace's condition, as

$$p_i - p_o = \sigma\left(\frac{1}{R_1} + \frac{1}{R_2}\right) \tag{6.1.8}$$

where $R_1$ and $R_2$ are the principal radii of curvature for the surface.

For very small deformations of the surface from its original spherical shape, reference [66] shows that the radii of curvature for the deformed surface are given, to a first-order approximation in $\eta$, by

$$\frac{1}{R_1} + \frac{1}{R_2} = \frac{2}{a} + \frac{(n-1)(n+2)}{a^2}\eta Y_n^m + O(\eta^2). \tag{6.1.9}$$

On the other hand, the pressure on either side of the volume can be evaluated, from equation (6.1.3), as

$$p_i = P_i(t) + \rho_i \left[ \frac{\partial \phi_i}{\partial t} - \frac{1}{2} (\nabla \phi_i)^2 \right]_{r=r_s}, \qquad (6.1.10a)$$

and

$$p_o = P_o(t) + \rho_o \left[ \frac{\partial \phi_o}{\partial t} - \frac{1}{2} (\nabla \phi_o)^2 \right]_{r=r_s}, \qquad (6.1.10b)$$

where $P_i(t)$ and $P_o(t)$ are constants of integration in Bernoulli's equation and specifically $P_0(t)$ represents the pressure outside the volume at infinity.

Substituting the above values into Laplace's equation (6.1.8), and using the solutions for the potential functions $\phi_i$ and $\phi_o$ from expressions (6.1.7), the following equations result for the new surface shape:

$$a \frac{d^2a}{dt^2} + \frac{3}{2} \left( \frac{da}{dt} \right)^2 = \frac{P_i - P_o - 2\sigma/a}{\rho_o - \rho_i}, \qquad (6.1.11)$$

$$\frac{d^2\eta}{dt^2} + \frac{3}{a} \frac{da}{dt} \frac{d\eta}{dt} - \lambda_n^2 \eta = 0, \qquad (6.1.12)$$

where

$$\lambda_n^2 = \frac{[n(n-1)\rho_o - (n+1)(n+2)\rho_i](d^2a/dt^2) - (n-1)n(n+1)(n+2)\sigma/a^2}{a[n\rho_o + (n+1)\rho_i]}.$$

For the general case of an oscillating sphere the volume of which is changing with time, both equations (6.1.11) and (6.1.12) must be solved simultaneously for $a(t)$ and $\eta(t)$. In the special case for which the volume of the sphere is constant (i.e., $da/dt = 0$), only equation (6.1.12) needs to be solved for the deformation amplitude $\eta$, where it reduces to the form

$$\frac{d^2\eta}{dt^2} + \omega_n^{*2}\eta = 0, \qquad (6.1.13)$$

with

$$\omega_n^{*2} = \frac{(n-1)n(n+1)(n+2)\sigma}{a^3[n\rho_o + (n+1)\rho_i]}. \qquad (6.1.14)$$

The general solution for the oscillation amplitude of the surface is obtained by solving equation (6.1.13). This solution can be expressed as

$$\eta_n = A_1 \cos(\omega_n^* t) + A_2 \sin(\omega_n^* t), \qquad (6.1.15)$$

where $\omega_n^*$ is the frequency of oscillation. For the case of an oscillating liquid drop in air for which $\rho_i \gg \rho_o$, the small amplitude oscillation frequency reduces to

$$\omega_n^2 = n(n-1)(n+2) \frac{\sigma}{\rho a^3}, \qquad (6.1.16)$$

where $\rho$ is the liquid drop density. The above solution for the oscillation frequency is commonly known as *Rayleigh's solution.* It should be stressed that this solution is for the linear problem where the oscillations are assumed to be infinitesimal.

Cases $n = 0$ and $n = 1$ correspond to rigid body motions. For the case $n = 2$, $\omega_2$ is known as the fundamental mode of oscillation. Its period $\tau_2$ is given by

$$\tau_2 = \pi\sqrt{\frac{\rho a^3}{2\sigma}}. \qquad (6.1.17)$$

For a 2.5-cm diameter water drop, for example, for which $\rho = 1$ gm/cm$^3$ and $\sigma = 75$ dyne/cm, the fundamental mode period would be $\tau_2 = 0.36$ sec, while for $n = 3$, $\tau_3 = 0.18$, and for $n = 4$, $\tau_4 = 0.11$. This example shows very clearly that Rayleigh's solution for small-amplitude oscillations gives reasonably good approximations for amplitude values for $\tau$ ranging from 0.375 to 0.625.

Further calculations include the effects of liquid viscosity on the oscillation frequency. For small values of viscosity, it was found that the only effect on an already oscillating spherical drop is the gradual damping of the oscillation amplitude. The decay of the oscillation amplitude is given in this case by the following expression:

$$\eta_n \propto A_0 e^{-\beta_n t}, \qquad (6.1.18)$$

where $A_0$ is the initial amplitude of the oscillation for the drop and $\beta_n$ is given by

$$\beta_n = \frac{(n-1)(2n+1)\nu}{a^2}, \qquad (6.1.19)$$

where $\nu$ is the kinematic viscosity for the liquid. For example, a drop of water 2.5 cm in diameter with $\nu = 0.014$ cm$^2$/sec oscillates with the fundamental frequency $n = 2$, $\beta_2 = 0.045$. The oscillation will thus decay to 1% of its initial amplitude in 102 sec.

A more complete analysis for the shape of an oscillating drop, for small oscillations that also includes viscous effects, is given in [67]. Treating the problem as an initial value problem, [67] provides a theoretical prediction for the behavior of a freely oscillating drop in the early transient period of oscillation. The following expression is derived for the $n$th resonant mode frequency, $\omega_n$, of a driven oscillating drop:

$$\omega_n = \omega_n^* - \frac{1}{2}\delta\sqrt{\omega_n^*} + \frac{1}{4}\delta^2, \qquad (6.1.20)$$

where $\omega_n^*$ is given by (6.1.14) and $\delta$ is given by

$$\delta = \frac{(2n+1)^2\sqrt{\mu_i\mu_o\rho_i\rho_o}}{\sqrt{2}r[n\rho_o + (n+1)\rho_i][\sqrt{\mu_i\rho_i} + \sqrt{\mu_o\rho_o}]}, \qquad (6.1.21)$$

where $\mu_o$ and $\mu_i$ are the dynamic viscosities of the inner and outer fluids, respectively.

The free decay of an oscillating drop is characterized by a damping constant $\tau_n^{-1}$ given by

$$\tau^{-1} = (1/2)\delta\sqrt{\omega_n^*} - (1/2)\delta^2 + (1/2)\gamma, \qquad (6.1.22)$$

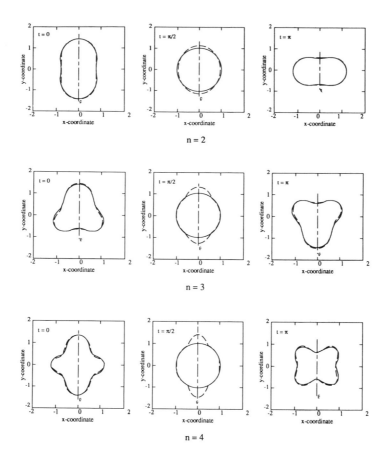

**Figure 6.1**    Analytical shapes of drop oscillations for the axisymmetric modes $n = 2, 3$, and 4, from [73]. Solid line: shapes obtained using only first-order correction; dotted line: shapes obtained using second-order correction.

where $y$ is a parameter given by

$$y = \frac{(2n + 1)\{2(n^2 - 1)\mu_i^2\rho_i + 2n(n + 1)\mu_o^2\rho_o + \mu_i\mu_o[(n + 2)\rho_i - (n - 1)\rho_o]\}}{a^3[\sqrt{\mu_i\rho_i} + \sqrt{\mu_o\rho_o}]^2[n\rho_o + (n + 1)\rho_i]}.$$

These results are valid only for small-amplitude oscillations in which the tangential stress at the drop boundary is vanishingly small and there is no circulation within the drop. Furthermore, the analysis does not account for the dependence of the resonant frequency on the oscillation amplitude. In other words, the various resonant modes are assumed to be uncoupled.

Figure 6.1 shows representative shapes of drops calculated in [73] through a half period of oscillations, for the lowest three modes $n = 2, 3$, and 4. These results are produced using asymptotic analysis for small amplitude axisymmetric oscillations of inviscid and incompressible liquid spheres. Figure 6.1 shows the calculated drop shapes according to both first- and second-order corrections, respectively.

Acoustic, electrostatic, and electromagnetic techniques can be used to levitate and manipulate liquid drops for experimental purposes. Acoustic and ultrasonic methods are the most widely used, showing versatility in positioning, oscillating, and rotating liquid samples, particularly at room temperature. Ultrasonic and electrostatic levitators can be used effectively for ground-based experimental investigations of levitated droplets ranging in size from 0.5 to 8 mm in diameter in a gas medium. In general, the power levels required for levitating samples in one-$g$ are such that varying degrees of drop distortion always exist, and the effects of the levitation stresses on the oscillation shapes cannot be neglected. The minimization of this interference is usually accomplished by reducing the sample size and using high surface-tension liquids. The influence of a nonuniform distribution of surface charges on a droplet behavior has not yet been studied consistently, and the effects of electrostatic levitation on the sample dynamics are yet to be assessed. The availability of long-duration low-gravity environment such as earth-orbiting spacecrafts can afford the ideal medium for levitating large-diameter droplets for longer periods of time.

The stability of droplet shapes within a distributed stress field, such as that found in an acoustic standing wave, is being actively investigated at the present time. Acoustic levitation methods are motivated by experimental observations revealing the onset of macroscopic shape oscillations for low-viscosity liquids when a threshold for the intensity level of the stress field is exceeded. It is observed that the various modes of shape oscillations could be excited in turn with the higher-lobed modes appearing at increasingly higher amplitudes of the sound field. Figure 2 from [71] shows the shape mode oscillations of low-viscosity liquid drops subject to acoustical field disturbances. The figure shows the lowest axisymmetric modes of deformation for $n = 2, 3$, and 4.

The oscillations shapes depicted in Figure 6.2 were induced using a low-frequency modulation of the acoustic radiation force. Two types of excitations were possible with the experimental configuration of [71]. One excitation corresponds to a periodic elongation of the drop at the poles and the other to a periodic compression of the drop at the poles. The restoring force is normally provided by the surface tension, which tends to drive the drop back to its equilibrium shape. For low-amplitude vibrations, both modes yield the same results for the resonance frequency. The shapes are set into oscillation for Figure 6.2 by acoustically driving the drop near its resonant frequency. The driver is subsequently turned off and the drop motion evolves into the free oscillations observed.

Figure 6.3 shows the effect of increasing the oscillation amplitude as measured by $(H/L)$ on the fundamental oscillation frequency. Here, $H$ is the length of the vertical axis and $L$ is the length of the horizontal axis. A comparison between the calculated effect of [73] and the measured effect of [70] is also shown in Figure 6.3.

The experimental results shown in both Figures 6.2 and 6.3 are for drops ranging in volume from 1.9 to 1.2 cm$^3$ of silicone/CCl$_4$ in water or phenetol in water/methanol. This drop size appears to be in the upper range of possible sizes in terrestrial environment. It is speculated that larger free drop sizes are possible

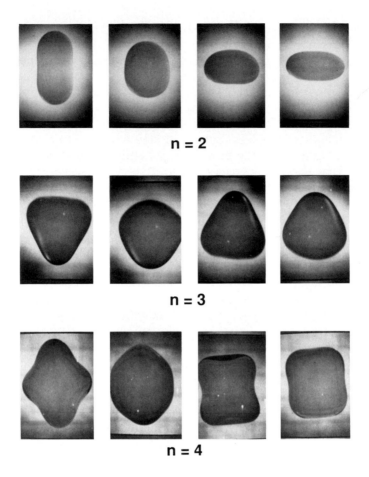

**n = 2**

**n = 3**

**n = 4**

**Figure 6.2** Photographs of the shape modes of oscillations for liquid drops suspended in another liquid for modes $n = 2$, 3, and 4, from [71]. *Courtesy E. H. Trinh.*

in the low-gravity environment of an earth-orbiting vehicle. Experimental work on larger size free drops can yield better results.

Measurements of the resonant frequencies of shape oscillations and of the free decay rate of such oscillations provides information on the values of both the surface tension and the viscosity of a pure liquid. Application of this technique to melts at very high temperatures has now become one of the goals of future microgravity flight experiments. Some preliminary results obtained in one-$g$ already exist in the case of electromagnetically levitated melts of ferrous alloys.

Drop oscillations can also be induced in nonrotating drops by applying an electrical charge to a conducting drop. Calculations by [72] reveal resonant motion of drops for specific values of the electrical charge in which two axisymmetric modes possessed the same linear frequency. Resonance occurred for a charged inviscid drop between the four- and six-lobed oscillations for a dimensionless charge given by

$$Q = Q_1 = \sqrt{32\pi/3}, \qquad (6.1.23)$$

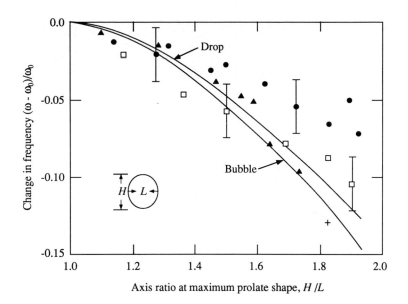

**Figure 6.3**  Change in the frequency of fundamental mode of oscillation $n = 2$ with increasing amplitude of oscillation as measured by $H/L$ for a prolate drop. The solid lines are the analytical predictions of [73] and the symbols are from the experiments of [70].

or an equivalent dimensional charge value of

$$Q^* = 2\sqrt{\frac{2}{3\sigma\epsilon_m a^3}} \, , \tag{6.1.24}$$

where $\epsilon_m$ is the electrical permittivity of the surrounding medium. The evolution of the oscillations of a charged drop with $Q = Q_1$ is shown in Figure 6.4 for an initially 4-lobed disturbance. The time scale for these calculations is the reciprocal of the frequency of the linear oscillation. In less than three cycles, Figure 6.4 shows that the oscillations have evolved from a 4-lobed to a 6-lobed form. Reference [72] shows that the oscillations will continuously exchange energy between these two modes whenever the components of both modes are present in the initial conditions.

Another application of drop shape oscillations is in assessing the effects of chemical surface agents on interfacial phenomena. Initial studies have revealed that the evolution of the concentration of the surface contaminants at the drop surface can be monitored through variations of the fundamental frequency period. In a terrestrial environment, two immiscible liquids are used in such experiments, making the theoretical predictions intractable. The possiblity of suspending liquid drops in a gas host can greatly simplify such analysis. Such experiments can provide much needed data on complex physical parameters, such as interfacial viscosity and elasticity, that are part of interfacial phenomena. Reference [69] provides an excellent review on free drop oscillations including low-gravity experimental prospects, advantages, and objectives.

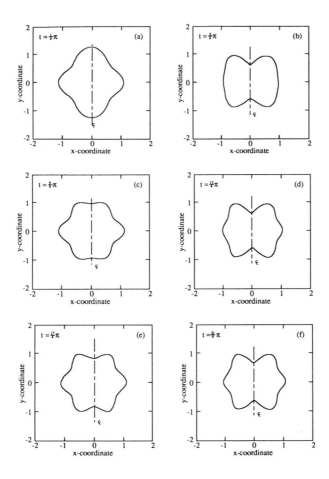

**Figure 6.4**   Analytical predictions of shape oscillations for conducting drops initially perturbed by a 4-lobed deformation as a function of time, from [72].

## 6.2 *ROTATING FREE DROPS*

The axisymmetric equilibrium shapes of rotating drops can be conveniently described in terms of a dimensionless *rotation parameter Ro* defined by

$$Ro = \frac{\Omega^2 \rho a^3}{2\sigma} \tag{6.2.1}$$

where $\Omega$ is the rotational angular velocity, $\sigma$ the surface tension, and $a$ the equatorial radius.

Reference [74] classifies the resulting oscillation shapes for rotating drops as a function of *Ro*. When *Ro* < 0.5, there are always two equilibrium shapes: the one of lower energy is simply connected, the other is torus-like and doubly connected. For *Ro* = 0.5, an additional collapsed shape appears with zero thickness at the center yielding a figure-eight cross section (peanut shape). For 0.5 < *Ro* < 0.533, there are two torus-like shapes and two simply connected shapes. When *Ro* = 0.533, there is only one torus-like shape but still two simply

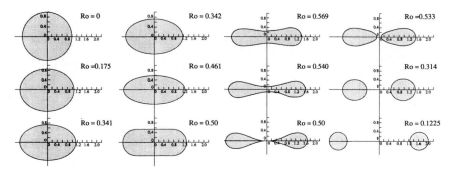

**Figure 6.5** Schematic interpretation of the axisymmeric equilibrium shapes of a rotating drop as a function of the rotation parameter *Ro*, from [74].

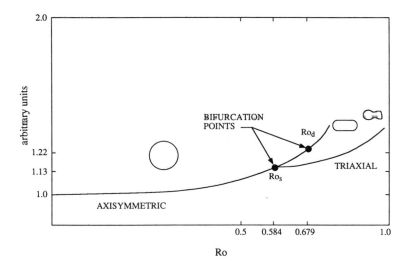

**Figure 6.6** Calculated shape bifurcation points as function of the rotation parameter, from [74].

connected shapes. The torus-like shapes are lost once *Ro* > 0.539, and when *Ro* = .5685 there remains only one simply connected shape. The angular velocity corresponding to this value of *Ro* is also the greatest angular velocity that an axisymmetric equilibrium shape can have. The various shape oscillations described above are shown in Figure 6.5.

For simply connected rotating drops, the only detailed stability analysis for the equilibrium shapes show that when *Ro* = 0.584, the drop can deform without changing its energy to another shape having rigid body rotation. The shape bifurcations described above as a function of the rotation parameter *Ro* are shown schematically in Figure 6.6. Figure 6.6 also shows the bifurcation points at which the secular stability passes from the sequence of axisymmetric shapes to triaxial shapes.

The exact shape evolution and bifurcations shown in Figure 6.5 have been confirmed analytically in [63] through detailed numerical calculations; the equilibrium shapes and stability are determined for rotating drops held together

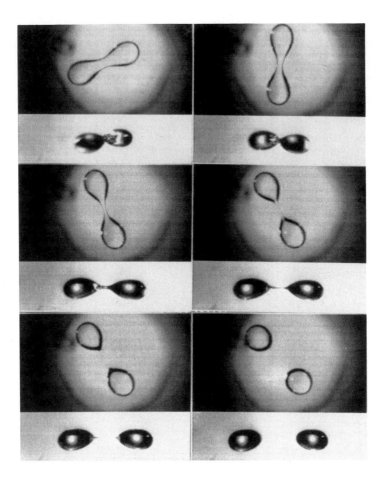

**Figure 6.7**   Photograph of a drop undergoing fission due to rotation, from [69]. The drop is levitated in a terrestrial laboratory using an ultrasonic device and rotated by standing waves at audio frequencies.

by surface tension only. The fluid inside the drop was assumed to be inviscid and incompressible, but the drop was allowed to take any three-dimensional shape configuration. In order to track the drop surface deformation, the finite element method was used for the numerical computations.

The stable and bifurcating shapes postulated by [63] have been tested experimentally; the critical rotational parameter at bifurcation was also measured. These experiments were conducted under terrestrial conditions in which the drops were suspended using ultrasonic levitation techniques. The drop sizes in this experiment were in the millimeter range. In terrestrial environment, the ultrasonic levitation methods induce a significant static distortion in ordinary liquid drops with diameters larger than 1 mm. Figure 6.7 shows the fission process of an ultrasonically levitated uncharged drop that is rotated by an acoustic torque.

The ultimate test for the various theories on the shape stability and bifurcation of rotating liquid drops must take place in low-gravity environment. It is in

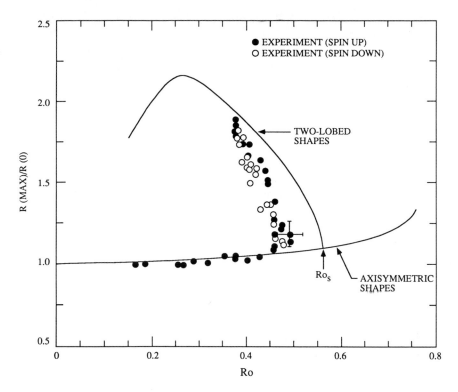

**Figure 6.8** Variation of the drop shape as function of the drop rotation rate as measured by [74] in low-gravity environment. The continuous curve depicts the computational results of [73].

such an environment that the secondary effects due to torque inducing acoustic and electrostatic stress fields can be minimized. Figure 6.8, which is reproduced from [74], shows experimental results on the stability and bifurcation of a rotating drop in low-gravity environment. This experiment was conducted in the Drop Dynamic Module aboard Spacelab-3 in 1985, where the drops were held within a small region inside a chamber by three orthogonal acoustic standing waves generating steady-state forces on the order of 1 dyne. These drops were driven into rotation by acoustic torques. This facility was designed to handle drops ranging in volume from 0.5 to 10 cm$^3$. The results, shown in Figure 6.8, are for a 3-cm$^3$, 100-cSt water-glycerin drop. The experiment involved the spin-up of the drop, its transition to a 2-lobed shape, and finally its relaxation to the axisymmetric shape as the acoustic torque is turned off.

Reference [74] provides a comprehensive survey of the experimental efforts concerning the shape oscillations and bifurcations of free and rotating drops. The survey also includes much of the recent results on low-gravity experiments on levitated drops.

## *6.3* DIMENSIONAL CONSIDERATIONS FOR DROPS AND BUBBLES

Consider a gas bubble in an unbounded pure liquid whose surface tension against the gas is $\sigma$, which is assumed to be constant. The bubble will rise steadily whenever its motion is stable with respect to random small disturbances, and also if the time taken to approach the terminal velocity is much less than the time required for the bubble size to change considerably. Changes in size may be due to evaporation or condensation, changes in ambient pressure, or gases moving in and out of solutions (as in a glass of beer).

Consider the steady motion of a bubble with the same constant volume as a sphere of diameter $2a$ rising at a speed $U$ in a liquid whose density is $\rho$ and dynamic viscosity $\mu$. The rise velocity $U$ can be represented functionally by

$$U = f(a, \rho, \mu, \sigma, g), \tag{6.3.1}$$

where $g$ is the acceleration due to gravity.

To make the calculations independent of a particular unit of measurement and specific fluids, a dimensionless product of the above parameters is sought. A number of these dimensionless parameters are commonly in use, but it has been shown that only one, the *Morton number, Mo*, can be formed from the given physical properties of the liquid. It is defined as

$$Mo = g\mu^4 / \rho\sigma^3. \tag{6.3.2}$$

The Morton number takes a wide range of values for different liquids. In highly viscous oils it can exceed $10^5$, and in liquid metals it can be less than $10^{-13}$.

Other dimensionless parameters must depend on $U$ or $a$ or both; these include the *Reynolds number Re*, the *drag coefficient* $C_D$, the *Weber number We*, the *Froude number Fr*, the *Eötvös number E*, and the *Bond number Bo*, which may be defined as

$$Re = 2Ua\rho/\mu = 2Ua/\nu, \tag{6.3.3}$$

$$C_D = \frac{\text{force on bubble}}{\rho U^2 \pi a^2 / 2} = \frac{8ga}{3U^2}, \tag{6.3.4}$$

$$We = 2\rho U^2 a/\sigma, \tag{6.3.5}$$

$$Fr = U^2/2ga, \tag{6.3.6}$$

$$E = Bo^2 = 4ga^2\rho/\sigma. \tag{6.3.7}$$

In a steady flow, any independent pair of these dimensionless numbers can determine all the rest of the parameters. It is convenient to use $Mo$ to specify the fluid and $Re$ to determine the dynamics. The basic problem for bubbles in liquids and liquid drops in liquids is the calculation of the flow pattern and hence $C_D$ as a function of both $Re$ and $Mo$ from the equations of motion.

The theory for moving drops is in most respects very much like that for gas bubbles, except more complicated. Instead of a single nondimensional number for the fluid properties $Mo$, in the case of bubbles there are now three independent dimensionless parameters that characterize the fluids inside and outside the drop; these are $\rho_o/\rho_i$, $\mu_o/\mu_i$ and

$$Mo_o = \frac{g\mu_o^4|\rho_o - \rho_i|}{\rho_o^2\sigma^3},$$  (6.3.8)

where the subscripts $o$ and $i$ stand for outside the drop and inside the drop, respectively.

Similarly, it is customary to define the Reynolds number $Re_o$, the drag coefficient $C_{Do}$, and the Weber number $We_o$ in the following manner:

$$Re_o = 2Ua\rho_o/\mu_o = 2Ua/\nu_o,$$  (6.3.9)

$$C_{Do} = \frac{8ga|\rho_o - \rho_i|}{3\rho_oU^2},$$  (6.3.10)

$$We_o = 2\rho_oU^2a/\sigma,$$  (6.3.11)

where $2a$ is the diameter of the *equivalent sphere* that has the same volume as the drop. $|\rho_o - \rho_i|$ appears in the formulas because it is convenient to have a positive drag coefficient whether the drop moves up or down.

Alternate dimensionless numbers $Mo_i$, $Re_i$, $C_{Di}$, and $We_i$ may also be defined by interchanging the subscripts $o$ and $i$ in the above definitions, but the external fluid parameters $Mo_o$, $Re_o$, $C_{Do}$, and $We_o$ are used more often.

## 6.4 HYDRODYNAMIC THEORY OF BUBBLES

The rise of bubbles in liquids is characterized by its slowness. Normally, when either the fluid motion or the body's movement is very small, the Reynolds number is also very small. For such a case, the spatial acceleration terms in the equation of motion also decline, while the viscous terms become more dominant. For vanishingly small velocities, the equations of motion for steady, incompressible flow, for instance, reduce to the following form:

$$\nabla \cdot \mathbf{v} = 0,$$  (6.4.1)

$$\nabla p = \mu\nabla^2\mathbf{v}.$$  (6.4.2)

Under these assumptions the equations of motion thus become linear and their solution can be readily obtained. Slow motions of bodies in fluids are commonly called *creeping flows*.

Consider the creeping motion, $U$, of a solid gas sphere of radius $a$ that is totally immersed in a fluid. Due to the spherical symmetry of this geometry it is convenient to use spherical polar coordinates $(r, \theta)$ to describe the flow field with $\theta = 0$ in the direction of the motion of the sphere. For such a spherically symmetric flow field, we can use the Stokes stream function $\psi$, which is related to the radial and tangential velocity components $u_r$ and $u_\theta$ in the following manner:

$$u_r = \frac{1}{r^2 \sin\theta}\frac{\partial\psi}{\partial\theta}, \qquad u_\theta = -\frac{1}{r \sin\theta}\frac{\partial\psi}{\partial r}.$$  (6.4.3)

Using the definition for the stream function, the momentum equation for this geometry takes the following form:

$$\left(\frac{\partial^2}{\partial r^2} + \frac{1}{r^2}\frac{\partial^2}{\partial \theta^2} - \frac{\cot\theta}{r^2}\frac{\partial}{\partial \theta}\right)^2 \psi = 0. \tag{6.4.4}$$

Equation (6.4.4) for the fluid motion can be solved easily for a solid sphere when the following no-slip boundary conditions on the surface of the sphere are imposed:

$$\frac{\partial \psi}{\partial \theta} = \frac{\partial \psi}{\partial r} = 0 \quad \text{at } r = a. \tag{6.4.5}$$

Meanwhile, at some distance from the sphere, the following condition must hold:

$$\psi \rightarrow \frac{1}{2}Ur^2 \sin^2\theta + const. \quad \text{as } r \rightarrow \infty. \tag{6.4.6}$$

Although this problem appears intractable, due to its linearity a solution can be obtained using the separation of variables method. In this method the stream function $\psi(r, \theta)$ may be represented in the form of a product of the two functions $f(r)$ and $g(\theta)$. Substituting such a representation for $\psi$ in the momentum equation results in the following solution for $\psi$ (see [62]):

$$\psi = \frac{1}{4}Ua^2 \sin^2\theta \left(\frac{a}{r} - \frac{3r}{a} + \frac{2r^2}{a^2}\right). \tag{6.4.7}$$

The two components of the velocity can be computed from (6.4.7):

$$u_r = U\cos\theta \left(1 - \frac{2a}{2r} + \frac{a^2}{2r^3}\right) \tag{6.4.8}$$

$$u_\theta = U\sin\theta \left(-1 + \frac{3a}{4r} + \frac{a^3}{4r^3}\right). \tag{6.4.9}$$

It can be seen that the velocity field given by this solution is totally independent of the fluid viscosity. This is true for all creeping flows. The solution for the stream function shows that the streamlines past the body possess perfect fore-and-aft symmetry. Also, solutions (6.4.8) and (6.4.9) show that the local fluid velocity is everywhere slower than its free stream value, indicating that the effect of the sphere can be felt at large distances from the body. At a distance of $r = 10a$, for example, the velocity is still about 10% below the free stream value.

With the velocity field determined above, the pressure on the surface of the sphere may be found by integrating the momentum equation (6.4.2) to yield

$$p = p_\infty - \frac{3\mu a U}{2r^2}\cos\theta, \tag{6.4.10}$$

where $p_\infty$ is the uniform and constant free stream pressure. Solution (6.4.10) shows that the pressure deviation is proportional to $\mu$, in which case it is positive at the front side and negative at the back side of the sphere. This imbalance in the pressure creates a drag force on the sphere, as might be expected.

The viscous shear stress $\tau_{r\theta}$ for this problem can also be evaluated as follows:

$$\tau_{r\theta} = \mu \left( \frac{1}{r} \frac{\partial u_r}{\partial \theta} + \frac{\partial u_\theta}{\partial r} \right) = -\frac{U\mu \sin\theta}{r} \left( 1 - \frac{3a}{4r} + \frac{5a^3}{4r^3} \right). \tag{6.4.11}$$

The total drag force $F_D$ on the sphere is found by integrating the pressure and the shear stress around the surface of the sphere in the following manner:

$$F_D = -\int_0^\pi \tau_{r\theta}|_{r=a} \sin\theta \, dA - \int_0^\pi p|_{r=a} \cos\theta \, dA, \tag{6.4.12}$$

where $dA = 2\pi a^2 \sin\theta \, d\theta$. Performing the integration yields the celebrated *Stokes formula* for the drag on an immersed sphere:

$$F_D = 4\pi\mu U a + 2\pi\mu U a = 6\pi\mu U a. \tag{6.4.13}$$

Expression (6.4.13) shows that the friction contribution is two thirds of the total drag, while the pressure term contributes only one third. Since this formula is for creeping flows, it is strictly valid for $Re \ll 1$.

The drag force is usually expressed in terms of the coefficient of friction $C_D$ in the following manner

$$C_D = \frac{2F_D}{\rho U^2 \pi a^2} = \frac{24}{Re}. \tag{6.4.14}$$

The Reynolds number in (6.4.14) is based on the diameter of the sphere, i.e., $Re = 2a\rho U/\mu$.

The above formula was derived for a solid sphere where the no-slip boundary condition was applied at its surface. For a sphere with a non-solid skin, such as a bubble or a drop, the condition of the continuity of the tangential stress across the surface must be used instead. For the case of a constant surface tension, the surface shear stress component $\tau_{r\theta}$ must vanish, leading to the following solution for the velocity distribution:

$$u_r|_{r=a} = 0, \qquad u_\theta|_{r=a} = -\frac{1}{2}U \sin\theta, \tag{6.4.15}$$

which gives the following value for the drag coefficient:

$$C_D = \frac{16}{Re}. \tag{6.4.16}$$

The method outlined above for evaluating the drag coefficient is identical for free and rigid surfaces; the only difference between the two is in the boundary conditions which result in different numerical values. Figure 6.9 shows a comparison between theory and experiments of the drag coefficient as a function of the Reynolds number, with good agreement between the two up to $Re \approx 1$.

For $Re \gg 1$, the theories on the fluid motion past solid spheres and drops or bubbles are quite different. In both cases the inviscid theory can be used outside a thin layer next to the surface, while boundary layer theory is used within this layer. In the case of the solid sphere, the no-slip boundary condition must be satisfied at the surface of the sphere. In either case, vorticity is generated at

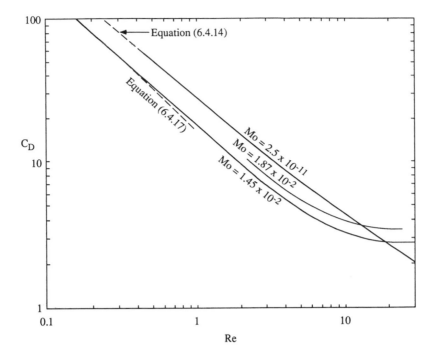

**Figure 6.9** $C_D$ vs. Reynolds number, from [64], for both theoretical predictions (dotted line) and experiments (solid line). The numbers on the experimental curves represent the respective values for the Morton number.

the surface because the irrotational flow does not satisfy all of the boundary conditions. Normally, this vorticity remains confined within the boundary layer next to the surface, the thickness of which, $\delta$, depends on the surface of the rigid sphere or drop or bubble. For a solid sphere the velocity must change by an order of $U$ across the layer, whereas for a gas bubble the velocity changes by an order of $U\delta/a = U Re^{-1/2}$.

The value of the fluid velocity in the boundary layer surrounding the bubble is only slightly different from that for the irrotational flow, while the velocity derivatives are of the same order. To a first-order approximation, the viscous dissipation integral, expression (6.4.12), has the same value as that for irrotational flow. This is due to the fact that the total volume of the boundary layer, which is proportional to $a^2\delta$, is much less than the volume of the region in which the velocity derivatives are of order $U/a$. The volume of the wake is not small, but the velocity derivatives in it are, and thus its contribution to the dissipation is of higher order significance. A first approximation to the coefficient of drag for a bubble at high $Re$ is obtained by [64] in the form of

$$C_D = \frac{48}{Re}, \qquad (6.4.17)$$

by evaluating the dissipation rate for irrotational flow past a sphere. The bubble in this case must be spherical and the surface tension force given by $\sigma/a$ is much greater than the dynamic pressure given by $\approx \rho U^2$, i.e., for $We \ll 1$.

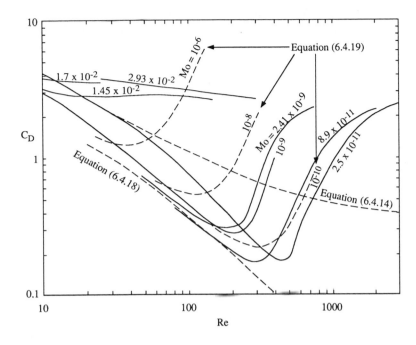

**Figure 6.10** $C_D$ as a function of the Reynolds number, from [64], for both theoretical predictions (dotted line) and experiments (solid line). The numbers on the experimental curves represent the respective values for the Morton number.

Reference [64] states that higher order approximation for the drag coefficient can be calculated efficiently using energy arguments. In a steady flow, the rate of the gravity work on the bubble, which is the drag force times the velocity $U$, is balanced by the rate of work against the surface stress plus the rate of viscous dissipation of energy throughout the fluid. Using the energy formulation with complicated perturbation analysis, which also accounts for the boundary layer, a final result for $C_D$ is obtained by [64] in the following form:

$$C_D = \frac{48}{Re}\left[1 - \frac{2.211}{\sqrt{Re}} + O(Re^{-5/6})\right]. \qquad (6.4.18)$$

The leading term in expression (6.4.18) represents the irrotational dissipation, the next term is due to the effects of the boundary layer and the wake, while the error term comes principally from neglecting viscous dissipation and surface stress in the rear stagnation region. Figure 6.10 shows the variation of the drag coefficient as a function of the Reynolds number according to (6.4.18).

Equation (6.4.18) gives a good correlation with the experimental data for bubbles that satisfy its conditions of validity, i.e., for high $Re$ ($\approx Re > 50$), low enough $Mo$ for the bubble to be almost spherical for some range of $Re$ over 50 ($M < 10^{-8}$ is sufficient), and fluids pure enough for $\sigma$ to be effectively constant. Figure 6.10 shows that the experimental value for $C_D$ does not continue to decrease in the way indicated by expression (6.4.18) for any fluid, but begins to rise steeply after a certain value of $Re$ is reached especially if $Mo$ is small.

In the above theory, the bubbles were assumed to be perfect spheres, while in practice they are frequently of very different shapes. In general, oblate spheroids are found to be a fair approximation to the true shapes of bubbles for quite large values of the Weber number. Reference [64] shows that the analysis for the flow around oblate spheroids has found the drag coefficient to be fairly well approximated by

$$C_D = \frac{48}{Re} G(\chi) \left[1 + \frac{H(\chi)}{\sqrt{Re}}\right] \tag{6.4.19}$$

where $\chi$ is the ratio of the longest to the shortest spheroid diameter. $G(\chi)$ and $H(\chi)$ are functions tabulated in [64]. Satisfying the pressure requirements only at the equator and the poles gives results that agree better with experiments and with the predicted values for $We(\chi)$. Using the tabulated values of $We(\chi)$, the drag coefficient may be calculated as a function of the Reynolds and Morton numbers with the aid of the following identity:

$$C_D = \frac{4}{3} Mo Re^4 We^{-3}. \tag{6.4.20}$$

Curves for $M = 10^{-6}, 10^{-8}$, and $10^{-10}$ are shown in Figure 6.10. Each curve extends from $\chi = 1.0$, where $C_D$ is close to its value for spherical bubbles, $\chi = 4.0$, where $We$ has become nearly constant making $C_D \propto Re^4$, approximately. Experiments indicate that the Weber number $We$ is not far from the calculated value when it becomes a slowly varying function of $\chi$. The shape of the bubble is then very sensitive to small changes in $We$, which may be caused by small currents in the surrounding fluid. It is therefore not surprising that bubbles in pure, low-$Mo$ liquids become unstable for values of $We > 3$. The type of motion that appears when a steady rise of bubbles in a straight line becomes unstable, is either a steady motion relative to the bubble up, a helix with a vertical axis, or a zigzag motion in a vertical plane on either side of a vertical line. In spite of the many experiments, there is no agreement on the conditions that give rise to either of these two modes of instability.

## 6.5 DROPS WITH CONSTANT SURFACE TENSION MOVING UNDER GRAVITY

The theory for moving drops is in most respects very much like that for gas bubbles, except more complicated. For drops, instead of a single nondimensional numbr $Mo$ there will be two additional independent dimensionless parameters that characterize the fluids both inside and outside the drop. These are $\rho_o/\rho_i$ and $\mu_o/\mu_i$. Figures 6.9 and 6.10, which illustrate the behavior of bubbles, can also be used for drops for which the values of both $\rho_o/\rho_i$ and $\mu_o/\mu_i$ are very large.

For a perfectly spherical drop it is possible to derive an analytical expression describing the fluid motion both inside and outside the drop in the limit of small values of the Reynolds number. The tangential velocity component for this case takes the form

$$u_\theta = \frac{U\mu_o \sin\theta}{2(\mu_o + \mu_i)}. \tag{6.5.1}$$

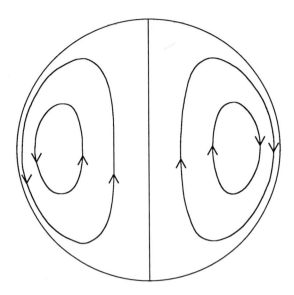

**Figure 6.11**  Streamline pattern from equation (6.5.3) for the flow field inside a drop.

Expression (6.5.1) can then be used to determine the drag coefficient, yielding the following expression:

$$C_{Do} = \frac{8(2\mu_o + 3\mu_i)}{Re_o(\mu_o + \mu_i)} \cdot \qquad (6.5.2)$$

Inside the drop the expression for the stream function is found to take the following form

$$\psi_i = \frac{U\mu_o[(r^4/a^2) - r^2]\sin\theta}{8(\mu_o + \mu_i)}, \qquad (6.5.3)$$

where the flow goes around the surface of the drop and back up through the middle. The maximum fluid velocity inside the drop is given by $U\mu_o/2(\mu_o + \mu_i)$, which occurs at both the equator and the center. There is a stagnation ring for this flow at $r = a/\sqrt{2}$ and $\theta = \pi/2$. The streamline pattern for this flow is shown in Figure 6.11. Such a circulation is commonly known as the *Hill vortex flow*. As $\mu_o/\mu_i \to \infty$, the results derived in the previous section for bubbles are recovered. In this case $C_{Do} \to 16/Re_o$ and $u_\theta \to 1/2U\sin\theta$. On the other hand, as $\mu_o/\mu_i \to 0$ the results for the Stokes flow over a rigid sphere are recovered, i.e., $C_{Do} \to 24/Re_o$ and $u_\theta \to 0$. For intermediate values of $\mu_o/\mu_i$, both $C_{Do}$ and $u_\theta$ assume intermediate values.

Expression (6.5.2) for the drag coefficient is valid in the limiting case of very small Reynolds numbers, i.e., $Re_o \to 0$. The expression for the drag coefficient at finite values of $Re_o$ is given in [64], in the form of a series expansion in terms of $Re_o$, in the following manner:

$$C_{Do} = \frac{8(2\mu_o + 3\mu_i)}{Re_o(\mu_o + \mu_i)}\left[1 + \frac{Re_o(2\mu_o + 3\mu_i)}{16(\mu_o + \mu_i)} + O(Re_o^2 \ln Re_o) + O(We_o)\right]. \quad (6.5.4)$$

The shape of the drop has also been investigated analytically for finite but small values of the Reynolds number. The axis ratio $\chi$ for a drop in this case, to a leading order, is given by the expression

$$\chi = 1 + \frac{3We_o}{16(\mu+1)^3} \left[ \left( \frac{81}{80}\mu^3 + \frac{57}{20}\mu^2 + \frac{103}{40}\mu + \frac{3}{4} \right) - \frac{\rho_i - \rho_o}{12\rho_o}(\mu+1) \right]$$
$$+ O(We_oRe_o) + O(We_o^2/Re_o), \tag{6.5.5}$$

where $\mu = \mu_i/\mu_o$. The drop shape, to a first-order approximation, is a spheroid, but it could be either prolate or oblate, unlike a bubble. Prolate (elongated) shapes require the inner fluid to be denser and either less viscous than the outer fluid or else slightly more viscous. Liquid drops in gases, for example, are normally oblate; so are mercury drops in water.

The limit $\mu \to \infty$ represents a drop behaving like a solid sphere; the drag coefficient in this case is given by $C_{Do} \to 24(1 + 3Re_o/8)/Re_o$ and $\chi \to 1 + (243/1280)We_o \approx 1 + 0.19We_o$.

The drag coefficient for a spherical drop at large values of the Reynolds number has been estimated by [64] to be

$$C_{Do} = \frac{48}{Re_o} \left\{ 1 + \frac{3\mu}{2} + \frac{\mu\lambda_b(1/\beta+1)}{\sqrt{Re_o}} \right. \tag{6.5.6}$$
$$\times \left[ \mu\lambda_b c_1(\lambda_a)\ln Re_o + \mu\lambda_b[1 + (1/\beta)]c_2(\lambda_a) + \left( 1 + \frac{3\mu}{2} \right) c_3(\lambda_a) \right] \right\},$$

where

$$\lambda_a = \frac{1-\beta}{1+\beta}, \qquad \lambda_b = \frac{2/\mu+3}{2+3/\beta}.$$

$c_1$, $c_2$, and $c_3$ in equation (6.5.6) are all functions of $\lambda_a$ and are tabulated in [64], and $\beta = \sqrt{\mu\rho_i/\rho_o}$. The above formula has been found to agree fairly well in carefully purified liquids up to values of the Reynolds number in several hundreds, provided that $Mo_o$ is low enough for the drops to remain nearly spherical. The values of $C_{Do}$ in these instances come to within 20% of the experimental data, whereas $C_{Do}$ for rigid spheres is too high by a factor of 2. Figure 6.12 shows a comparison between theory and various experiments for $C_{Do}$ as a function of $Re_o$. Agreement between experiment and theory is generally worse for drops than for bubbles.

## 6.6 EFFECTS OF VARIABLE SURFACE TENSION

When a gas bubble is placed in a liquid possessing a temperature gradient, the interfacial tension at the bubble surface will vary with position, due the strong dependence of the latter on temperature. Typically, the surface tension is smallest at the warm region and greatest at the cold region. The resulting gradient in surface tension leads to a tangential stress at the interface in the direction of the cold region. Consequently, the neighboring fluid is dragged in this direction by the stress, driving the bubble in the opposite direction, namely, in the direction of the temperature gradient. Therefore, bubbles experience an attraction toward the hot liquid regions, and their behavior under gravity could be affected by this phenomenon whenever the motion is parallel to gravity.

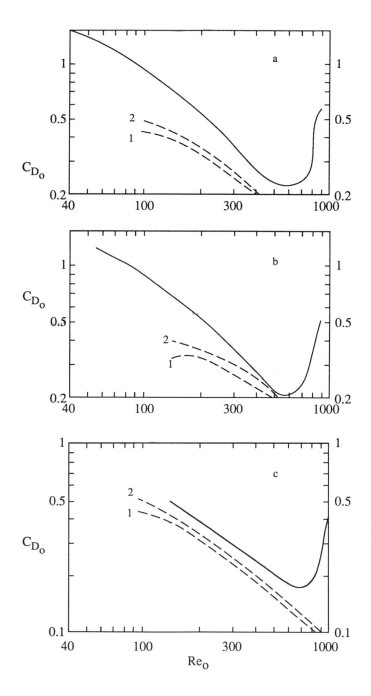

**Figure 6.12**  $C_D$ vs. Reynolds number, from [64], for both theoretical predictions (dotted line) and experiments (solid line). Curve 1 is equation (6.5.6) without the logarithmic term; curve 2 is equation (6.5.6). (a) $CCl_3$ drop in water, $Mo = 5.1 \times 10^{-11}$; (b) $o$-dichlorobenzene drop in water, $Mo = 3.1 \times 10^{-11}$; and (c) bromobenzene drop in water, $Mo = 1.7 \times 10^{-11}$.

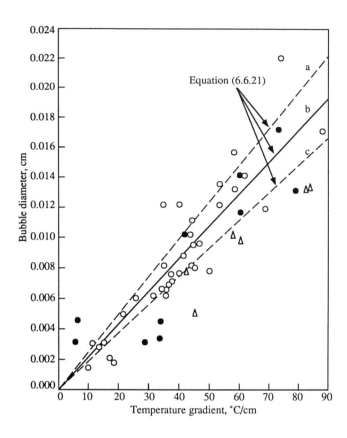

**Figure 6.13**   The variation of bubble diameter with applied temperature gradient in the liquid (see [75]). The different lines denote different values for the surface tension variation with temperature:  (a) $\partial\sigma/\partial T = -0.06$, (b) $\partial\sigma/\partial T = -0.07$, and (c) $\partial\sigma/\partial T = -0.08$ dyne/cm°C.

The effect of a vertical temperature gradient on bubbles in a liquid host has been demonstrated experimentally in [75] by confining a small column of liquid between the anvils of a machinist's micrometer.  Several Dow-Corning DC-200 series silicone oils were used for the experiments, together with a set of control experiments in n-hexadecane.  In this experiment, the lower anvil was kept at a higher temperature than the upper.  The temperature gradients were kept sufficiently small, however, to avoid buoyancy effects due to the unstable density stratification in the liquid.  Air bubbles were injected into the liquid column and were found to collect at the lower anvil. This abnormal behavior confirmed qualitatively the expectations regarding the action of surface tension gradients. The vertical temperature gradient was then slowly reduced until a relatively large bubble was seen to detach from the lower anvil and rise slowly. Another adjustment to the temperature gradient allowed the bubble to remain nearly motionless midway between the anvils. The bubble diameter was measured with a traveling microscope. The temperature gradient necessary to hold bubbles of various sizes from rising under Earth's gravity is shown in Figure 6.13.

While these experiments were for gas bubbles, the phenomenon is equally applicable to liquid drops. Motion driven by interfacial tension gradients is usually referred to as *capillary driven motion* or simply *capillary motion*. When the gradients are the result of temperature variations, the term *thermocapillary motion* is used. When composition variations cause the gradients, the term *diffusocapillary motion* has been suggested.

For the analytical description of the problem, reference [75] considers the motion of a fluid sphere under the combined action of gravity and a downward temperature gradient. Assuming the motion of the bubble to be very small, the fluid velocity both inside and outside the bubble is described by equations (6.4.1) and (6.4.2) for creeping motion. For the present problem, these equations take the following form:

$$\nabla(p_o + \rho_o g z) = \mu_o \nabla^2 \mathbf{v}_o, \quad \nabla \cdot \mathbf{v}_o = 0, \tag{6.6.1}$$

$$\nabla(p_i + \rho_i g z) = \mu_i \nabla^2 \mathbf{v}_i, \quad \nabla \cdot \mathbf{v}_i = 0. \tag{6.6.2}$$

The subscript $o$ and $i$ denote outside and inside of the bubble, respectively. Note that the pressure term in (6.6.1) and (6.6.2) has been replaced by the modified pressure, which includes the hydrostatic head (see Section 2.4).

The solution for the velocity field outside the bubble can be obtained by solving equation (6.6.1), resulting in a solution that is analogous to the one given by (6.4.8) and (6.4.9). The tangential and radial velocity components and the pressure for a spherically symmetric bubble of radius $a$ and coordinates $r$ and $\theta$ are given by (see [62]):

$$u_{ro} = \left[ \frac{A_o}{\mu_o} \left( \frac{1}{r} - \frac{a^2}{r^3} \right) + U \left( 1 - \frac{a^3}{r^3} \right) \right] \cos\theta, \tag{6.6.3}$$

$$u_{\theta o} = \left[ \frac{A_o}{\mu_o} \left( \frac{1}{r} - \frac{a^2}{r^3} \right) + U \left( 1 + \frac{a^3}{2r^3} \right) \right] \sin\theta, \tag{6.6.4}$$

$$p_o = \left( \frac{A_o}{r^2} - \rho_o g r \right) \cos\theta. \tag{6.6.5}$$

The solutions given by (6.6.3)–(6.6.5) are consistant with the following far field conditions:

$$\mathbf{v}_o \to (0, 0, U), \quad p_o \to \rho_o g z, \quad \text{as } |\mathbf{r}| \to \infty. \tag{6.6.6}$$

$A_o$ and $U$ are a constant of integration and the bubble rise velocity, respectively.

The velocity field and the fluid pressure inside the bubble are determined by solving equations (6.6.2), which result in (see [62]):

$$u_{ri} = \frac{A_i}{10\mu_i}(r^2 - a^2) \cos\theta, \tag{6.6.7}$$

$$u_{\theta i} = -\frac{A_i}{10\mu_i}(2r^2 - a^2) \sin\theta, \tag{6.6.8}$$

$$p_i = (A_i - \rho_i g)r \cos\theta + B_i. \tag{6.6.9}$$

The solutions in (6.6.7)–(6.6.9) are valid for $|\mathbf{r}| \to 0$. $A_i$ and $B_i$ are constants of integration.

The temperature distributions both inside and outside the bubble, $T_i$ and $T_o$, are given by the steady-state energy conservation equation for both inside and outside the bubble, i.e.,

$$\mathbf{v}_o \cdot \nabla T_o = \kappa_o \nabla^2 T_o, \tag{6.6.10}$$

$$\mathbf{v}_i \cdot \nabla T_i = \kappa_i \nabla^2 T_i, \tag{6.6.11}$$

where $\kappa = \rho c_p / k$ is the coefficient of thermal diffusion and $k$ is the thermal conductivity for the fluid. Consistent with the creeping flow assumption, equations (6.6.10) and (6.6.11) can be further reduced to the following form in the limit of very slow fluid motion:

$$\nabla^2 T_{o,i} = 0. \tag{6.6.12}$$

The solutions to equations (6.6.12) are

$$T_o = T_1 + |\nabla T_\infty|(r + C_o/r^2)\cos\theta, \tag{6.6.13}$$

$$T_i = T_1 + |\nabla T_\infty| C_i r \cos\theta, \tag{6.6.14}$$

where $T_1$, $C_o$, and $C_i$ are also constants of integration and $|\nabla T_\infty|$ is the imposed vertical temperature gradient in the fluid outside the bubble. The solutions given in (6.6.13) and (6.6.14) are consistent with the following far field temperature condition:

$$T_o = T_1 + |\nabla T_\infty|z \quad \text{as} \quad |\mathbf{r}| \to \infty. \tag{6.6.15}$$

The velocity fields, temperature, and pressure for both inside and outside the bubble must satisfy the following interface conditions at the bubble surface, $r = a$:

1. The tangential velocity components must be the same on both sides of the bubble, i.e.,

$$u_{\theta o}(a,\theta) = u_{\theta i}(a,\theta). \tag{6.6.16}$$

2. The shear stress at the bubble surface must be balanced, i.e.,

$$\mu_i\left(\frac{\partial u_{\theta i}}{\partial r} - \frac{u_{\theta i}}{r}\right)\bigg|_{r=a} - \mu_o\left(\frac{\partial u_{\theta o}}{\partial r} - \frac{u_{\theta o}}{r}\right)\bigg|_{r=a} = \frac{1}{r}\frac{\partial\sigma}{\partial\theta}\bigg|_{r=a}. \tag{6.6.17}$$

3. The normal stress at the bubble surface must be balanced, i.e.,

$$\left(p_i - 2\mu_i\frac{\partial u_{ri}}{\partial r}\right)\bigg|_{r=a} - \left(p_o - 2\mu_o\frac{\partial u_{ro}}{\partial r}\right)\bigg|_{r=a} = \frac{2\sigma}{a}. \tag{6.6.18}$$

4. The temperature on both sides of the bubble surface must be continuous, i.e.,

$$T_i(a, \theta) = T_o(a, \theta). \qquad (6.6.19)$$

5. The heat flux at the bubble surface must be continuous, i.e.,

$$k_o \frac{\partial T_o}{\partial r}\bigg|_{r=a} = k_i \frac{\partial T_i}{\partial r}\bigg|_{r=a}. \qquad (6.6.20)$$

Substituting the solutions to (6.6.3)–(6.6.5), (6.6.7)–(6.6.9), and (6.6.12)–(6.6.13) into the conditions at the bubble surface, (6.6.16)–(6.6.20), and solving for the constants of integration, [75] obtains the following result for the bubble rise velocity $U$:

$$U = \frac{2}{3\mu_o(2 + 3\mu)} \left[ \frac{3|\nabla T_\infty||\partial\sigma/\partial T}{2 + k} a + (\rho_o - \rho_i)g(1 + \mu)a^2 \right]. \qquad (6.6.21)$$

$\mu = \mu_i/\mu_o$ is the ratio of the dynamic viscosities, $g$ is the magnitude of the acceleration due to gravity, $\partial\sigma/\partial T$ is the change of interfacial tension with temperature, and $|\nabla T_\infty|$ is the magnitude of the downward temperature gradient imposed in the continuous phase fluid. $k = k_i/k_o$ represents the ratio of the thermal conductivities. A negative value for $U$ in the above expression implies downward motion. Note that when the velocity on the left-hand side of (6.6.21) is set equal to zero, the temperature gradient needed to hold the bubble motionless is a linear function of the bubble radius and independent of the viscosity of the continuous phase. Figure 6.13 shows a comparison between expression (6.6.21) and the experiments of [75].

For a gas bubble, the viscosity and thermal conductivity of the gas are both quite negligible compared to the properties of the surrounding liquid. Thus, it is possible to set the property ratios $\mu$ and $k$ to zero in expression (6.6.21) as well as assume that the density of the gas is negligible compared to that of the continuous phase. Reference [68] shows that a solution for the problem under these conditions may constitute a first step in a perturbation expansion for small values of both the Reynolds and Peclet numbers. The *Peclet number* is defined by

$$Pe = \frac{aU}{\kappa}.$$

Such a solution should be a good approximation when the values for these parameters are small.

The experiments described in [75] are remarkable in their simplicity and yet produced results of substantial impact. For instance, the original experiments verified that the temperature gradient required to hold a bubble stationary increased with the bubble size and was independent of viscosity, as predicted from expression (6.6.21). These experiments opened up a substantial research area that thrives at the present time stimulated by applications in reduced gravity. A steady temperature gradient is relatively straightforward to induce and maintain in a liquid. This has led to a number of experiments that already have been conducted on ground, and several are planned for space flights in the next decade.

According to the survey of [68], there is clear evidence of surface tension-driven motion of bubbles in various liquids, based on ground-based experiments. This fact allows low-gravity experiments to go beyond the range of parameters attainable on ground. Nondimensionalization of the governing equations for bubble motion in a temperature gradient reveals that the scaled bubble speed depends on two key parameters, namely the Reynolds number and the Peclet number. When the motion is caused by purely thermocapillary effects, the Marangoni number is usually used in place of the Peclet number. The Reynolds number gives the relative importance of convective transport of momentum compared to molecular transport, while the Peclet number gives the relative importance of convective transport of energy compared to molecular transport.

In practical situations appropriate for space applications, a wide range of values for each of these numbers may be encountered depending on the physical properties of the fluids involved. However, virtually all of the ground-based experimental effort has been concerned with small or negligible values for these two numbers. Reference [68] suggests examining these definitions in order to understand the reasons:

The Reynolds number $Re$ and the Marangoni number $M$ can be alternatively defined, according to [68], in the following manner:

$$Re = \frac{aU_1}{\nu_o} , \qquad (6.6.22)$$

$$M = \frac{aU_1}{\kappa_o} . \qquad (6.6.23)$$

Here, $\nu_o$ is the kinematic viscosity of the continuous phase, $\kappa_o$ is its thermal diffusivity, and $a$ is the radius of the bubble. $U_1$ in the above definitions is a characteristic velocity defined in terms of the surface tension gradients in the melt:

$$U_1 = \frac{|\partial \sigma / \partial T||\nabla T_\infty|a}{\mu_o} . \qquad (6.6.24)$$

In the limit $g \rightarrow 0$, the drop migration velocity given by expression (6.6.21) reduces to

$$U = \frac{2a|\nabla T_\infty||\partial \sigma / \partial T|}{\mu_o(2 + 3\mu)(2 + k)}. \qquad (6.6.25)$$

For bubbles in the limit as $\mu$, $k \rightarrow 0$, expression (6.6.24) reduces to

$$U = U_1/2, \qquad (6.6.26)$$

which states that the bubble migration velocity is one-half the characteristic velocity $U_1$.

Reference [68] argues that the problem with using experiments on the ground under conditions, appropriate for a large Peclet number, is that they will not necessarily be representative of experiments in reduced gravity under conditions appropriate for a large Marangoni number. There are fundamental differences in the resulting flow fields, depending on whether the motion is driven by a body force for which a nonzero hydrodynamic force is necessary to achieve steady migration, or whether the motion is driven by the interface under conditions of zero hydrodynamic force. This, in turn, will affect the temperature fields in the surrounding fluid and therefore the driving force for thermocapillary motion, which is a function of the temperature field on the drop surface. Therefore, experiments on ground, even if they can be conveniently performed, are of limited utility in predicting the behavior of drops and bubbles in reduced gravity. In addition, there are difficulties in performing these experiments even on ground.

The most immediate application for this phenomenon in low-gravity environment is related to materials processing in space. In a terrestrial environment, bubbles in glass melts tend to migrate toward the upper regions of the melt because of buoyancy effects. In a low-gravity environment this will not happen, and consequently another mechanism is required to drive the bubbles to a specific location in the melt. Thermocapillary migration of bubbles has been suggested as one such mechanism.

Very few low-gravity experiments on thermocapillary motion of gas bubbles have been conducted to date. Some early experiments were performed aboard NASA's sounding rocket flights. In these experiments a temperature gradient was applied to a sodium borate melt, contained in a low narrow cell of rectangular cross section, in which the motion of small gas bubbles suspended in the melt was photographed through a microscope. The objective of the experiments was to demonstrate that the bubbles would move in the glass melt due to the temperature gradient. The bubbles were indeed observed to move toward warmer regions, and larger bubbles moved more rapidly than smaller ones. The bubbles accelerated as they moved into warmer melt because of reduced viscosity. Precise quantitative comparisons were not attempted due to experimental complications, such as interactions among bubbles and neighboring surfaces.

Recent low-gravity experiments on bubble migration in a liquid possessing a temperature gradient are discussed in [65]. In this reference, a summary is presented on the recent experimental efforts by the European Space Agency on thermocapillary effects on bubbles. The reference cites experimets already performed on the D-1 (Deutsche Spacelab 1) mission as well as the proposed work to be performed in the D-2 mission. The D-1 mission experiments were conducted with air bubbles in silicone oil. These experiments demonstrated bubble motion under the action of thermocapillary forces for a range of Marangoni numbers up to $M = 300$ and for Reynolds numbers up to $Re = 0.4$.

Three types of silicone oil were used in these experiments: AK100, AS100, and AP100. Figure 6.14 summarizes the results for thermocapillary bubble migration from these low-gravity experiments. Bubbles in silicone oils AK100 and AS100 were found to move with velocities very close to the propagation velocities predicted by expression (6.6.24). However, the bubbles in silicone oil AP100 did not move at all during the flight experiments. These results led [65] to suggest a minor modification to expression (6.6.24) to account for an additional dissipation term at the liquid gas interface. Reference [65] indicated that all of the results of the tests on this flight experiment correlated very well with the modified expression

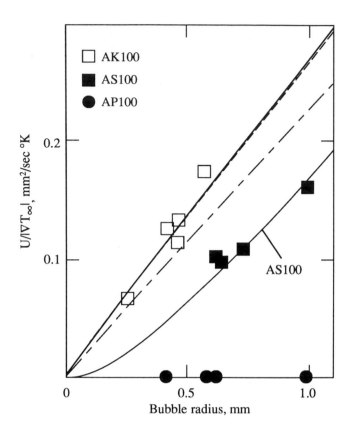

**Figure 6.14**  Measured bubble diameter as a function of normalized speed in the low-gravity experiments of [65]. Dotted line: equation (6.6.25); solid line: equation (6.6.27) with $\lambda = 6.0 \times 10^{-5}$ kg/(msec).

for the bubble migration speed

$$U = \frac{2a|\nabla T_\infty|\partial\sigma/\partial T}{2\mu_o(2+k)[2+3\mu+2\lambda/(a\mu_o)]},\qquad(6.6.27)$$

where $\lambda$ is the surface dilation viscosity.

*Chapter 7*

# Liquid-Vapor Two-Phase Flows

T he future trend in spacecrafts and space systems design is toward ever increasing size, complexity, and power consumption. Two-phase flow technology has been proposed for meeting the increased requirements placed on certain space subsystems such as thermal control, propulsion, power generation, and life support. Operations involving multiphase flows are also required for a wide variety of manufacturing processes involving condensation and boiling and for separation and chemical reactions. However, the thermofluid dynamics of two-phase systems encompasses a wide range of complex phenomena that are not yet completely understood in the terrestrial environment. The fundamental conservation equations for multiphase systems are very complex and often very difficult to solve. Engineers have traditionally relied, to a very large extent, on empirical or mechanistic expressions for predicting realistic design criteria for systems involving two-phase flows.

Low gravity in space presents additional problems and new challenges for two-phase systems engineering design. The validity of existing analytical and empirical formulations is limited since all were developed under terrestrial conditions. Numerous low-gravity experiments designed for gathering the necessary data for formulating such empirical expressions for space must be performed before we can define realistic design criteria. A limited amount of two-phase, low-gravity research has been performed to date, yet there are still many more questions than answers. There still remains a considerable need for reliable design guidelines for two-phase, low-gravity components and systems.

This chapter begins with a review of the fundamentals of liquid-vapor two-phase flows without phase change, i.e., for adiabatic conditions. Next, the basic principles for pool boiling are presented, followed by flow boiling formulation. Finally, the process of vapor condensation is reviewed.

## 7.1 *CHARACTERISTICS OF LIQUID-VAPOR TWO-PHASE FLOWS*

The intrinsic complexity of two-phase flows can be effectively illustrated using one of the simplest flows for which a great deal is known. One of the most striking areas, where the difference between single- and two-phase flows can be demonstrated, is in evaluating the skin friction for pipe flows. The *skin friction coefficient*, $C_f$, for a single-phase, fully developed laminar pipe flow is given by

$$C_f = \frac{16}{Re}, \tag{7.1.1}$$

where $Re$ is the Reynolds number defined in this case by

$$Re = d\rho U/\mu,$$

$d$ is the pipe diameter, and $U$ is the average velocity of the fluid. For turbulent flows, however, the skin friction on a smooth wall can be shown to be related to the Reynolds number in the following manner (see [89]):

$$C_f = 0.079 Re^{-1/4}. \tag{7.1.2}$$

Formula (7.1.2) is an empirical relation, while a semi-empirical formula for the turbulent pipe flow, which is applicable in the logarithmic region, can be shown to have the following form:

$$\sqrt{2/C_f} = 2.5 \ln\left(Re\sqrt{C_f/2}/2\right) + 1.5. \tag{7.1.3}$$

Expressions (7.1.1)–(7.1.3) all have the common characteristic that the skin friction coefficient is a function of a single nondimensional parameter, the Reynolds number. This is due to the fact that for a simple pipe flow the only parameter appearing in the governing conservation equations is the Reynolds number.

When two phases of fluid are flowing simultaneously in a pipe (say, a liquid and a gas), the problem of pipe flow must be reformulated. The skin friction in this case depends very strongly on the specific fluid phase adjacent to the wall. Since in most two-phase flows the ratio of the two phases as well as their speeds vary with both time and position from the wall, the skin friction similarly will be a function of time and position. Thus any rational formulation for the skin friction coefficient requires a thorough knowledge of the specific flow pattern.

These arguments may be illustrated by examining the equations governing two-phase flows. In two-phase flows, the conservation equations for both the liquid and the gas must be solved simultaneously. For the simple adiabatic case, these equations take the following nondimensional form:

$$\nabla \cdot \mathbf{v}_l = 0, \quad \nabla \cdot \mathbf{v}_g = 0, \tag{7.1.4}$$

$$\frac{\partial \mathbf{v}_l}{\partial t} + \mathbf{v}_l \cdot \nabla \mathbf{v}_l = -\nabla p_l + \frac{1}{Re_l}\nabla^2 \mathbf{v}_l + \frac{1}{Fr_l}\mathbf{F}_B, \tag{7.1.5}$$

$$\sqrt{Fr_l/Fr_g}\frac{\partial \mathbf{v}_g}{\partial t} + \mathbf{v}_g \cdot \nabla \mathbf{v}_g = -\nabla p_g + \frac{1}{Re_g}\nabla^2 \mathbf{v}_g + \frac{1}{Fr_g}\mathbf{F}_B. \tag{7.1.6}$$

The subscripts $g$ and $l$ in equations (7.1.4)–(7.1.6) denote gas and liquid, respectively. $\mathbf{v}$, $p$, and $\mathbf{F}_B$ are the velocity vector of the specific fluid, the pressure, and the appropriate body force. Equations (7.1.4)–(7.1.6) must be solved subject to the no-slip boundary conditions at the walls and the appropriate interface conditions at all liquid-gas surfaces. These conditions can be written as

$$\mathbf{v}_l = \mathbf{v}_g = 0, \tag{7.1.7}$$

at the solid wall, and

$$\mathbf{v}_l = \mathbf{v}_g, \tag{7.1.8}$$

$$\left( p_l - \frac{\rho_g}{\rho_l}\frac{Fr_g}{Fr_l} p_g \right) \mathbf{n} - 4 \left( \frac{\mathbf{e}_l}{Re_l} - \frac{\rho_g}{\rho_l}\frac{Fr_g}{Fr_l}\frac{\mathbf{e}_g}{Re_g} \right) \mathbf{n} = -\frac{4}{We_l}\mathbf{n}\left( \frac{1}{R_1} + \frac{1}{R_2} \right), \tag{7.1.9}$$

at all liquid-gas interfaces. $\mathbf{e}$ in condition (7.1.9) is the rate of strain tensor; $\mathbf{n}$, $R_1$, and $R_2$ are the unit normal to, and the principal radii of curvature of, the liquid-gas interface, respectively.

The difficulty with two-phase flow modeling, whether empirical or analytical, is immediately recognized as due to the increase in the nondimensional numbers in equations (7.1.4)–(7.1.9). In the single-phase flow only the Reynolds number was used, but in two-phase flows six nondimensional numbers are needed. These nondimensional numbers for this problem are

(a)  The Reynolds numbers for both the liquid and the gas:

$$Re_l = \frac{v_{sl}d}{\nu_l}, \qquad Re_g = \frac{v_{sg}d}{\nu_g}.$$

(b)  The Froude numbers for the liquid and the gas:

$$Fr_l = \frac{v_{sl}^2}{gd}, \qquad Fr_g = \frac{v_{sg}^2}{gd}.$$

(c)  The Weber number for the liquid:

$$We_l = \frac{\rho_l v_{sl}^2 d}{\sigma}.$$

(d)  The density ratio:

$$\frac{\rho_g}{\rho_l}.$$

In the definitions given above, $v_{sl}$ and $v_{sg}$ denote the superficial velocities in the liquid and gas, respectively. The *superficial velocity* $v_s$ is usually defined for pipe flows as the volumetric flow rate of the particular phase divided by the cross-sectional area of the tube. Equations (7.1.4)–(7.1.6) show that regardless of the type of flow pattern (i.e., whether bubbly, mist, or stratified), the skin friction coefficient $C_f$ must be formulated in terms of all the nondimensional parameters appearing in the equations in the following manner

$$C_f = \frac{2\tau_w}{\rho_m v_m^2} = f(Re_l, Re_g, Fr_l, Fr_g, We_l, \rho_l/\rho_g), \tag{7.1.10}$$

where

$$\rho_m = \alpha_l \rho_l + \alpha_g \rho_g, \quad v_m = v_{sl} + v_{sg},$$

$\alpha_l$ is the liquid fraction, and $\alpha_g$ is the void fraction that satisfy the mass conservation requirement

$$\alpha_l + \alpha_g = 1.$$

It is interesting to note that the above derivation shows gravity appearing explicitly in the equations via the Froude numbers. Gravity does play an important role in two-phase flows by specifying the shape of the gas/liquid interfaces and by supplying the driving mechanism.

The derivation given above also shows that the complexity of the analysis has increased tremendously so that any attempt to solve two-phase flow problems must be achieved in terms of at least six nondimensional numbers. In terrestrial applications, empirical formulas have been devised to predict the pressure drop and hence the wall stress $\tau_w$ in terms of the void fraction $\alpha$, or vice versa, under specific flow conditions. Different sets of empirical relations exist depending on whether the pipe is horizontal, vertical with up-flow, or vertical with down-flow. Gravity is usually not included explicitly in such empirical formulation, but is implicitly through the geometry of the problem. The flow patterns for this type of formulation account for gravity through the different formulas that apply for each flow pattern. For instance, the pressure drop formula for plug flow is different from that for annular flow.

## 7.2  FLOW PATTERNS IN LIQUID-VAPOR FLOWS

The empirical formulation approach is attractive only because of its simplicity, but it has a major drawback in that it cannot be generalized. These relations cannot be used, for instance, when gravity is lowered by two orders of magnitude. It is impossible to produce a comparable set of empirical formulas for application in low-gravity since such a task requires performing a large number of space-based experiments. If, however, the effect of gravity is only manifested by the alteration of the two-phase flow patterns for a specific flow configuration, then it may be possible to modify the terrestrially defined empirical relations for use in low-gravity applications.

All analytical methods for describing two-phase flows must rely on the proper identification of the specific gas-liquid flow patterns. Once these are identified, then each pattern is defined by an appropriate physical model. The concept of a flow pattern is intrinsically related to the gas-liquid interfacial structure. Flow pattern maps are essential regardless of the degree of sophistication of the analytical model used. These maps are just as important for developing simple empirical mechanistic models as they are for developing complex two-fluid models based on conservation laws.

Flow pattern maps represent a method of classifying two-phase flows in terms of measurable flow parameters. Simple flow pattern maps usually have the same coordinates for all flow patterns and their transitions. The identification of the various flow patterns result from visual observations and thus may be subjective. However, there exists a minimum number of generic liquid-vapor flow patterns that can be easily characterized. The classifications of [81] are used as an example

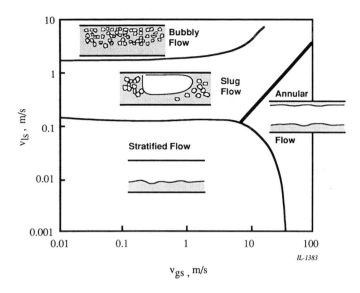

**Figure 7.1**  Typical flow pattern map for horizontal pipe flow, from [82].

here in which the flow pattern map axes are given in terms of the superficial velocities $v_{sg}$ and $v_{sl}$.

Figure 7.1 shows the flow pattern map from [82], which has been summarized by [81], that is usually observed in a *horizontal pipe flow.* The map is derived from water-air mixture, at atmospheric pressure and under adiabatic conditions, flowing in a 5-cm ID pipe. Four different patterns are generally identified for such a flow: stratified flow, annular flow, bubbly flow, and intermittent or slug flow. The specific characteristics of these flow patterns follow:

(a)  *Stratified flow* is normally observed at low liquid-flow rates. Gravity is responsible for the separation of the two phases, in which case the liquid flows at the bottom of the pipe while the gas flows at the top. The liquid is driven by the interfacial stress exerted at the interface by the gas and by the pressure gradient. At high gas-flow rates, waves will develop at the interface. This flow pattern may be further subdivided into *smooth* or *wavy* stratified flows.

(b)  *Annular flow* normally occurs at high gas-flow rates. In this flow pattern, the liquid takes the form of an annulus at the pipe wall. The thickness of the annular is generally greater at the bottom surface of the pipe than at the top. The liquid velocity in this flow pattern is much smaller than the gas velocity. Due to the high interfacial shear stress, waves form and propagate at the liquid-gas interface. Normally, some of the liquid is entrained in the gas core as a result of the wave action and forms small droplets that are dispersed and deposited on the liquid film.

(c)  *Bubbly flow* occurs at high liquid-flow rates. The gas is mainly in the form of small bubbles that are carried by the liquid. The bubbles are dispersed in the liquid phase by turbulence, which tends to counteract stratification due

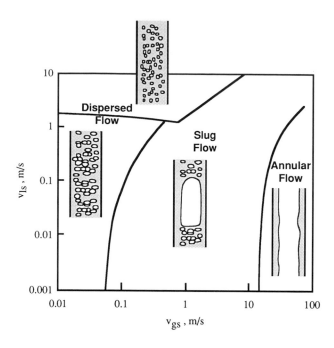

**Figure 7.2**   Typical flow pattern map for vertical pipe flow, from [82].

to gravity. Due to buoyancy effects, the bubbles in this flow tend to aggregate more at the top of the pipe.

(d) *Intermittent flow*, which is also called *slug flow*, is a complex flow pattern consisting of space-time sequences of stratified and bubbly flows. The gas in this pattern is mainly carried in long bubbles with a round nose and a flat inclined back. Inside the long bubbles, the flow appears to be stratified where the gas fills the upper part of the bubble next to the upper wall, while the liquid forms a film at the bottom. Between two consecutive long bubbles, liquid is observed to fill the whole cross-sectional area of the pipe; this is normally called a *slug* flow. For very low flow rates, only liquid is present in liquid slugs. As the flow rate increases, a large number of small bubbles appear to be entrained behind the long bubbles. These are usually overtaken by the larger bubbles, which move faster.

For *vertical pipe flow*, the two-phase liquid-vapor flow patterns are different from the patterns for the horizontal pipe flow due to the absence of stratification induced by gravity. These patterns depend on the flow direction in the tube, whether it is upward or downward. For upward flow, three main flow regimes are commonly observed under adiabatic conditions. The flow pattern map appropriate for vertical upflow conditions is shown in Figure 7.2. This flow is comprised of the following patterns:

(a) *Bubbly flow* is observed at low gas velocity. The liquid wets the pipe wall and the gas is transported in bubbles whose longest diameter is typically smaller than the pipe diameter. The motion of the bubbles for this flow is highly influenced by buoyancy. Two types of bubbly flows can be distinguished in this pattern: one consists of small dispersed bubbles whose motion and size are strongly influenced by turbulence at high gas velocities; the other has millimeter-size bubbles with an ellipsoidal shape that are commonly seen at low liquid and gas velocities.

(b) *Annular flow* pattern for this geometry is very similar to the annular pattern for the horizontal pipe configuration, with a minor difference. The flow in the liquid film, in this case, is very chaotic, and supports large-amplitude roll waves propagating at the film surface. The liquid is usually entrained through droplets in the gas core. Even though the liquid fraction in the droplets is much smaller than in the liquid film, the liquid flow rate in the droplet can be significant since the droplets move much faster than the liquid film.

(c) *Intermittent flow* pattern consists of a succession of long cylindrical bubbles and liquid slugs containing smaller bubbles. In this flow the bubbles are surrounded by a thin annular film so that an observer will see the flow as a sequence of annular flow and bubbly flow. At low pressure in upward flow the liquid film is moving downward such that its velocity with respect to the cylindrical bubbles is high enough to entrain a large amount of gas at the lower end. When the gas flow rate increases, the flow becomes more chaotic, with the shape of the bubble nose becoming less spherical and waves appearing on the liquid film that move upward and carry small bubbles.

The essential question in this chapter concerns the role of gravity in modifying any or all of the flow patterns observed in terrestrial conditions. A number of two-phase liquid-vapor experiments under low-gravity conditions have been performed recently with the objective of characterizing the flow patterns. References [78], [79], and [80] show that *basically the same flow patterns occur under low-gravity conditions as in the terrestrial upward flow in a vertical tube under adiabatic conditions.* The specific low-gravity patterns observed can be classified as bubbly flow, annular flow, and intermittent flow. The components of the flow pattern map appropriate for *low-gravity flows*, shown in Figure 7.3, are described in the following manner:

(a) *Bubbly flow* possesses some variations from its terrestrial counterpart due to the absence of gravity in the body force. The bubble motion in this case appears frozen in the sense that the bubbles move gently with the same velocity as the surrounding liquid. The bubbles do not appear to be deformed, as they would otherwise be in a vertical upflow. They are simply transported by the liquid. The size of the bubbles increases significantly with *void fraction*, unlike its upflow terrestrial counterpart.

(b) *Annular flow* is characterized by an annular liquid film flowing adjacent to the tube wall. Again, for this case, waves develop at the liquid-gas interface similar to the terrestrial upflow pattern; however, the waves appear to have smaller amplitudes. Traveling roll waves have been observed at almost all gas and liquid flow rates. The gas core may contain liquid droplets, but the entrainment process is not an important factor in this case.

(c) *Intermittent flow* occurs in low gravity at moderate gas-flow rates. The flow pattern consists of liquid slugs and long cylindrical bubbles surrounded by an annular film in which the liquid velocity is very small. The slug flow in

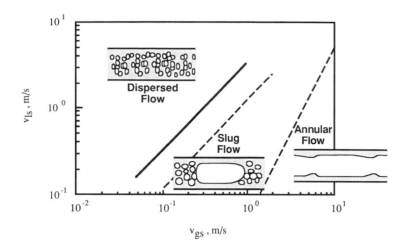

**Figure 7.3** Typical flow pattern map in low gravity conditions from [82].

this case contains fewer bubbles with larger sizes. The slugs move at almost the same velocity as the cylindrical bubbles with what appears to be a frozen structure. Again, their motion is mainly due to being transported by the liquid. No entrainment behind the long bubbles has been observed in low-gravity flows.

## 7.3 *ANALYTICAL MODELING OF TWO-PHASE FLOWS*

The modeling of two-phase flows is far more complex than its single-phase counterpart due to the simultaneous existence of the two phases. If, however, the phasic distributions in both time and space are known, say, from experimental data, it is possible to define suitable models from conservation principles. There are basically two approaches that are commonly used for two-phase flows: the *mechanistic models* and the *separated models*. In a mechanistic model, the two phases are represented by a single fluid whose physical properties are assumed based on experimetal data. Mechanistic models are very simplistic and require large amounts of empirical data.

In the separated models, the solution is obtained by applying the conservation principles for each phase separately and then matching the flow variables across all interfaces. The liquid-vapor interface motion in almost all two-phase flows appears to be random and thus complicates the analysis further. This difficulty can be overcome by averaging the equations of motion in both time and space. The averaging process eliminates the random behavior of the interfaces and allows for the equations to be solved for the averaged variables in a straightforward manner. The substantial penalty to be paid for such a simplification is discussed later.

The separated model approach is illustrated by considering the one-dimensional, cross-sectional area averaged equations for the conservation of mass, momentum, and energy in each phase. The one-dimensional area averaged mass conservation equations take the following form for each phase (see [83]):

$$\frac{\partial}{\partial t}(\rho_k \alpha_k) + \frac{\partial}{\partial x}(\rho_k \alpha_k U_k) = m_k, \qquad k = l, g \tag{7.3.1}$$

with

$$m_l + m_g = 0, \tag{7.3.2}$$

$$\alpha_l + \alpha_g = 1. \tag{7.3.3}$$

The momentum equations under the same conditions for each phase are

$$\frac{\partial}{\partial t}(\rho_k \alpha_k U_k) + \frac{\partial}{\partial x}(\rho_k \alpha_k U_k^2) + \frac{\partial}{\partial x}(\alpha_k P_k)$$

$$= \tau_{wk} + \tau_{ik} - \Pi_{ik} + I_{ik} - \alpha_k \rho_k g \sin \varphi, \qquad k = l, g \tag{7.3.4}$$

with

$$(\tau_{il} - \Pi_{il} + I_{il}) + (\tau_{ig} - \Pi_{ig} + I_{ig}) = 0, \tag{7.3.5}$$

while the energy conservation equation for each phase are

$$\frac{\partial}{\partial t}[\rho_k \alpha_k (H_k + U_k^2/2)] + \frac{\partial}{\partial x}[\rho_k \alpha_k (H_k + U_k^2/2)U_k] - \frac{\partial}{\partial t}(\alpha_k P_k)$$

$$= -q_{wk} - q_{ik} + w_{ik} - \rho_k \alpha_k U_k g \sin \varphi + e_{ik}, \qquad k = l, g \tag{7.3.6}$$

with

$$(-q_{il} + w_{il} + e_{il}) + (-q_{ig} + w_{ig} + e_{ig}) = 0. \tag{7.3.7}$$

In equations (7.3.1)–(7.3.7) the subscripts $l$ and $g$ denote the liquid and gas phases, respectively. The primary flow variables in the above equations are the averaged streamwise velocity $U_k$, the averaged phase fraction $\alpha_k$, the averaged pressure $P_k$, and the averaged enthalpy $H_k$. $\varphi$ is the pipe inclination to the horizontal, and $g$ is the gravitational acceleration. $m_k$ is the mass flux per unit volume into phase $k$, $\tau_{wk}$ is the shear stress at the wall of phase $k$, $\tau_{ik}$ is the interphasic stress, $\Pi_{ik}$ is the pressure force exerted across phase $k$, and $I_{ik}$ is the momentum flux across the interfaces. $q_{wk}$, $q_{ik}$, $w_{ik}$, and $e_{ik}$ are the energy flux at the wall for each phase, the energy flux between the phases, the work done by interfacial forces, and the energy source per phase $k$ due to interfacial mass transfer, respectively.

The one-dimensional conservation equations (7.3.1)–(7.3.7) can in principle be solved for the primary variables if all of the quantities $m_k$, $\tau_{ik}$, $\tau_{wk}$, $\Pi_{ik}$, $I_{ik}$, $q_{ik}$, $q_{wk}$, and $e_{ik}$ are defined in terms of these variables. These equations apply for all of the basic flow patterns enumerated in the previous section, including bubbly, stratified, and annular flows, provided that the interfacial transport terms and various fluxes are known. The fluxes for each flow pattern can be determined either from experimentally derived empirical relations or from simple models based on conservation principles.

It is possible to classify most of the flow patterns discussed in the previous section into two generic classifications, dispersed and separated. The *dispersed flow* consists of both mist and bubbly flow patterns, while the *separated flows* are those in which there is a distinct and clear separation between the phases, i.e., stratified, annular, or *plug flows*.

There exist numerous methods and procedures in the literature for solving the two-phase flow equations given above. One interesting method discussed in is outlined here. This method is also described in [82], and depends on using the statistical averages of the variables instead of the instantaneous averages. A new parameter, $\beta$, is defined with this method as the probability of occurrence of a separated flow in both space and time. For steady flow with periodic sequences of long bubbles and liquid slugs moving with the same velocity, $\beta$ would be the ratio of the bubble length to the observational length. $\beta = 0$ for a dispersed flow, $\beta = 1$ for a separated flow, and $0 < \beta < 1$ when the flow contains both separated and dispersed flows sequentially.

This method requires defining the new primary flow variables $\alpha_k^q$, $\rho_k^q$, $U_k^q$, $P_k^q$, $H_k^q$ where $q$ can be either $d$ for a dispersed flow or $s$ for a separated flow. The following definitions for the primary variables are introduced:

$$\alpha_k = \beta \alpha_k^s + (1 - \beta) \alpha_k^d, \tag{7.3.8}$$

$$\rho_k = \frac{\beta \rho_k^s \alpha_k^s + (1 - \beta) \rho_k^d \alpha_k^d}{\alpha_k}, \tag{7.3.9}$$

$$U_k = \frac{\beta \rho_k^s \alpha_k^s U_k^s + (1 - \beta) \rho_k^d \alpha_k^d U_k^d}{\rho_k \alpha_k}, \tag{7.3.10}$$

$$H_k = \frac{\beta \rho_k^s \alpha_k^s H_k^s + (1 - \beta) \rho_k^d \alpha_k^d H_k^d}{\rho_k \alpha_k}. \tag{7.3.11}$$

Reference [83] used this technique for a specific case in which the mass conservation was given by a separate equation for each phase while the momentum and energy conservation were each described by a single equation for the mixture of the two phases. Such a description for a two-phase flow covers a wide range of engineering applications, especially for adiabatic flow. With these conditions, the conservation of mass equations for each phase take the following form:

$$\frac{\partial}{\partial t}(\rho_k \alpha_k) + \frac{\partial}{\partial x}(\rho_k \alpha_k U_k) = m_k, \tag{7.3.12}$$

$$m_g + m_l = 0, \tag{7.3.13}$$

while the momentum equation for the mixture is given by

$$\frac{\partial}{\partial t}\left[\sum_{k=l,g}(\rho_k \alpha_k U_k)\right] + \frac{\partial}{\partial x}\left[\sum_{k=l,g}(\rho_k \alpha_k U_k^2)\right] + \frac{\partial P}{\partial x}$$

$$+ \frac{\partial}{\partial x}\left[\beta(1-\beta)\sum_{k=l,g}\left(\frac{\rho_k^s \rho_k^d \alpha_k^s \alpha_k^d (U_k^s - U_k^d)^2}{\rho_k \alpha_k}\right)\right]$$

$$= \beta(\tau_{wg}^s + \tau_{wl}^s) + (1-\beta)\tau_w^d - \sum_{k=l,g}(\rho_k \alpha_k)g \sin\varphi. \tag{7.3.14}$$

Similarly, the following single equation is used for the conservation of energy for the mixture:

$$\frac{\partial}{\partial t}\left[\sum_{k=l,g}\rho_k\alpha_k(H_k+U_k^2/2)\right]+\frac{\partial}{\partial x}\left[\sum_{k=l,g}\rho_k\alpha_k(H_k+U_k^2/2)U_k\right]$$

$$-\frac{\partial P}{\partial t}+\frac{\partial}{\partial t}\left[\beta(1-\beta)\sum_{k=l,g}\left(\frac{\rho_k^s\rho_k^d\alpha_k^s\alpha_k^d(U_k^s-U_k^d)^2}{2\rho_k\alpha_k}\right)\right]$$

$$+\frac{\partial}{\partial x}\left[\beta(1-\beta)\sum_{k=l,g}\left(\frac{\rho_k^s\rho_k^d\alpha_k^s\alpha_k^d}{\rho_k\alpha_k}(U_k^s-U_k^d)\right.\right.$$

$$\left.\left.\times\left\{(H_k^s-H_k^d)+\frac{(U_k^s-U_k^d)(U_k+U_k^d+U_k^s)}{2}\right\}\right)\right]$$

$$=-\beta(q_{wg}^s+q_{wl}^s)-(1-\beta)q_w^d-\sum_{k=l,g}(\rho_k\alpha_kU_k)g\sin\varphi. \qquad (7.3.15)$$

Reference [82] points out that $\beta$ is included in both the momentum and the energy equations because in slug flow, the liquid velocity is higher in the liquid slugs than in the film flowing under and around the long bubbles. Due to the averaging procedure, the nonlinear terms containing $U_l$ will give these extra terms. Also, the friction and heat flux terms in equations (7.3.14) and (7.3.15) are not given in an analogous manner for both the dispersed flow and the separated flow. In the derivation of these equations, it was assumed that both phases may contact the wall in the separated flow, whereas only the continuous phase will do so in the dispersed flow. In addition to equations (7.3.8)–(7.3.14), the following conditions must also be satisfied at all times:

$$\alpha_l+\alpha_g=1,\qquad \alpha_l^s+\alpha_g^s=1,\qquad \alpha_l^d+\alpha_g^d=1. \qquad (7.3.16)$$

The conservation equations given in (7.3.12)–(7.3.16) are referred to by [82] as the *drift flux model*. Setting $\beta=1$ in these equations leads to equations appropriate only for the separated flow, while setting $\beta=0$ results in equations for the dispersed flow.

For a steady, fully developed two-phase liquid-vapor flow without heat and mass transfer between the phases (i.e., for adiabatic conditions), the conservation equations simplify to the following form, when the density of each phase is assumed not to vary significantly:

$$\rho_k^s=\rho_k^d=\rho_k, \qquad (7.3.17)$$

$$v_{sk}=\alpha_kU_k, \qquad (7.3.18)$$

$$\frac{dP}{dx}=\beta(\tau_{wg}^s+\tau_{wl}^s)+(1-\beta)\tau_w^d-\sum_{k=l,g}(\rho_k\alpha_k)g\sin\varphi. \qquad (7.3.19)$$

Equation (7.3.18) for the conservation of mass defines a relationship between the superficial velocity and the averaged velocity of each phase. The momentum equation (7.3.19) in this case gives the pressure gradient.

The usefulness of this formulation is best illustrated with the following two simple examples from [82]. The first example is for the case of a bubbly flow, the second for a separated flow.

For a bubbly flow the parameter $\beta$ is set equal to zero, resulting in the following equations:

$$v_{sk}^d = \alpha_k U_k^d, \tag{7.3.20}$$

$$\frac{dP}{dx} = \tau_w^d - \sum_{k=l,g} (\rho_k \alpha_k) g \sin \varphi. \tag{7.3.21}$$

The problem in this example is that in order to solve for the pressure gradient, expressions for the phasic dispersed flow velocities $U_k^d$ and the wall friction $\tau_w^d$ need to be formulated.

Normally, the movement of the bubbles with respect to the mixture is given by a *drift flux expression.* In the vertical upflow, for example, the velocity of the bubbles can be determined with the aid of the following relationship (see [82]):

$$U_g^d = C_0 U^d + f(\alpha_g^d) U_B, \tag{7.3.22}$$

where $U_B$ is the rise velocity of a single bubble in still liquid, $C_0$ is a constant that accounts for the phase and velocity distributions in the pipe cross section, and is normally close to unity, and $U^d$ is the mixture fluid velocity defined by

$$U^d = \alpha_l^d U_l^d + \alpha_g^d U_g^d. \tag{7.3.23}$$

Expression (7.3.23) simply states that $U^d$ represents the sum of the superficial velocities of the liquid and the gas. These velocities are usually known from the inlet conditions for adiabatic flow. Furthermore, it is assumed that the distributions of the radial velocity and the void fraction $\alpha$ can be represented in terms of powers of the radial distance from the center of the tube as

$$\frac{u}{u_0} = 1 - \left(\frac{2r}{d}\right)^m, \qquad \frac{\alpha - \alpha_w}{\alpha_0 - \alpha_w} = 1 - \left(\frac{2r}{d}\right)^n, \tag{7.3.24}$$

where $m$ and $n$ are different powers, $d$ is the tube diameter, and the subscript 0 refers to the tube centerline. For such distributions the coefficient $C_0$ takes the following form:

$$C_0 = 1 + \frac{2}{m+n+2}\left(1 - \frac{\alpha_w}{\alpha_g}\right) = \frac{m+2}{m+n+2}\left(1 + \frac{\alpha_w}{\alpha_g}\frac{n}{m+2}\right). \tag{7.3.25}$$

If the maximum value of $\alpha$ is at the centerline, then $C_0 > 1$, while $C_0 < 1$, if $\alpha$ attains its maximum value at the wall. Of course, gravity plays a decisive role in the phasic distribution and the rise velocity.

The functions $f(\alpha_g^d)$ in expression (7.3.22) are empirical functions such as the following expression:

$$f\left(\alpha_g^d\right) = \left(1 - \alpha_g^d\right)^n. \tag{7.3.26}$$

$n = 3/2$ is a commonly used value here.

The wall friction term, on the other hand, may be extrapolated from the single-phase flow relationship, with appropriate modifications, in the following manner (see [89]):

$$\tau_w^d = -f_w^d \rho_m^d \frac{U^d |U^d|}{2} \frac{4}{d} , \tag{7.3.27}$$

where $f_w^d$ is a friction factor that must be known and $\rho_m^d$ is the mixture density given by

$$\rho_m^d = \alpha_l^d \rho_l^d + \alpha_g^d \rho_g^d. \tag{7.3.28}$$

Other relationships for the wall stress may also be used.

The conservation equations for the separated flow can now be derived from equations (7.3.17)–(7.3.19) by setting $\beta = 1$. However, using the drift flux model equations for either the annular or the stratified flows will degrade the accuracy of the model considerably. Thus, it is more accurate for either of these two flow patterns to separate the momentum equation for each phase.

Upon eliminating the pressure gradient between the momentum equation for the liquid and the gas, the following single equation results:

$$\frac{\tau_{wl}^s}{\alpha_l^s} - \tau_{ig}^s \left( \frac{1}{\alpha_l^s} - \frac{1}{\alpha_g^s} \right) - \frac{\tau_{wg}^s}{\alpha_g^s} - (\rho_l - \rho_g) g \sin \varphi = 0 \tag{7.3.29}$$

where $\tau_{ig}^s$ is the interfacial stress. The stress terms in this case may be modeled using single-phase relationships through the friction factor both at the wall and at the interface in the following manner:

$$\tau_{wg}^s = -f_{wg}^s \rho_g \frac{U_g^s |U_g^s|}{2} \frac{C_{wg}^s}{A} , \tag{7.3.30}$$

$$\tau_{wl}^s = -f_{wl}^s \rho_l \frac{U_l^s |U_l^s|}{2} \frac{C_{wl}^s}{A} , \tag{7.3.31}$$

$$\tau_{ig}^s = -f_{ig}^s \rho_g \frac{(U_g^s - U_l^s)|U_g^s - U_l^s|}{2} \frac{C_i^s}{A} , \tag{7.3.32}$$

where $C_{wg}^s$ and $C_{wl}^s$ are the segments of the wall circumference that are wetted by the gas and liquid, respectively, while $C_i^s$ and $A$ are the interfacial perimeter and the pipe cross-sectional area, respectively. These perimeters are functions of the pipe diameter and the void fraction, which can be different functions depending on whether the flow is annular or stratified. The functions $U_g^s$, $U_l^s$ are also related to $v_{sg}$, $v_{sl}$, and $\alpha_g^s$ through similar expressions as those derived for the dispersed flow.

The above solution illustrates very clearly the difficulty associated with determining solutions for simple two-phase flow configurations. Note the very strong dependence of the form of the solution on the character of individual flow pattern. This highlights the essential role played by the flow pattern on the development of any analytical model.

## 7.4 POOL BOILING

Boiling is characterized by the combined transport of energy, momentum, and mass in the two-phase flow boiling media. The physical transport phenomena are mathematically described by the conservation equations of mass, momentum, and energy. In general, boiling phenomena are very complicated owing to the interaction of a number of factors and effects. Because of the nonlinearity of the basic equations, hydrodynamic or thermal instabilities may occur under appropriate conditions. To avoid the difficulties associated with solving the basic equations, the connection between the single-phase heat transport, the temperature, and the flow field is generally expressed in the form of semi-empirical relations between the characteristic nondimensional numbers.

Liquid boiling can take place via different processes depending on the specific mechanism and the geometric configuration. Reference [89] identifies three main mechanisms of boiling:

1. *Nucleate boiling*, in which vapor bubbles are formed usually at a solid surface;
2. *Convective boiling*, in which heat is conducted through a thin liquid film where the liquid subsequently evaporates at the vapor-liquid interface without bubble formation; and
3. *Film boiling*, in which the heated surface is blanketed by a vapor film. Heat is transferred in this mode by conduction through the vapor. The liquid vaporization in this boiling mode occurs at the vapor-liquid interface.

Liquid boiling can also be classified by boiling forms:

1. *Pool boiling*, in which boiling occurs at a heated surface in a pool of liquid that is mainly stagnant apart from any convection induced by the boiling process itself, and
2. *Flow boiling*, where the liquid is pumped through a heated channel, typically a tube.

Nucleate boiling and film boiling occur in both pool boiling and flow boiling, while convective boiling occurs only in flow boiling. Nucleate boiling is an important mode of heat transport due to the fact that small temperature differences provide large heat transfer rates. This can result in significant economic and other benefits associated with the smaller heat transfer areas necessary to accomplish a given function. Figure 7.4 shows as an example the boiling curve for water at 1 bar of pressure. The figure shows very clearly the sharp rise in the heat flux during the nucleate boiling phase.

In this section the phenomenon of pool boiling is explored; flow boiling is covered in the following sections. The following salient characteristics of nucleate pool boiling are identified by [85]:

(a) The phase change from liquid to vapor occurs with the formation of discrete bubbles at individual sites.
(b) The bulk liquid in this boiling process is essentially stagnant, in contrast with flow boiling, where the liquid is set in motion by external means.
(c) The rates of energy transport for this boiling process are large with small temperature difference driving potentials.

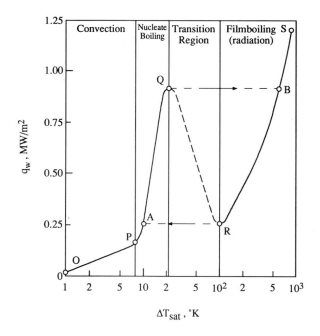

**Figure 7.4** Schematic of a typical boiling curve.

(d) The process is inherently transient, although quasicyclic repetitions are possible when vapor removal mechanisms, such as buoyancy, are acting.

Beginning with the transient heating of a liquid at a solid-liquid interface, reference [85] characterizes pool boiling with the following physical sequence of events.

*Conduction.* With an initially static liquid, heat is transported through conduction alone until buoyancy, thermophysical, or other forces set the liquid in motion. The rates of temperature rise and temperature distribution in this early stage depend on the nature of the heat source and the dynamic interaction with the system. The temperature distribution can be determined easily from the conduction equation. The onset of natural convection modifies this temperature distribution.

*Onset of natural convection.* Natural convection is driven by buoyancy, and its onset is described in terms of instabilities due to disturbances that are always present. Reducing the effects of buoyancy by reducing the body forces in low gravity delays the onset of convection and reduces the resulting convection velocities. Both of these effects serve to increase the temperature levels in the liquid adjacent to the heating surface for a given heating time, regardless of whether the bulk liquid is initially saturated or subcooled. The liquid temperature levels and distributions adjacent to the heater surface are influenced by buoyancy and can in turn influence nucleation and bubble growth rates.

*Nucleation.* Vaporization can take place only at an existing liquid-vapor interface, which then leads to the growth phase of nucleate boiling. If an interface does not exist, it must be formed. The formation of a vapor nucleus is called *nucleation* and is classified as either homogeneous or heterogeneous depending on the presence of other components or species in the vicinity of the nucleation site. Nucleation is affected by the following three factors:

(a) The heater surface geometry, which provides crevices and intergranular defects serving as preexisting nucleation sites.

(b) The solid-fluid properties, which govern not only the temperature distribution in both the heater and the fluid, but also the surface energy relationships between the solid, the liquid, and the vapor. This is often expressed in terms of a contact angle or wettability.

(c) The liquid temperature distribution, including the solid-liquid interface temperature. Once nucleation has occurred, the bubble growth rates that follow are governed by the bulk liquid temperature distribution.

*Vapor bubble growth and collapse.* Vapor bubble growth requires that the liquid-vapor interface be superheated with respect to the saturation temperature corresponding to the interfacial liquid pressure. The rate of vapor formation, and hence bubble growth, depends on the amount of superheat and on the liquid temperature gradient at the interface, and thus on the liquid temperature distribution at the onset of bubble growth. The interfacial liquid superheat governs the internal bubble vapor pressure, which acts to move the bulk liquid away from the heater surface. If the bulk liquid is supercooled, the pressure difference may reverse with subsequent collapse of the vapor bubble. Bubble growth or collapse is influenced primarily by the following five factors:

(a) Internal bubble pressure, which is governed by the liquid temperature distribution, which is in turn influenced by buoyancy.

(b) Liquid momentum or bulk liquid inertia.

(c) Buoyancy, including both the pressure differences associated with the liquid-vapor density differences and the temperature differences in the liquid.

(d) Surface tension, including both the tension at the liquid-vapor interface and the liquid-solid-vapor contact line.

(e) Liquid viscosity in the vicinity of the solid surface. Vapor viscosity could also be a factor during the early moments when surface rates of vapor formation are large. The interface shape during growth is governed by the net balance of the dynamic forces at each point on the interface.

*Departure.* The subsequent motion of the vapor bubble depends on the net balance of the forces listed above as well as molecular momentum effects associated with the simultaneous evaporation and condensation across a vapor bubble. The bulk liquid momentum induced by the rapid bubble growth can act to assist in the removal of the bubble from the heater surface. In low gravity, buoyancy effects are reduced significantly, which may alter the bubble removal rate.

*Motion following departure.* When a bubble departs from the surface, its subsequent motion will depend on the following factors:

(a) Buoyancy.

(b) Initial bubble velocity upon departure. This velocity induces momentum in the bulk liquid that must be conserved and can either accelerate the bubble collapse or decelerate the bubble growth.

(c) Amount and distribution of superheat or subcooling in the liquid. The bulk liquid temperature distribution due to surface tension or thermocapillary effects greatly influences the bubble motion.

During the transient events described above, heat is continually being transported between the solid surface and the liquid. Nucleation, growth, and departure of successive vapor bubbles may be taking place at or near the initial nucleation site. As a result of the liquid motion induced by the growth and departure of the bubbles, the behavior of the liquid appears to be random and chaotic and the detailed behavior of successive bubbles may be quite variable. Nevertheless, observations over a reasonable period of time reveal a periodic behavior for the bubbles which can be used to describe some features of pool boiling.

An adequate understanding of the fundamental mechanisms constituting nucleate pool boiling could lead to a description of this phenomena in terms of the governing parameters of the problem. Such descriptions will include the onset of boiling, the dynamic behavior of vapor bubbles—including number density of the active nucleation sites and frequency of formation—and the associated heat fluxes. The governing parameters include fluid and heater properties such as surface tension, surface roughness, wetting angles, and pressure. The degree of the initial subcooling, macrogeometry, and the body force angle are also parameters that govern this process.

## 7.5 *ANALYTICAL FORMULATION OF POOL BOILING*

Although a considerable amount of research has been devoted to nucleate boiling, the prediction capability for the process remains very limited. The present state of understanding of nucleate boiling requires separating the initial transients at the inception of boiling from the quasisteady periodic phase. This latter phase can take place only when vapor bubbles are removed from the vicinity of the heating surface. For pool boiling in a terrestrial environment, this will occur due to the buoyancy forces provided by the Earth's gravity in the absence of externally induced fluid motion. However, the minimum body force necessary to provide this removal mechanism is not precisely known.

For the case of steady-state pool boiling only empirical correlations can be used. The most well-known and prevalent pool boiling correlation for the heat flux for saturated liquids in a terrestrial environment is that given by [86]. Reference [86] suggests the following correlation for the heat flux $q_b$ during boiling:

$$q_b = B\mu_l h_L \left[\frac{g(\rho_l - \rho_g)}{\sigma}\right]^{1/2} (\phi\Delta T_{sat})^{m+1}, \tag{7.5.1}$$

$$\phi^{m+1} = \frac{k_l^{1/2}\rho_l^{17/8}c_{pl}^{19/8}h_L^{(m-23/8)}\rho_g^{(m-15/8)}}{\mu_l(\rho_l - \rho_g)^{9/8}\sigma^{(m-11/8)}T_{sat}^{(m-15/8)}}, \tag{7.5.2}$$

where $h_L$ is the latent heat of vaporization, $k_l$ is the thermal conductivity for the liquid, $T_{sat}$ is the saturation temperature, and $\Delta T_{sat} = T_w - T_{sat}$, where $T_w$ is the wall temperature. *B* in equation (7.5.1) is given in [86] as a dimensional constant depending on the magnitude of the gravity and the boiling surface properties as

$$B = \left(\frac{r_c}{2}\right)^m \frac{2C_1 C_2^{5/3}}{g^{9/8}} \sqrt{C_3/\pi}, \qquad (7.5.3)$$

where $r_c$ is the nucleation site cavity radius. The empirical constants in the above expressions (namely, $m$, $C_1$, $C_2$, and $C_3$) are defined as follows: $m$ sets the slope of $q_b$ versus $\Delta T_{sat}$ curve with a value of 2 used for most surfaces. $C_3$ arises from an empirical correlation relating the product of the frequency of departure and bubble size at departure. $C_1$ represents the smallest active cavity size that corresponds to a given heater surface superheat. $C_2$ describes the departure size of a vapor bubble with mechanical equilibrium balance between buoyancy and surface tension.

Even though gravity appears explicitly in equation (7.5.1), this expression is not valid for applications in environments other than terrestrial. Due to the fact that it is empirical in nature, (7.5.1) is valid only for a specific pressure, fluid properties, and heating surface. The difficulty with adapting terrestrially derived empirical expressions similar to (7.5.1) to low-gravity applications is demonstrated by the following example from [85].

Expression (7.5.1) for the boiling heat flux can be rearranged as

$$q_b = K g^{-5/8} \Delta T_{sat}^3, \qquad (7.5.4)$$

where $m = 2$, and all of the fluid properties and constants are lumped into a single coefficient $K$. For a given heat flux, expression (7.5.4) predicts a value of $\Delta T_{sat}$ of 1.62 times and 2.61 times its value at $g_0$ for $g/g_0$ of 10 and 100, respectively. Reference [85] shows that experimentally measured values for $\Delta T_{sat}$ are only 1.03 and 1.24 times as great, respectively, for water and for the corresponding gravity levels. This example clearly illustrates that empirical correlations obtained under terrestrial conditions cannot be extended to low-gravity environment even when the correlation functions contain gravity explicitly.

For the transient nucleate boiling process that occurs at the onset of boiling, reference [85] suggests modeling each element separately. In the first stage of the transient boiling process, heat is transferred to the liquid from the heating surface through conduction only. Gravity does not play any role in conduction heat transfer as long as the liquid remains attached to the heating surface. This has been verified with low-gravity experiments with the conclusion that terrestrial correlations for this stage of boiling are appropriate for low-gravity applications.

The next stage in the transient boiling process is the onset of natural convection in the liquid. In this stage, the strength and the structure of buoyancy-driven convection depends entirely on the magnitude and direction of the gravity vector. Buoyancy-driven convection was discussed in Chapter 4. It was shown there that if the liquid layer is normal to the gravity vector, then gravity is directly proportional to the critical Rayleigh number, while for a liquid layer aligned in the same direction as the gravity vector, the strength of the convective motion is proportional to $\sqrt{g}$. Once liquid motion has begun, the process can be characterized as transient natural convection with the strength of the motion depending on the geometry, temperature distribution, body force, and fluid properties.

Vapor bubble nucleation takes place when the local temperature of the liquid in the vicinity of the heating surface exceeds the saturation temperature. The degree of superheat necessary for nucleation depends on the microgeometry of the solid surface, the solid/fluid properties, the surface temperature of the solid, and the temperature distribution in the liquid. A bubble critical radius, $r_c$, can be calculated as a function of the degree of superheat using thermodynamical homogeneous equilibrium between a spherical bubble and pure liquid in the following manner (see [85]):

$$(T_w - T_{sat}) = \frac{2\sigma T_{sat}}{\rho_g h_L r_c}. \tag{7.5.5}$$

A liquid superheat greater than this amount results in the growth of the vapor bubble, whereas for a smaller degree of superheat the bubble will collapse.

Another method for determining the onset of nucleation is to use homogeneous nucleation theory to describe heterogeneous nucleation (see [87]). In that formulation, the nucleation delay time $t$ is given in terms of the heater surface superheat as

$$T_l = \left\{ 1 - \left( \frac{p_l}{\rho_l h_L} \right)_{T=T_{sat}} \ln \left[ \left( \frac{16\pi\sigma^3 F}{3KT_l p_l^2 \ln(NKT_l t/\lambda)} \right)^{1/2} + 1 \right] \right\}^{-1} T_{sat} \tag{7.5.6}$$

where $K$ and $\lambda$ are the Boltzmann and Planck constants, respectively, and $N$ is the number density of the molecules. $F$ is a nucleation factor defined in [85] as the ratio of the free energy for heterogeneous nucleation to that for homogenous nucleation:

$$F = \frac{A_{het}}{A_{hom}}. \tag{7.5.7}$$

This factor $F$ must be determined experimentally.

Once a particular nucleation site has become activated, the subsequent rate of growth and later departure and/or collapse of the bubbles depends on the transient temperature distribution in the vicinity of the bubble interface. This phenomenon also depends on the liquid inertia effects. The temperature distribution for bubble growth is different for an initial site than for the succeeding bubbles. The bubble growth rate is affected by reduced gravity only if the temperature distribution in the liquid is affected. Bubble growth rates from a semi-infinite solid surface with a step change of the interface temperature for a spherical bubble is given in [85] by the following correlation:

$$r_c(t) = (2/\sqrt{\pi}) \left( \frac{\rho_l c_{pl}(T - T_{sat})}{\rho_g h_L} \right) \sqrt{Ct}. \tag{7.5.8}$$

Expression (7.5.8) gives a value of 1.13 for the coefficient $C$. Different analytical formulations give different values for $C$ ranging from 1.13 to 2. Bubble growth rates have been found experimentally to be relatively insensitive to gravity, whether high or low. However, this rate is dependent on the temperature distribution within the liquid, which is in turn affected by the level of the body force due to gravity.

The departure of a grown bubble from the nucleation site is determined from the balance of forces acting on the bubble. These forces include buoyancy, momentum of the liquid, and surface tension effects at the liquid-vapor interface and the liquid-solid-vapor interface. It also depends on phase change momentum effects, viscous effects at the solid surface, and the pressure difference between the inside and the outside of the bubble. The most recent and widely used empirical formula for the maximum diameter at departure $D_d$ of a spherical bubble is given by (see [85]):

$$D_d = C \left( \frac{\sigma}{g \Delta \rho} \right)^{1/2} \left( 1 + C_1 \frac{dD_d}{dt} \right), \qquad (7.5.9)$$

where $C_1$ is an empirical constant that depends on the heating surface/fluid properties, the heat flux, and bulk supercooling. The constant $C$ in expression (7.5.9) includes the contact angle.

*Departure diameter* measurements were made in low-gravity environment using drop towers for both saturated distilled water vapor bubbles and water-sucrose solution vapor bubbles. According to [85], the results from these experiments show good agreement between expression (7.5.9) and the boiling water-sucrose solution. No differences in departure diameters were observed over a range of gravity levels of $0.014 < g/g_0 \leq 1.00$. The results of these experiments lead [85] to conclude that the departure process is inertia dominated.

Nucleate pool boiling heat flux data is considered an extremely useful design tool; it has been the subject of intensive research in terrestrial environment. The nucleate pool boiling curve, which correlates the heat flux in terms of the heating surface superheat $\Delta T_{sat}$, is considered the most significant and repeatable boiling data in terrestrial applications. It has been conjectured, based on numerous terrestrial experiments, that the nucleate boiling heat flux increases as buoyancy is reduced for a given heater surface superheat $\Delta T_{sat}$. Reference [85] summarizes the results from numerous experiments on pool boiling in both low- and high-$g$'s. In these experiments the boiling curve, under the appropriate conditions, was compared with the known curve in terrestrial environment. Due to the strong influence of buoyancy on the characteristics of the whole boiling curve, three different types of experiments were conducted in which the orientation of the heating surface was varied.

Figure 7.5 shows the boiling curve for an upward-facing heating surface (HU) for both 1-$g$, and low-$g$ for $g/g_0 \approx 0.008$. The data shown in Figure 7.5 are for saturated liquid nitrogen at atmospheric pressure covering the full range of film, transition, critical, and nucleate boiling. Figures 7.6 and 7.7 show similar data for a downward-facing (HD) and vertical heating surfaces, respectively.

Reference [85] presents the following conclusion on pool boiling in low-gravity environment based on these experiments. For the initial HU orientation in Figure 7.5, a somewhat significant decrease in the heat flux appears to occur in the film and transition boiling regions, when compared with the initial HD data of Figure 7.6. This difference implies that the residual liquid velocities in the vicinity of the heater persist in the latter case. The data of Figure 7.5 suggest that the phenomena of a critical heat flux can exist at drastically reduced levels of buoyancy. The change in behavior in the nucleate boiling region with reduced gravity in Figure 7.5 is subtle, but a tendency for an increase in the heat flux can be discerned. This is in contrast to the initial HD orientation where a reasonably large decrease in the heat flux takes place, representing a degradation in the nucleate boiling

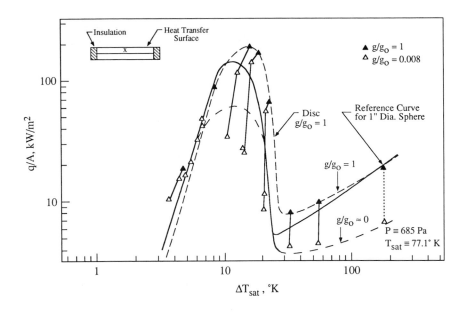

**Figure 7.5**   Boiling curve with a face-up heater, from [85].

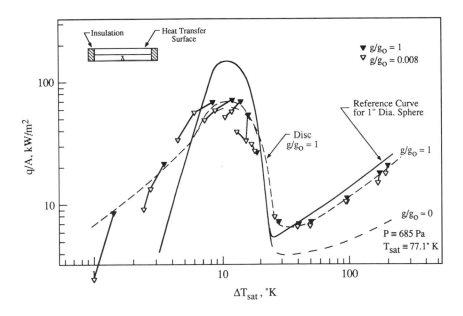

**Figure 7.6**   Boiling curve with face-down heater, from [85].

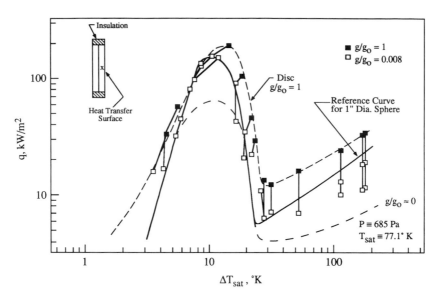

**Figure 7.7**   Boiling curve with side-mounted heater, from [85].

process as the body force changes from 1-$g$ to low-$g$. It may be concluded from these experiments that the nucleate boiling process is enhanced when buoyancy acts to hold the vapor bubbles at the heater surface.

## 7.6 FLOW BOILING

Power and other systems that use a working fluid to extract energy for providing useful work are characterized under the broad heading of thermohydraulic systems. This usually means forcing the working fluid to undergo phase change as it cycles through the loop. The thermohydraulic process takes advantage of the high heat-transfer rates available in flow boiling, generally orders of magnitude more efficient than single-phase heat transfer. Nuclear reactors, solar power plants, environmental heating and cooling systems, and refrigeration systems for cryogenic fluid storage and handling are a few space application examples of thermohydraulic systems. The thermohydraulic systems in most of these applications need to be as efficient as possible due to size and weight constraints imposed by the high cost of lifting objects into orbit.

Correlations and constitutive models for two-phase flow and heat transfer derived in terrestrial applications cannot a priori be assumed to apply in low-gravity environment. The gravity force, $g$, which is implicit in all terrestrial correlations as a fixed constant, is transformed to a fundamental variable in space applications. A whole new data base, derived from experiments over a wide range of design conditions, is needed to accurately characterize the behavior of flow boiling in reduced gravity. For purely economic reasons, these correlations need to be as accurate as possible so that space systems can be designed without waste. They must be capable of performing consistently within the design parameters.

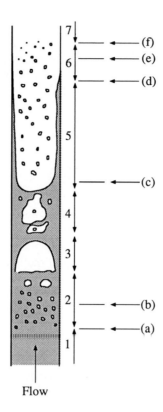

**Figure 7.8** Typical flow boiling patterns for up flow in a vertical tube.

At the present time the existing flow boiling data base in reduced gravity is not sufficient to develop useful correlations and constitutive models for designing the required systems.

It is useful to first review the phenomenon of flow boiling under terrestrial conditions. The liquid in vertical boiler tubes typically undergoes the flow patterns shown in Figure 7.8 as the coolant moves up the tube. The flow boiling phenomenon consist of the following patterns:

1. *Single phase liquid* is observed as the coolant is injected into the lower end of the tube boiler.
2. *Bubbly flow* is observed as the coolant undergoes nucleate boiling in the vicinity of the hot boiler tube wall.
3. *Plug flow* will develop as the bubbles generated at the tube wall are dragged by the coolant and coalesce.
4. *Churn flow* is usually observed during transition from the plug flow.
5. *Annular flow* is the next stage, in which the churn flow transitions to annular flow.
6. *Dispersed flow* will develop due to the combined effects of film boiling and liquid entrainment from the liquid film surface.

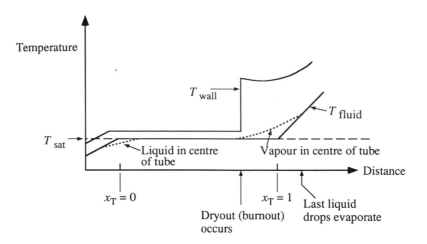

**Figure 7.9**  Tube wall temperature variation with distance. $x_T$ denotes the thermodynamic quality.

7. *Single-phase vapor* is the final stage of the flow boiling process.

The important physical phenomena involved in the flow boiling process can be summarized for each region in the following manner:

1. Bubble nucleation begins when the thermodynamic quality $x_T$ is less than zero. The *thermodynamic quality* is defined here (see [89]) as

$$x_T = \frac{h - h_{l,sat}}{h_L},\qquad(7.6.1)$$

where $h$ is the enthalpy of the flowing mixture and $h_{l,sat}$ is the saturation enthalpy of the liquid phase. Thus, the nucleation process begins when the liquid bulk flow away from the walls is still subcooled.

2. $x_T = 0$.

3. Bubble nucleation ceases and the boiling process becomes convective in character.

4. The liquid film dries out. This is the dryout, burnout, or the critical heat flux condition.

5. $x_T = 1$.

6. The last liquid drops evaporate.

The fact that bubble nucleation begins when $x_T < 0$ and liquid drops persist when $x_T > 1$ demonstrates that there is thermodynamic nonequilibrium in these regions. Both the liquid and the vapor are not saturated and consequently not in equilibrium with each other.

During the flow boiling process the temperature distribution in the tube wall and the distribution of the fluid inside the tube can take the form shown in Figure 7.9. It can be seen that there is a large increase in the wall temperature when dryout occurs, which may be comprised of many hundreds of degrees. Also,

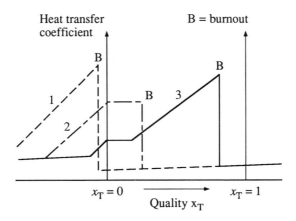

**Figure 7.10**   Heat transfer coefficient at the tube wall as a function of the thermodynamic quality for flow boiling.

before dryout occurs, the difference between the wall temperature and the saturation temperature decreases gradually. Therefore, the boiling heat transfer coefficient increases due to the thinning of the liquid film in the annular flow region as the quality increases.

A convenient method for expressing the heat transfer coefficient variation along the tube is by graphing its value against the thermodynamic quality $x_T$, as shown in Figure 7.10. At the highest heat flux, curve 1, burnout occurs before $x_T = 0$. At the lower heat flux, curve 2, burnout occurs after $x_T = 0$. In the region of positive quality, nucleate boiling occurs that ceases at burnout. At the lowest heat flux, curve 3, there are five distinct regions, starting from $x_T < 0$, which can be described in the following manner:

1. Single-phase liquid-forced convection in which the heat transfer coefficient is almost constant.

2. Subcooled boiling, in which the heat transfer coefficient increases as the bulk liquid approaches the saturation temperature.

3. Saturated nucleate boiling, in which the heat transfer coefficient is almost constant.

4. Saturated convective boiling, in which the heat transfer coefficient increases slowly.

5. Post-burnout heat transfer, in which the heat transfer coefficient is low. This regime gradually merges into a single-phase vapor-forced convection regime.

The sequence of flow patterns shown in Figure 7.8 for flow boiling is just a typical sequence of events that may occur. Recent work on flow boiling relevant to the flooding problem during the loss of coolant accident in nuclear reactor safety analyses have demonstrated that the specific pattern depends very strongly on the state of the injected coolant. If the injected coolant temperature is considerably below the saturation temperature (i.e., the liquid is subcooled), the flow pattern during the boiling process is slightly different, as downstream of the nucleate boiling regime the liquid forms an axial jet at the center of the tube that is surrounded by a vapor film, as shown in Figure 7.11. The dispersed regime for

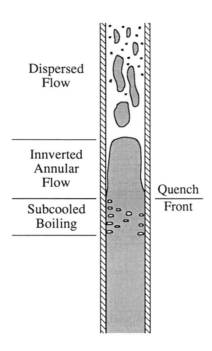

**Figure 7.11**   Flow pattern for flow boiling using subooled coolant.

that case is brought about by instabilities of the liquid jet, which eventually lead to its breakup into liquid drops.

The above two flow regions are somewhat different from the patterns arising when the injected liquid is at its saturation temperature, as shown in Figure 7.8. Figure 7.12 shows typical temperature values measured as a function of time during quenching of a straight vertical copper tube at terrestrial conditions. The tube is open to the atmosphere at the upper end. The coolant in that experiment is liquid nitrogen at atmospheric pressure and thus at saturation conditions.

Low-gravity flow boiling experiments have been conducted recently on board the NASA/KC-135 airplane with the aim of classifying the flow patterns during the boiling process. These experiments were conducted to study problems associated with cryogen fuel transfer line chilldown in space environment. Two sets of low-gravity experiments were conducted, one of which used Freon 113 for the coolant to study the flow boiling patterns arising from quenching with a subcooled liquid. Liquid nitrogen at 1 bar was used as the coolant for the second set of experiments to model the quenching process with a saturated liquid.

For the quenching experiments with subcooled liquid, the results indicated very little difference in the flow patterns between terrestrial and low-gravity environments. The flow boiling was characterized by an inverted annular region just downstream of the nucleate boiling region. Figure 7.13, from [84], shows a typical inverted annular region for low-gravity quenching. Detailed descriptions of each of these experiments are given in [84].

For the quenching experiments with saturated liquid, reference [76] indicates a substantial difference between the terrestrial and low-gravity experiments. The

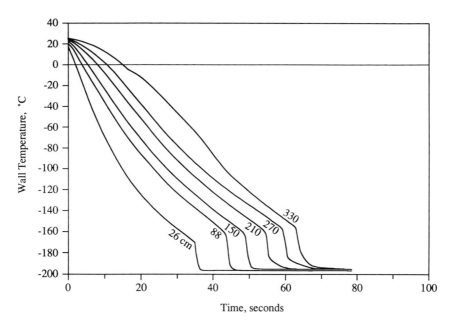

**Figure 7.12** Tube wall temperature variation with time during tube quenching with liquid nitrogen from [76].

major difference in the flow patterns is the existence of a *filamentry flow* just ahead of the quench front at the commencement of quenching in low gravity. The filamentary flow consisted of large and elongated liquid drops coalescing and flowing freely at the center of the tube, giving the impression of long meandering liquid filaments. This flow persisted for a few seconds after opening the coolant inlet valve to the test section. However, after it subsided the regular annular flow downstream of the quench front was observed. The boiling process after that time can be characterized as very similar to its terrestrial counterpart. No filamentary flow regime observed has been observed in terrestrial flow boiling. Figure 7.14 shows a typical filamentary flow regime observed in low-gravity saturated quenching. Much more low-gravity research needs be conducted on this problem before a suitable analytical model can be formulated.

## 7.7 *CONDENSATION*

The process of vapor condensation results physically from the interaction of high-energy vapor phase molecules with the interface of a bulk liquid phase, and either joining it or rebounding from it. High-energy molecules may also escape into the vapor region. It is commonly thought that fundamentally the process of condensation does not depend on the gravity level. This will be assumed to be correct even though no experiments have directly investigated its validity. Condensation may be classified into two different processes: *homogeneous condensation* and *inhomogeneous condensation*.

Homogeneous condensation occurs in the bulk phase of a vapor when molecules of sufficiently low energy coalesce to form microscopic liquid phase

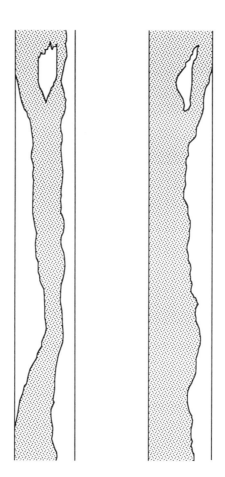

**Figure 7.13**   Interpretation of the flow pattern observed during quenching of a hot tube with subcooled liquid from the experiments of [84].

particles. Although this process is of interest in certain technology applications, homogenous nucleation does not play any role in condensation occurring in equipment and therefore is not very relevant here.

Inhomogeneous condensation occurs when the bulk vapor phase molecules interact with the bulk liquid/solid phase molecules, exchanging momentum and energy and becoming bulk liquid phase molecules. Condensation from the vapor phase directly to the solid phase (the opposite of sublimation) is possible but is not of importance in heat transfer process equipment.

Inhomogeneous condensation occurs either as *filmwise* or *dropwise*, as shown schematically in Figure 7.15. Filmwise condensation occurs when vapor at saturation temperature comes into contact with a liquid or solid below the working fluid saturation temperature. Energy is transferred due to this temperature difference, resulting in vapor condensing into liquid at the interface.

Momentum transfer, in this case, occurs due to two effects: one is due to the shear forces produced by the velocity difference between the vapor and the liquid; in the other, the convective momentum transport is due to momentum formerly

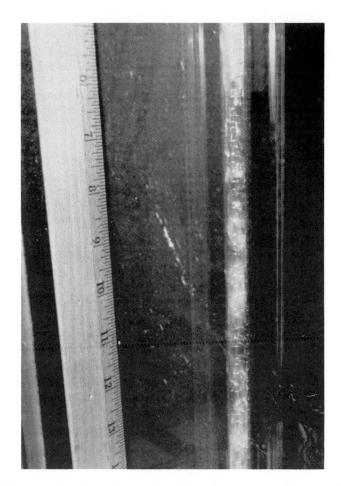

**Figure 7.14** Photograph of the filamentary flow during quenching of a hot tube with saturated liquid, from [76].

carried by the condensed vapor, which now resides in the liquid bulk. Energy and momentum are transported from the liquid bulk to the solid due to temperature and velocity differences, respectively.

Superheated vapor must be cooled to saturation temperature in order to condense. This can occur by convective cooling of the bulk vapor without condensation, where the mixed mean temperature of the vapor is brought to the state of saturation. Alternatively, the cooling to saturation conditions occurs in the vapor layer near the liquid, while the bulk vapor core remains superheated.

For both filmwise and dropwise condensation, the local temperature difference between the liquid and the vapor at the interface is normally small and usually neglected in heat transfer calculations. Energy transfer away from the condensation interface is controlled by conduction and convection through the associated liquid phase to the cooled solid surface. The condensation heat-transfer coefficient in the filmwise process depends on the thickness of the liquid layer. All models of filmwise condensation basically deal with calculating the film thickness.

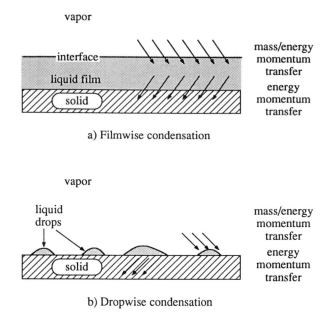

**Figure 7.15**   Filmwise and dropwise condensation.

Terrestrial condensing equipment relies principally on gravity to remove the liquid film, e.g., to drain liquids from the outside of condenser tubes. Both processes rely on gravity to settle the liquid droplets out of the bulk vapor.

Filmwise condensation occurs when the liquid phase wets the solid surface, and dropwise condensation occurs when the liquid phase does not wet the solid surface. Filmwise condensation occurs with most metal surfaces and working fluids. Dropwise condensation occurs only for specially treated surfaces, although any non-wetting surface/fluid combination will produce this phenomena. Glass/mercury is one such non-wetting pair.

One of the earliest filmwise condensation models was developed by Nusselt for a vertical plate geometry. This model has been subsequently extended to many geometries and conditions, as shown in Figure 7.16 in which the plate is at an angle $\varphi$ to the direction of the gravity vector. Neglecting momentum changes in the fluid, interfacial condensation resistance, and interfacial shear, the following correlations for the film thickness $\delta$ and the average heat transfer coefficient $h$ are given in [77]:

$$\delta(x) = \left[ \frac{4k\mu x(T_{sat} - T_w)}{g \cos \varphi \rho_l (\rho_l - \rho_g) h_L'} \right]^{1/4}, \tag{7.7.1}$$

$$h = 0.943 \left[ \frac{g \cos \varphi \rho_l (\rho_l - \rho_g) k^3 h_L'}{L\mu(T_{sat} - T_w)} \right]^{1/4}, \tag{7.7.2}$$

where the latent heat of vaporization is given by

$$h_L' = h_L + 0.68 c_p (T_{sat} - T_w). \tag{7.7.3}$$

$h_L'$ corrects for the liquid film subcooling and $L$ is the plate length. $k$ and $\mu$ are the coefficients of thermal conductivity and viscosity of the liquid, respectively.

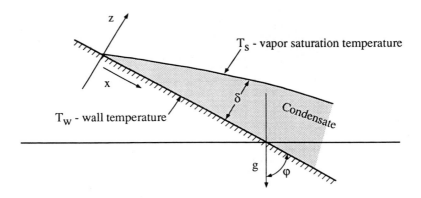

**Figure 7.16**  Schematic of filmwise condensation over an inclined flat plate, from [77].

Equations (7.7.1) and (7.7.2) show that the film thickness at any position, $\delta(x)$, varies with $g^{-1/4}$, while the overall heat transfer coefficient varies as $g^{1/4}$. These results indicate that the heat transfer coefficient is a slowly varying function of the local gravity vector, decreasing with decreasing $g$. In the limit as $g \to 0$, the condensate does not drain at all, and the stagnant film thickness and heat transfer coefficient for this case are given in [77] by

$$\delta = \sqrt{\frac{2k(T_{sat} - T_w)t}{\rho h_L}}, \tag{7.7.4}$$

$$h = 1/\sqrt{\frac{2(T_{sat} - T_w)t}{\rho h_L k}}. \tag{7.7.5}$$

The above expressions were derived with the assumption that the liquid film is laminar.

Since many industrial condensers rely on tubular design for increasing the efficiency of the condensation process, relevant empirical formulas have been derived for such configurations. For condensation on the external surface of a tube that is cooled internally, an interesting approach was suggested by [77]; the liquid film thickness $\delta$ and the heat transfer coefficient $h$ for condensation at the external surface of a tube, as shown in Figure 7.17, were calculated consequently by [88] in the limit $g/g_0 \to 0$ and for stagnant vapor:

$$\left[\left(1 + \frac{\delta}{r_i}\right)^2 + \frac{1}{2}\right]\left[\ln\left(1 + \frac{\delta}{r_i}\right) - \frac{1}{2}\right] = \frac{2k(T_{sat} - T_w)t}{\rho h_L r_i^2} \tag{7.7.6}$$

$$h = \frac{k}{r_i \ln(1 + \delta/r_i)} \tag{7.7.7}$$

where $r_i$ is the tube inner radius.

Shear driven film condensation within an externally cooled constant diameter tube can be considered also as a form of film condensation. As shown in Figure 7.18, vapor enters the tube at saturation conditions, condenses on the cool

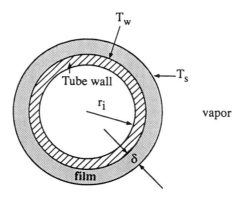

**Figure 7.17** Filmwise condensation on the outside of a tube, from [77].

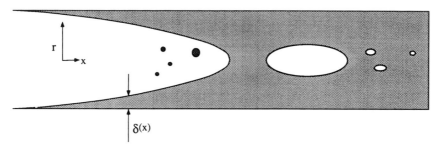

**Figure 7.18** Schematic of filmwise condensation on the inside of a tube, from [77].

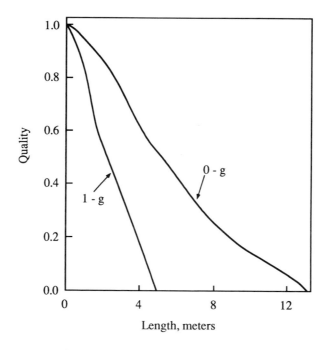

**Figure 7.19** Variation of quality $x$ during condensation in $g/g_0 = 1$ and 0, from [77].

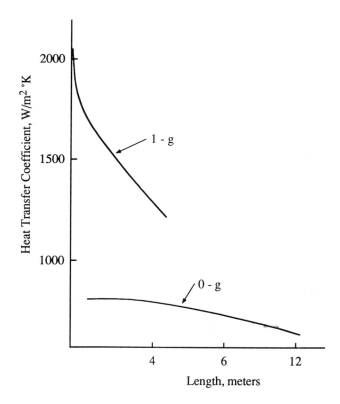

**Figure 7.20** Heat transfer variation with tube length for $g/g_0 = 1, 0$, from [77].

wall, wets the surface, and forms an annular liquid film. This type of condensation process is basically inertia dominated, in which the effects of gravity are minimal.

Figure 7.19 shows the calculated quality as a function of tube length for $g/g_0 = 1$ and $g/g_0 \approx 0$ models. Figure 7.20 shows the heat transfer coefficient variation as a function of tube length, again for $g/g_0 = 1$ and $g/g_0 = 0$ for the same model.

Dropwise condensation occurs when the condensing liquid phase does not wet the cooled surface. The heat transfer coefficient for this case is usually ten or more times greater than those for filmwise condensation. The major heat transfer effect is due to the small drops, whereas the large drops are relatively inert. For instance, for water at 100°C, this effect is due to drops on the order 100 $\mu$m or less. The active, small condensation drops nucleate in cracks and pits on the surface where liquid remains from previous drops. Small drops grow until they touch other drops. These drops coalesce, continuing to grow by merging with smaller drops. Under low-gravity conditions the large droplet population is depleted by entrainment into the vapor or it is removed by mechanical means.

Dropwise condensation depends on the large droplet coalescence and removal rate, the effective constriction of the wall conduction heat transfer path due to highly localized condensation nuclei, and the resistance of drop promoting coating and noncondensable gases. No generally applicable quantitative heat

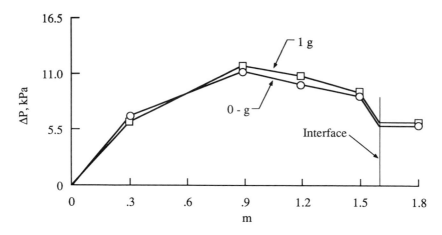

**Figure 7.21** Pressure drop as function of tube length for a constant diameter tube for $g/g_0 = 1$ and $g/g_0 \approx 0$, from [77].

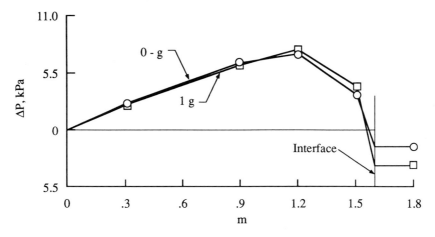

**Figure 7.22** Pressure drop as function of tube length for a tapered tube for $g/g_0 = 1$ and $g/g_0 \approx 0$, from [77].

transfer models have been developed, although empirical relationships do exist. Figures 7.21 and 7.22 show experimental results for dropwise condensation in a constant diameter and tapered diameter tubes for $g/g_0 = 1$ and low-$g$ conditions.

# Chapter *8*

# *Combustion and Flame Propagation*

hat distinguishes combustion from other processes in fluid dynamics are the large temperature variations that exist in combusting flows. These are caused by the exothermic heat release that results from the chemical reactions that characterize combustion processes. The temperatures can range from about 300 K (the unreacted ambient state) to as high as 2500 K at the height of the reaction. These large variations lead to correspondingly large density differences which in turn can cause strong flow currents to arise from buoyancy-driven convection.

In addition to causing buoyancy-driven convection, the presence of gravity in combusting two-phase flows causes particles and droplets to settle, leading to stratification in the mix. Furthermore, the presence of buoyancy-driven flows together with the hydrostatic pressure can distort the effects of surface tension, resulting in variations in the interfacial contours and thermocapillary motion in a body of liquid fuel. Gravity can also induce a degree of asymmetry in an otherwise symmetrical phenomenon. For example, a combustion gaseous jet, injected in the direction normal to the gravity vector, will quickly lose its axial symmetry as the hot flame plume eventually turns upward due to buoyancy. This departure from symmetry leads to highly complex multidimensional fluid flows.

The conventional practice in most analytical formulations of combustion processes is to neglect buoyancy-driven convection. This is done in order to make the analysis simpler. Such assumptions can render direct comparison between theory and terrestrial experiments very difficult, making the feedback process between the two meaningless. Consequently, data from low-gravity experiments become more valuable for comparison with theoretical results from such an analysis. Thus low-gravity experimental endeavors can lead to a significant improvement in the understanding of combustion science.

In a habitable low-gravity environment, important fire hazard and fire safety issues include flame front spread rates, smoldering fires, and flammability limits. At this time, our data base for the necessary safety design procedures is meager, requiring a considerable amount of further study. Tests already performed reveal considerable new information on the effects of gravity on the combustion process itself. These issues and related findings are the content of this chapter.

The intent of this chapter is not to present a comprehensive theory of the combustion process, but rather to give the necessary tools for analyzing the effects of gravity on the process. The theory itself is extremely complex, and the reader is advised to consult the many excellent texts that have been written on this subject, such as reference [99].

## 8.1 BASIC CONSIDERATIONS IN COMBUSTION

Combustion is a process in which strongly exothermic chemical reactions take place between atomic or molecular reactants in the form of fuel and oxidant, resulting in product molecules with a significant emission of light. Chemical reactions in combustion can only take place under specific conditions of pressure, temperature, and fuel-oxidant ratio mixture. Many other processes, in addition to chemical reactions, are involved in combustion. Combustion can only persist when the energy generated by the chemical reactions is sufficient to preheat the reactants to the combustion condition. Depending on the state of the reactants prior to admission to the combustion zone, whether solid, liquid, or gaseous, and on the injection conditions, whether the fuel and oxidant are premixed or not prior to entering the combustion environment, the phenomenon of combustion can involve all or some of the following processes: (a) heating of the condensed reactants, (b) vaporization or sublimation of the reactant to the gaseous phase, (c) mixing of the oxidant and the fuel in the gaseous phase, (d) heating of the oxidant and the fuel in the gaseous phase, and (e) chemical reaction in the zone in which the conditions for the temperature, pressure, and mixture ratio are correct. Each of these processes is essential to combustion, but only the slower ones control it. The relative importance of each process can only be appreciated when its characteristic time scale is evaluated in terms of readily identifiable parameters of the system.

The modeling of the combustion processes is extremely complex when all of the phenomena listed above are included. This becomes especially obvious when the conservation equations governing combustion are examined. These equations for combustion are the same as those for fluid mechanics, supplemented by equations for the conservation of chemical species including the effects of chemical kinetics.

The full set of equations necessary for defining the combustion process can be written in a nondimensional form in the following manner (see [98]):

$$\varpi \frac{\partial \rho}{\partial t} + \nabla \cdot (\rho \mathbf{v}) = 0, \tag{8.1.1}$$

$$\varpi \frac{\partial \rho \mathbf{v}}{\partial t} + \nabla : (\rho \mathbf{v}\mathbf{v}) = -\frac{1}{\mathcal{M}^2} \nabla p + \frac{1}{Fr} \rho \mathbf{F}_b + \frac{1}{Re} \nabla : \tau, \tag{8.1.2}$$

$$\left[ \varpi \rho \frac{\partial}{\partial t} + \rho \mathbf{v} \cdot \boldsymbol{\nabla} \right] \left[ \int c_p dT + \alpha \sum_{i=1}^{N} h_i Y_i + \frac{\gamma - 1}{\gamma} \mathcal{M}^2 \left( \frac{\mathbf{v} \cdot \mathbf{v}}{2} \right) \right]$$

$$= \frac{\gamma - 1}{\gamma} \varpi \frac{\partial p}{\partial t} + \frac{1}{PrRe} \boldsymbol{\nabla}$$

$$\cdot k \boldsymbol{\nabla} T - \alpha \sum_{i=1}^{N} h_i [\boldsymbol{\nabla} \cdot (\rho Y_i \mathbf{U}_i)] + \frac{1}{B} \boldsymbol{\nabla}$$

$$\cdot \mathbf{q}_r + \frac{\gamma - 1}{\gamma} \frac{\mathcal{M}^2}{Re} \boldsymbol{\nabla} \cdot \mathbf{v} : \tau + \frac{\gamma - 1}{\gamma} \frac{\mathcal{M}^2}{Fr} \left( \rho \mathbf{v} \cdot \mathbf{f} - \rho \sum_{i=1}^{N} Y_i \mathbf{U}_i \cdot \mathbf{f}_i \right), \quad (8.1.3)$$

$$\varpi \frac{\partial (\rho Y_i)}{\partial t} + \boldsymbol{\nabla} \cdot (\rho \mathbf{v} Y_i) = -\boldsymbol{\nabla} \cdot (\rho Y_i \mathbf{U}_i) + P_c, \quad i = 1, \dots, N \quad (8.1.4)$$

where $P_c$ is the chemical production term which can be a very complex function of the specific chemical reactions and the thermodynamic state of each species, $N$ is the number of different chemical species involved, and $\mathbf{U}_i$ is the individual species diffusion velocity. $\rho$, $p$, $T$, $\mathbf{v}$, $Y_i$, and $h_i$ are the density, pressure, temperature, flow velocity vector, mass fraction, and enthalpy of species $i$, respectively. $\gamma$ is the ratio of the specific heats defined as $\gamma = c_p/c_v$. Other symbols are explained later in the proper context.

Equations (8.1.1)–(8.1.3) are the usual equations for the conservation of mass, momentum, and energy for a compressible fluid, respectively. Equation (8.1.4) is the species conservation equation for each species $i$ that must be used when several are present in the flow. The equations are to be solved for the primary variables: $\rho(\mathbf{x}, t)$, $\mathbf{v}(\mathbf{x}, t)$, $T(\mathbf{x}, t)$, $p(\mathbf{x}, t)$, and $Y_i(\mathbf{x}, t)$, subject to the appropriate initial and boundary conditions.

Equations (8.1.1)–(8.1.4) are not complete and must be augmented with an equation of state and the equation for the viscous stress tensor. The simplest equation of state that is commonly used is that for an ideal gas, for which

$$p = \rho R_0 T / \left( \sum_{i=1}^{N} X_i M_i \right), \quad (8.1.5)$$

where $R_0$ is the universal gas constant, and $X_i$ and $M_i$ are the mole fraction and the molecular weight of species $i$, respectively. Also, it is common to assume that the fluid is Newtonian, in which case the stress tensor is given by

$$\tau_{ij} = 2\mu \left( \frac{\partial v_i}{\partial x_j} + \frac{\partial v_j}{\partial x_i} - \frac{1}{3} \frac{\partial v_i}{\partial x_i} \delta_{ij} \right), \quad (8.1.6)$$

where $\mu$ is the mixture viscosity and $\delta_{ij}$ is the Kronecker delta.

The four different terms that appear in the species conservation equations (8.1.4) lie at the heart of combustion science. The first term describes the accumulation of chemical species within a volume element, the second represents the convection of this species out of the volume element, the third is the diffusion of the species into the volume element, and the last is the production of the species within the element by chemical reactions. This accumulation-convection-diffusion-production balance is the most general conservation statement for molecules undergoing chemical reaction.

The energy conservation equation (8.1.3) has an additional term in the total energy balance besides the usual kinetic and internal energy. This equation for combustion conserves the following three forms of energy:

(a) the internal energy given by $\int c_p dT$,

(b) the kinetic energy, given by $\mathcal{M}^2 \mathbf{v} \cdot \mathbf{v}[(\gamma - 1)/\gamma]/2$, and

(c) the chemical potential energy in the form $\alpha \sum_{i=1}^{N} h_i Y_i$, where $\alpha$ is a nondimensional heat release function (see [98]).

The latter form of energy is special to combustion; it is associated with the binding energies of the different molecules in their ground states. The chemical energy term provides a chemical-kinetic contribution to the change of thermal energy of the material through the species conservation equation.

The energy conservation equaticn (8.1.3) includes additional balance terms that are unique to combusting fluids. These are the transient energy accumulation through the pressure increase term $((\gamma - 1)/\gamma)\varpi \partial p/\partial t$ and the radiant energy input term $\mathbf{q}_r$ (measured by the Boltzman number $B$). The radiation term is extremely important due to the high temperatures involved in a typical combustion process.

Equations (8.1.1)–(8.1.6) are appropriate for compressible flow, in which the density $\rho$ is one of the unknowns that must determined as a function of space and time. The Boussinesq approximation becomes redundant in this case. It is not valid for combustion processes since the density changes range from a factor of three to one order of magnitude in a typical combustion process. Thus, unlike buoyancy-driven convection problems, the full compressible flow equations must be used in the analysis of combustion processes.

Equations (8.1.1)–(8.1.6) can only be solved if all the necessary auxiliary relationships are given in terms of the primary variables. For chemically reacting ideal gas mixtures, the kinetic theory of gases provides viscosity and thermal conductivity as functions of $T$ and $X_i$. The caloric equations of state gives $c_p$ as $c_p = \sum_{i=1}^{N} Y_i c_{pi}(T)$ for ideal gas mixtures. An equation for the radiative transport must be added to define $\mathbf{q}_r$. If absorption and scattering is neglected, then $\nabla \cdot \mathbf{q}_r \approx \epsilon T^4$, where $\epsilon$ is the Stefan-Boltzman constant. The species diffusion velocities $\mathbf{U}_i$ that appear in the energy and species conservation equations can be obtained from a diffusion equation whose form is given by the kinetic theory of gases.

The unsteady terms in equations (8.1.1)–(8.1.4) contain a nondimensional ratio of the time scales, $\varpi$, representing the ratio of the convective time scale to the reaction time scale. This parameter is unique to combustion. The chemical reaction time scale is typically on the order of a few milliseconds, depending on the system temperature and pressure. On the other hand, the convective time scale is typically on the order of one tenth to one hundredth of a second. For many combustion processes it is assumed that the chemical reaction rates are small compared with the fluid mixing rates. This shows that the overall combustion phenomenon is often primarily driven by the regular fluid dynamic processes, while at the same time instantaneous reactions are assumed. In other words, the gaseous flow field has an important influence on the combustion process.

Equations (8.1.1)–(8.1.4) contain several nondimensional numbers, the ranges of which must be specified for each application for which a solution is desired. The most significant of these numbers are the Reynolds number $Re$, representing the ratio of the inertia force to the viscous force, the Prandtl number $Pr$, the Mach number $\mathcal{M}$, representing the ratio of the fluid velocity to the local speed of sound, and the Froude number $Fr$, representing the ratio of the inertia force to the buoyancy force. There are many other nondimensional numbers in these equations, but those listed above are the most relevant for low-gravity applications.

Low-gravity combustion studies can be classified into different regimes according to the ranges of the nondimensional parameters. There are certain parameter ranges in which no benefit can be anticipated by performing experiments in low gravity. One is the regime of large $\mathcal{M}$, in which the dynamic forces dominate over buoyancy. This can be illustrated by taking $g_0 \approx 10^3$ cm/s$^2$ and the speed of sound $c \approx 10^5$ cm/s, resulting in $\mathcal{M}^2/Fr \approx 10^{-7}L$. These conditions require typical length scales, $L$, on the order of $10^4$ cm for the ratio $\mathcal{M}^2/Fr$ to be of order unity. Thus as long as $L \ll 10^4$ cm, the buoyancy force is much less than the inertia force.

The Reynolds number $Re$ in combustion processes can vary from very small to very large values. Usually turbulent flows occur when the Reynolds number is moderate. It is still unknown whether inertia forces dominate buoyancy forces in flows with the Reynolds number large enough for the flow to be turbulent. In general, the smaller the convective velocity, in the absence of buoyancy, the larger the influence of buoyancy. Thus, low-speed combustion processes are typically those in which buoyancy is of greatest importance. Typical velocities in laminar combustion range from 1 to 100 cm/s. This makes the Froude number $Fr < 1$ for length scales $L$ on the order of $10^{-3}$ to $10^{-1}$ cm at normal Earth gravity causing buoyancy to become dominant. By reducing $g/g_0$, low gravity enables laminar combustion to be investigated with negligible buoyancy effects. Thus a major reason for conducting combustion science experiments in low-gravity environment is the need to analyze certain combustion processes in a manner unencumbered by buoyancy.

Low-speed combustion processes in the laboratory often tend to be buoyancy-dominated, i.e., $Fr \to 0$. In low-gravity environment, however, where buoyancy forces are negligible, the value of the Froude number can be very large and thus for most low-gravity applications $Fr \to \infty$. Another parameter that is usually encountered in combustion processes is the Lewis number $Le$, defined as the ratio of the thermal to mass diffusivities. Almost all of the solutions in combustion modeling are normally obtained for the limit of $Le \approx 1$.

## 8.2 CHEMICAL KINETICS IN COMBUSTION

The chemical reactions involved in realistic combustion processes can be very complex when all the individual elementary reaction steps are considered. Such considerations make the solution for the species conservation equations untenable. The value of $N$ in equation (8.1.4) can be on the order of 500 or more for realistic processes. For example, in hydrogen air combustion, there could be up to 50 relevant elementary reaction steps. This makes performing realistic combustion calculations for such a simple reaction quite extensive. A substantial effort has been devoted recently to developing reduced chemical-kinetic mechanisms for describing combustion processes. First, all elementary reactions that do not play any role in the specific combustion process are discarded. Second, steady-state and partial equilibrium approximations are assumed.

For example, reference [98] shows that an excellent description for methane-air flames has been achieved with only the following four overall reactions:

$$CH_4 + 2H + H_2O \rightleftharpoons CO + 4H_2, \tag{8.2.1}$$

$$CO + H_2O \rightleftharpoons CO_2 + H_2, \tag{8.2.2}$$

$$2H + O_2 \rightarrow H_2 + M, \tag{8.2.3}$$

$$3H_2 + O_2 \rightleftharpoons 2H_2O + 2H, \tag{8.2.4}$$

with rates related to the following elementary reaction steps respectively:

$$CH_4 + H \rightleftharpoons CH_3 + H_2, \tag{8.2.5}$$

$$CO + OH \rightleftharpoons CO_2 + H, \tag{8.2.6}$$

$$H + O_2 + M \rightarrow HO_2 + M, \tag{8.2.7}$$

$$3H + O_2 \rightleftharpoons OH + O. \tag{8.2.8}$$

The simplest approximation to combustion chemistry is the *one-step approximation*. For the methane flame, for example, this can be represented by the following reaction

$$CH_4 + 2O_2 \rightarrow CO_2 + 2H_2O. \tag{8.2.9}$$

The combustion of methane involves the four elementary reactions indicated above, while the one step reaction described by equation (8.2.9) does not occur at the elementary level even though it describes the overall process. The one-step approximation is a very poor approximation for methane combustion; however, many aspects of the dynamics of methane flames have been adequately studied using the single reaction alone. This equation is treated as an overall step with a rate derived empirically and not related to the rates of the elementary reaction steps actually occurring. The empirical approach has been used extensively in the past but now, with high-speed computational facilities, more than one-step reactions can be used without much difficulty.

## *8.3* FLAMMABILITY LIMITS

Some of the most important combustion phenomena related to fire safety and fire hazard control are the processes of ignition, extinction, and flammability limits. Ignition is the process in which a material capable of reacting exothermically is brought to a state of rapid combustion. This process has long been of practical interest from numerous viewpoints, ranging from the desire to prevent unwanted explosions in transportation and storage of combustibles to the need to initiate controlled combustion in furnaces, motors, and guns. Methods for achieving ignition include exposure of the combustibles to a sufficiently hot surface such as an electrically heated wire, to a radiant energy source, etc. For additional discussion, see reference [99].

The process of fire extinction is of interest in problems of safety and of fire suppression, as well as in various aspects controlling combustion processes in industry and propulsion. Extinction (quenching) of deflagrations or diffusion flames may be achieved in various ways, including passing the flame through tubes of small diameters (a design principle of flame arrestors), removing an essential reactant from the system (an approach especially suitable for diffusion flames), adding sufficiently large quantities of a material that slows combustion (such as water or a chemical fire extinguisher), or physically removing the flame from the reactant mixture by inducing high gas velocities (blowing the flame away, as accomplished for instance by explosives in large production-well fires). General strategies for extinguishment may be listed as cooling, flame removal, reactant removal, and flame inhibition. Reference [99] suggests that since flame extinction usually occurs relatively rapidly if at all, more emphasis should be placed on ascertaining critical conditions for extinction than on extinction times.

Flammability limits are limits of composition or pressure beyond which a fuel-oxidizer combination cannot be made to burn. These limits are of practical interest especially in connection with safety considerations because mixtures outside the flammability limits can be handled without concern about ignition. For this reason, extensive tabulations of flammability limits have been prepared.

In spite of decades of studies, flammability limits are not completely understood. Many theories have been proposed including conductive and/or radiative heat losses, flame curvature effects, fuel chemistry, and buoyancy. Theories based on energy conservation requirements and heat losses have served very well in deriving important results. These simple theories serve to produce reasonable practical results without including the finer details of chemical kinetics, such as radical diffusion or surface reactions. This is due to the fact that energy budgeting is essential to combustion, without which it does not take place.

An example for ascertaining flame propagation speed and flammability limits with the aid of the energy conservation equation alone is given in reference [99] and is outlined here. In that model, the flammability limits are derived as eigenvalues of the combustion problem. The burning velocity for a simplified model of a plane, one-dimensional, non-adiabatic flame is derived with the proper heat loss considerations. Reference [99] shows that under certain conditions involving finite heat losses, two burning velocities are obtained. One of these velocities presumably corresponds to an unstable flame. Under different conditions, no solution for the burning velocity can be found, which indicates the existence of flammability limits.

The one-dimensional, steady-state energy equation describing the temperature distribution as a function of position and the strength of the heat sources can be written as

$$k\frac{d^2T}{dx^2} - mc_p\frac{dT}{dx} = -f(T)q_0 + q_L(T). \qquad (8.3.1)$$

Equation (8.3.1) is the simplest one-dimensional version of equation (8.1.3) in which the flame is at rest with respect to the origin of the coordinate system. $x$ is the coordinate direction normal to the flame, $k$ is the constant mean thermal conductivity, $c_p$ is the constant average specific heat at constant pressure, $m = \rho U_f$ is the constant mass flow per unit area in the positive $x$-direction, and $U_f$ is the flame propagation speed in the $x$-direction.

The terms on the right-hand side of equation (8.3.1) are the source terms. $f$ is the chemical reaction rate, which is defined in terms of the mass of fuel converted per unit volume per second; $q_0$ is the heat release in the chemical reaction per unit mass of converted fuel; and $q_L$ is the heat loss per unit volume per second by radiation or conduction to the walls. This equation preserves the balance between heat conduction, thermal-enthalpy convection, heat production, and heat loss at every point in the flame. Equation (8.3.1) is an example of the thermal theory for flame propagation.

The chemical reaction rate function $f$ is normally a very strong function of the temperature $T$, making the governing equation, the solution of which is dependent on the specific form chosen for the function $f$, highly nonlinear. A reasonably simple solution to equation (8.3.1) can be obtained if the burning rate is known and there is no heat loss, i.e., when $q_L = 0$. Also, the adiabatic mass burning rate $m_a$ is assumed to be known and is used in the nondimensionalization process. A length scale may be defined using this known quantity given by $L = k/m_a c_p$. The nondimensional temperature $\theta$, the burning velocity $U$, and the streamwise coordinate may be defined in the following manner:

$$\theta = c_p(T - T_0)/(q_0 Y_{F0}), \qquad U = m/m_a, \qquad z = x m_a c_p/k. \qquad (8.3.2)$$

Substituting these nondimensional variables into equation (8.3.1) results in the following equation:

$$\frac{d^2\theta}{dz^2} - U\frac{d\theta}{dz} = -\Lambda_a \omega(\theta) + \phi(\theta), \qquad (8.3.3)$$

where $U$ is known as the burning velocity ratio, $T_0$ is the initial combustible mixture temperature, and $Y_{F0}$ is the initial mass fraction of the fuel. The reaction rate $f$ in equation (8.3.3) has been nondimensionalized as

$$\Lambda_a \omega(\theta) = kf/(Y_{F0} m_a^2 c_p), \qquad (8.3.4)$$

where $\Lambda_a$ is known as the burning rate eigenvalue for the adiabatic problem defined in [99]. Finally, the nondimensional heat loss rate is given by

$$\phi(\theta) = kq_L/(q_0 Y_{F0} m_a^2 c_p). \qquad (8.3.5)$$

Assuming that all of the fuel is consumed in the reaction zone, an expression for $Y_{F0}$ can be derived from the fuel conservation equation. The heat loss distribution function $\phi(\theta)$ plays an important role in the structure of the solution for the non-adiabatic flame. A solution to equation (8.3.3) is derived in [99] using asymptotic expansion techniques in terms of an expansion parameter $\beta$ defined by

$$\beta = E(T_{af} - T_0)/(R_0 T_{af}^2), \qquad (8.3.6)$$

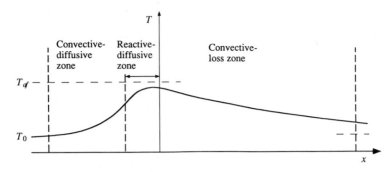

**Figure 8.1**   Sketch of the three reaction zones, from [99].

where $T_{af} = T_0 + q_0 Y_{F0}/c_p$ is the adiabatic flame temperature and $E$ represents the overall activation energy.

The solution for equation (8.3.3), given in [99], is valid for three distinct regions of the flame that depend on the specific forms of both the reaction rate function $\omega$ and the heat loss function $\phi$ in these regions. These three specific combustion zones used in deriving the solution are (1) a region upstream of the reaction zone, in which there exists a balance between heat convection and diffusion; (2) the reaction region itself; and finally (3) a convective loss region downstream of the reaction zone.

Reference [99] formulates a solution for the temperature distribution valid within each of the three regions. In the upstream convection region, reference [99] defines the following solution for the temperature:

$$\theta = \theta_f e^{Uz},\qquad(8.3.7)$$

while in the downstream convective loss zone, the following solution is obtained:

$$\theta = \theta_f e^{cz/(U\beta)},\qquad(8.3.8)$$

where $c$ is a constant of order unity and the subscript $f$ identifies conditions just downstream from the reaction zone (see [99]).

In the reaction zone itself a solution is constructed by ensuring a smooth matching of the temperature between the upstream and the downstream zones. The temperature distribution in the three zone structure for the non-adiabatic flame is shown schematically in Figure 8.1.

The burning velocity ratio $U$ is determined as an eigenvalue for the problem in the reaction zone. Reference [99] shows that $U$ in the reaction zone must satisfy the following condition

$$U^2 = e^{-\lambda/U^2},\qquad(8.3.9)$$

where the heat loss parameter $\lambda$ is defined in the following manner:

$$\lambda = [\phi(\theta_f) + \int_0^{\theta_f} (\phi/\theta)d\theta]\beta.\qquad(8.3.10)$$

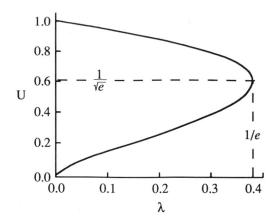

**Figure 8.2**    A sketch of the flammability limits, from [99].

Equation (8.3.9) can be inverted to give the heat loss parameter $\lambda$ as a function of $U$ in the following form

$$\lambda = U^2 ln(1/U^2).\qquad(8.3.11)$$

Figure 8.2 shows the variation of $U$ as a function of $\lambda$. Expression (8.3.11) shows that

$$d\lambda/dU = 0 \quad\text{at}\quad U = 1/\sqrt{e}\qquad(8.3.12)$$

at $\lambda = 1/e$, and that there is no solution for $\lambda > 1/e$, which thus represents a flammability limit. For $\lambda < 1/e$ there exist two values for the burning velocity ratio for each value of $\lambda$, giving upper and lower limits for the burning velocity. These two solutions may not exist experimentally depending on the stability of the steady state. These stability criteria are likely to be influenced by the experimental configuration and by parameters that have not been included in the solution.

Static stability arguments have suggested that the upper branch is stable and the lower is unstable. However, under certain conditions, a small portion of the lower branch has been stabilized near the maximum value of $\lambda$ for a flat-flame burner.

Reference [94] cites several experimental studies that are performed with the purpose of determining quantitatively the values for the flammability limits as well as the mechanisms responsible for their existence. Flammability limits determined under normal terrestrial conditions suggest significant influence of buoyancy effects. This is deduced from the fact that the flammability limit is much wider for upwardly propagating flames than for downwardly propagating flames. Recent experiments on flammability limits of combustible mixtures in low gravity have been conducted in drop towers. Figure 8.3 shows the flammability limits measured experimentally for upwardly propagating, downwardly propagating, and zero-$g$ flames. Figure 8.3 shows that in this case the zero-$g$ limit lies between the terrestrial upward and downward limits.

The fact that flammability limits do seem to exist in low gravity resolves a controversy regarding the role of buoyancy in the limit phenomena. Previous theories that related such limits to buoyancy-driven convection predicted that no limits should exist in the absence of gravity. The drop tower experiments have ruled out buoyancy as a controlling mechanism for flammability limits. These

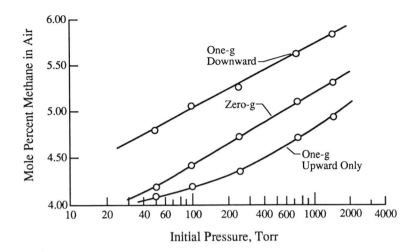

**Figure 8.3**   Flammability limits for upwardly propagating, downwardly propagating, and zero-gravity conditions, from [97].

experiments also show that since the low-$g$ flammability limit is always bounded between the upward and downward terrestrial limits, a conservative estimate for low-$g$ flammability limits would be the upward normal-$g$ values. This will allow for the direct use of flammability limit results from terrestrial conditions for space applications.

Recent extensive numerical calculations by reference [96] clearly illustrate the various flame stability criteria. Figure 8.4 shows the instability and eventual breakup of an upward propagating hydrogen-air flame. Also shown in Figure 8.4 are the downward propagating flame, which is stable, and the zero-gravity flame.

## *8.4* FLAME SPREADING

Flames spreading over fuel surfaces is a crucial issue in fires both on earth and in space. This involves the understanding of the mechanisms governing spreading, the determination of the spreading rate as a function of the fuel properties and environment parameters, and the identification of the limiting conditions under which spreading is not possible. Flame spreading can be classified into two broad categories: *opposed flow spreading*, as shown in Figure 8.5 in which the spreading direction is opposite to that of the external flow, and *concurrent flow spreading*, as shown in Figure 8.6, in which the spreading direction is the same as that for the external flow.

For a vertically oriented surface in the presence of gravity, the flame induces buoyant flow and these two spreading modes respectively correspond to downward and upward spread. In concurrent spread, the hot combustion products form a plume that bathes the unburnt fuel causing the spread to be generally acceleratory and rapid. Opposed spread is slower, in general, in which case a steady state can be achieved more readily.

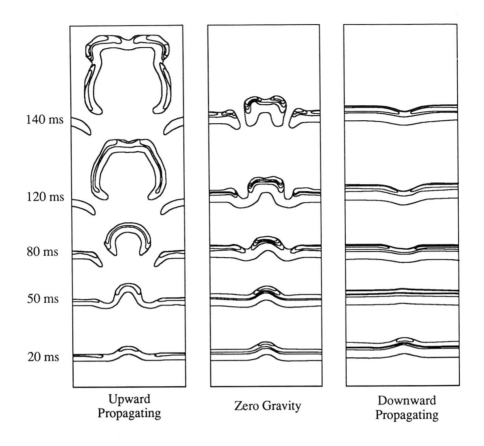

**Figure 8.4**  Effects of gravity on cellular flame formation, from [96].

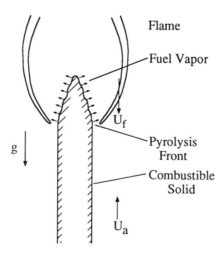

**Figure 8.5**  Sketch of opposed flow flame spread mechanism.

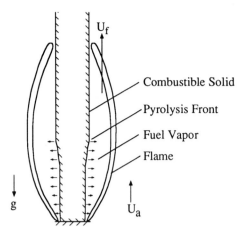

**Figure 8.6**  Sketch of concurrent flow flame spread mechanism.

Figure 8.7 shows a more detailed schematic for the opposed spreading configuration. For a finite flame speed, the figure shows the flame experiencing different upstream velocities against its motion in a coordinate system that is moving with the flame speed. This relative motion depends on whether the flame is situated in (a) a zero-$g$ stagnant environment, (b) a zero-$g$ forced convective environment, or (c) a stagnant buoyant environment. Thus spreading is accomplished through highly coupled convective-diffusive heat and mass processes around the leading segment of the flame.

One of the earliest calculations for determining the flame spread velocity was that of reference [91], in which the mechanism of flame spread over a stationary fuel bed was investigated. The model used was very simple in that it does not account for buoyancy effects and only the energy and species conservation equations are used. The analysis was simplified by assuming that the gas is moving with a uniform velocity $U_a$ parallel to the fuel bed surface, as shown schematically in Figure 8.8. Also, the gas phase is assumed to have constant density $\rho$, pressure $p$, conductivity $k$, specific heat $c_p$, and coefficient of mass diffusion $D$. The Lewis number $Le = k/\rho c_p D$ is assumed to be unity as is commonly done in most combustion models. The mass diffusion for the different species is assumed to be driven by the species concentration gradients only.

The energy conservation equation for this model in a steady, two-dimensional domain $(x, z)$, can be written as

$$\rho c_p U_a \frac{\partial T}{\partial x} = k \left[ \frac{\partial^2 T}{\partial x^2} + \frac{\partial^2 T}{\partial z^2} \right] + q_{chem} - q_r, \qquad (8.4.1)$$

where $q_{chem}$ and $q_r$ are the net rates of chemical heat release and radiative heat loss per unit volume, respectively. If $Y_i$ is the mass fraction of species $i$, and $m_i$

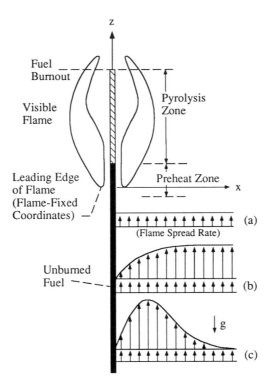

**Figure 8.7**   Detailed schematic of opposed flame spread.

is the net rate of mass generation per unit volume of species $i$, then the species concentration equation can be written as

$$\rho U_a \frac{\partial Y_i}{\partial x} = \rho D \left[ \frac{\partial^2 Y_i}{\partial x^2} + \frac{\partial^2 Y_i}{\partial z^2} \right] + m_i, \qquad i = F, O \qquad (8.4.2)$$

where the subscripts $F$ and $O$ represent the fuel and oxidizer, respectively.

Assuming that the reaction can be represented by a single global reaction equation in which the fuel reacts with oxygen to produce the combustion products and heat, gives:

$$n'_F(\text{fuel}) + n'_O(\text{oxidant}) \rightarrow n''_{p1}(\text{product1}) + n''_{p2}(\text{product2})$$
$$+ Q(\text{heat of combustion}), \qquad (8.4.3)$$

where $n$ represents the stoichiometric coefficients. In addition, reference [91] assumes that the heat released by combustion is in the form of radiation only, i.e., $q_r = \chi q_{chem}$.

Reference [91] introduces a nondimensional function for the temperature and oxygen species defined as

$$\theta(x, z) = c_p \frac{T - T_\infty}{h_L} + \frac{(Y_O - Y_{O\infty})(1 - \chi)Q}{M_O n'_O h_L}, \qquad (8.4.4)$$

where $h_L$ is the latent heat of vaporization for the fuel, $T_\infty$ and $Y_{O\infty}$ are the ambient temperature and the oxygen concentration at infinity, respectively, and $M_O$ is the

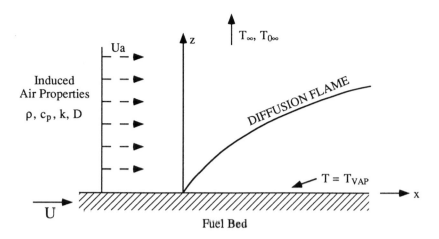

**Figure 8.8** Detailed schematic of opposed flame spread.

molecular weight of the oxidant. Substituting this nondimensional function into the energy and species equations gives:

$$\left(\frac{U_a}{\kappa}\right)\frac{\partial\theta}{\partial x} - \frac{\partial^2\theta}{\partial x^2} - \frac{\partial^2\theta}{\partial z^2} = \frac{q_{chem}(1-\chi)}{kh_L} + m_O Q\frac{(1-\chi)}{M_O n'_O kh_L}. \tag{8.4.5}$$

The right-hand side of equation (8.4.5) is zero for $-\infty < x < \infty$ and $z \geq 0$. This is due to the assumption that heat in the amount $Q(1-\chi)/M_O n'_O$ is released into the temperature field when a unit mass of oxygen is consumed. Another nondimensional function $\zeta$ is defined by

$$\zeta(x,z) = \frac{Y_F Q(1-\chi)}{M_F n'_F h_L} + \frac{(Y_O - Y_{O\infty})(1-\chi)Q}{M_O n'_O h_L}, \tag{8.4.6}$$

where $M_F$ is the molecular weight of the fuel. The fuel and oxygen species conservation equations can then be written in terms of $\zeta$ in the following manner:

$$\left(\frac{U_a}{\kappa}\right)\frac{\partial\zeta}{\partial x} - \frac{\partial^2\zeta}{\partial x^2} - \frac{\partial^2\zeta}{\partial z^2} = \frac{q_F(1-\chi)}{M_F n_F kh_L} + m_O Q\frac{(1-\chi)}{M_O n'_O kh_L}. \tag{8.4.7}$$

Again, for $-\infty < x < \infty$ and $z \geq 0$ the right-hand side of equation (8.4.7) is zero since $M_F n'_F$ grams of fuel are consumed with $M_O n'_O$ grams of oxygen. The infinite reaction rate assumption is used in [91] to solve equations (8.4.5) and (8.4.7) and also to formulate the boundary conditions.

Outside the downstream flame wake it is assumed that

$$\theta, \zeta \to 0 \quad \text{as } z \to \infty, \text{ or } x \to -\infty. \tag{8.4.8}$$

Assuming the fuel bed is very thin, and neglecting radiative effects by setting $\chi = 0$, the energy equation at the surface of the fuel bed takes the form

$$\delta\rho_w c_{pw} U_f \partial T/\partial x = k\partial T/\partial z \quad \text{at } z = 0 \text{ and for } x \leq 0, \tag{8.4.9}$$

where $\rho_w$, $c_{pw}$, and $\delta$ are the fuel bed density, specific heat, and thickness, respectively. The subscript $w$ denotes wall values.

The fuel bed is assumed to vaporize at a constant temperature, $T_{vap}$. Also, the fuel mass transfer at the surface must be related to the heat flux in the following manner:

$$Y_F = (k/h_L)\partial T/\partial z \quad \text{at} \quad z = 0 \text{ and } x \geq 0. \tag{8.4.10}$$

The problem defined above was solved analytically in [91] using the Fourier transform method. The following solution was obtained for the case of fast burning rate; i.e., for $\rho_w c_{pw} U_f \delta / k \gg 1$,

$$\frac{\rho_w c_{pw} U_f \delta}{k} \simeq \sqrt{2} \frac{T_f - T_{vap}}{T_{vap} - T_\infty}, \tag{8.4.11}$$

where

$$T_f = T_{vap} + \frac{Bh_L}{c_p}\left(1 - \frac{1}{K}\right) - (T_{vap} - T_\infty) \tag{8.4.12}$$

is the downstream asymptotic flame temperature. $U_f$, in the solution given in (8.4.11), is the flame spread velocity, and $B$ in equation (8.4.12) is known as the mass transfer driving force, which is defined as

$$B = -\theta(x \geq 0, z = 0) = c_p \frac{T_\infty - T_{vap}}{h_L} + \frac{Y_{O\infty}Q}{M_O n'_O h_L}, \tag{8.4.13}$$

while $K$ is given by $K = [B/ln(1 + B)]^{-1}Q/(M_F n'_F h_L)$.

A physical interpretation of the flame speed $U_f$ as derived in expression (8.4.11) may be obtained when the solution is rearranged in the following manner:

$$\rho_w c_{pw} U_f \delta (T_{vap} - T_\infty) \simeq \sqrt{2}k(T_f - T_{vap}). \tag{8.4.14}$$

The left-hand side of equation (8.4.14) represents the heat flux necessary for raising the fuel bed temperature to its vaporization temperature. The right-hand side represents the gas phase conductive heat transfer rate from the flame forward to the unburned fuel bed.

Figure 8.9 shows the approximate flame shape curve as calculated with the assumption that both the oxygen and fuel concentrations are zero at the flame. The simple formula derived for the flame spread rate is very useful since it can be easily written in terms of the applied external variables. However, this solution is limited due to the severe assumptions imposed on it including a prescribed vaporization temperature $T_{vap}$, the uniform gas velocity $U_a$, and the infinite rate gas phase kinetics.

The flame spread model analyzed in here included a forced gas phase flow, while other studies included buoyancy-driven gas phase flow; in both models the gas phase velocity was large compared to the spread rate. This assumption resulted in a decoupling between the two phases to the extent that the gas phase was treated without the necessary information on the spread rate. In a quiescent, gravity-free environment, this simplification is not possible. In a flame-fixed coordinate system, both the gas and the solid approach the flame at the same speed, which is the desired spread rate in this case.

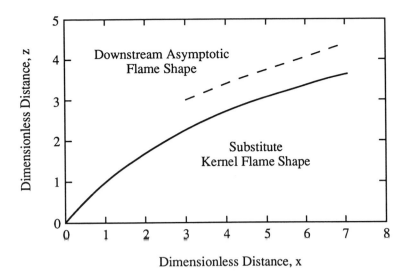

**Figure 8.9**   Calculated flame front position as a function of time, from [91].

## 8.5 *LAMINAR FLAME OVER A VERTICAL FUEL SURFACE*

Under terrestrial conditions, buoyancy-driven convection caused by the combustion process represents an important driving mechanism. This section explores the coupling between buoyancy-driven convection and the dynamics of the combustion process. This coupling is illustrated by one of the simplest examples in combustion available, namely the laminar diffusion flame. The effects of buoyancy can be clearly demonstrated for a flat, vertical surface along which fire is burning, as shown schematically in Figure 8.10. Such a fire may be divided into two regions. In the *pyrolysis region*, the wall material is gasified and partially burned in the gas phase adjacent to the surface. In the *overfire region*, the combustion process has been completed and the thermal plume generated by the fire rises freely along the surface if a noncombusting wall exists downstream. Buoyancy effects on the fire decay downstream due to entrainment of ambient fluid and heat loss to the surface.

The pyrolysis region is the more interesting of the two regions. The flow in this region is dominated by the thermal boundary layer next to the vertical fuel bed. Consider a two-dimensional, steady-state, laminar boundary layer flame adjacent to a pyrolyzing fuel slab, as shown in Figure 8.11. The combustion zone is assumed to be thin so that the reactants do not coexist. In this case they are separated by the flame sheet. It is assumed that boundary layer type equations are valid for the flow next to the fuel surface, which are given by

$$\frac{\partial(\rho u)}{\partial x} + \frac{\partial(\rho w)}{\partial z} = 0, \tag{8.5.1}$$

$$\rho u \frac{\partial w}{\partial x} + \rho w \frac{\partial w}{\partial z} = \frac{\partial}{\partial x}\left(\mu \frac{\partial w}{\partial x}\right) + g(\rho_\infty - \rho), \tag{8.5.2}$$

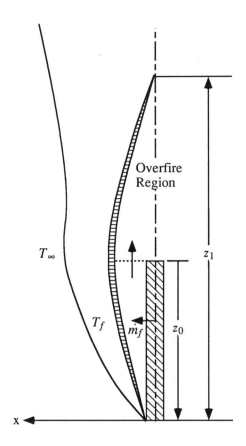

**Figure 8.10** Characteristics of vertical diffusion flame.

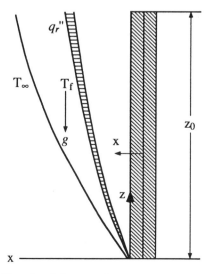

**Figure 8.11** Sketch of the pyrolysis region for a vertical flame.

$$\rho u \frac{\partial h}{\partial x} + \rho w \frac{\partial h}{\partial z} = \frac{\partial}{\partial x}\left(\frac{k}{c_p}\frac{\partial h}{\partial x}\right) + q_{chem} - q_r, \qquad (8.5.3)$$

$$\rho u \frac{\partial Y_i}{\partial x} + \rho w \frac{\partial Y_i}{\partial z} = \frac{\partial}{\partial x}\left(\rho D\frac{\partial Y_i}{\partial x}\right) + m_i, \qquad i = F,O \qquad (8.5.4)$$

where $u$ and $w$ in equations (8.5.1)–(8.5.4) are the $x$- and $z$-components of the fluid velocity vector, respectively. $h = \int c_p dT$ is the total enthalpy, and $q_{chem}$ and $m$ are the volumetric combustion heat and mass generation rates, respectively. $q_r$ is the thermal radiative flux in the $x$-direction. If the fuel surface is inclined at angle $\varphi$ to the vertical direction then $g$ must be replaced by $(g\cos\varphi)$ in equation (8.5.2).

Consistent with the usual boundary layer approximation, only the $z$-component of the momentum equation is included in the conservation equations. In other words, it is assumed that the velocity component $w$ is a function of the horizontal coordinate $x$ only, i.e., $w(x)$. In addition to the conservation equations an equation of state is needed, which for an ideal gas, assuming no change in the averaged molecular weight, is given by

$$\rho T = \rho_\infty T_\infty. \qquad (8.5.5)$$

The chemical reaction may also be approximated in this case by a single reaction equation in the form

$$n'_F(\text{fuel}) + n'_O(\text{oxidant}) \rightarrow n''_p(\text{products}) + Q(\text{heat of combustion}), \qquad (8.5.6)$$

where $n$ are the stoichiometric coefficients. If gaseous methane is taken as the fuel, and assuming nitrogen does not participate in the reaction but is involved in the heat transfer and energy storage processes, the reaction equation may be written as

$$n'_F(CH_4) + n'_O(O_2) + n'_p(H_2O + CO_2) \rightarrow n''_F(CH_4) + n''_O(O_2)$$
$$+ n''_p(H_2O + CO_2) + Q. \qquad (8.5.7)$$

From stoichiometry, the molecular coefficients for the above reaction are given by

$$n'_F = 1, \ n'_O = 2, \ n'_p = 0, \ n''_F = 0, \ n''_O = 0, \ n''_p = n''_c + n''_H = 1+2 = 3, \ (8.5.8)$$

where $n''_c$ and $n''_H$ represent the coefficients for $CO_2$ and $H_2O$, respectively. The species production rate is related to the energy generation by the expressions

$$\frac{m_F}{M_F(n''_F - n'_F)} = \frac{m_O}{M_O(n''_O - n'_O)} = \frac{m_p}{M_p(n''_p - n'_p)} = \frac{q_{chem}}{Q}, \qquad (8.5.9)$$

where $M$ is the corresponding molecular weight. The solution for the conservation equations must be determined in this case subject to the following boundary conditions:

$$w = h = Y_F = 0, \qquad Y_O = Y_{O\infty}, \qquad \text{as} \ \ x \rightarrow \infty \qquad (8.5.10)$$

where $Y_{O\infty}$ is the mass fraction of oxygen in the ambient medium. At the wall, $x = 0$, the application of the no-slip boundary condition on the velocity together with the use of the diffusion flame approximation, leads to the wall conditions

$$w = Y_O = 0, \quad Y_w = \rho u = \frac{k}{h_L c_p}\frac{\partial h}{\partial x}, \quad h = h_w,$$

$$\rho u(Y_{FT} - Y_F) = -\frac{k}{c_p}\frac{\partial Y_F}{\partial x}, \quad \text{at } x = 0 \qquad (8.5.11)$$

where $Y_w$ is the wall mass flux, $h_L$ is the effective latent heat of vaporization for the fuel, $h_w$ is the enthalpy at the wall, and $Y_{FT}$ is the fuel fraction in the gas due to the transfer of fuel into the gas phase. The heat flux to the wall is given by

$$q_w = (k/c_p)\partial h/\partial x, \quad \text{at } x = 0. \qquad (8.5.12)$$

In addition to the boundary conditions, an expression for the radiative heat flux $q_r$ is needed.

The solution for this problem, given in [93], is outlined here for illustrative purposes. Assuming the fuel surface is infinite in length (i.e., $z \to \infty$), then, as is common in boundary layer theory, a similarity solution is sought for this system by first defining a similarity variable $\eta$ with the corresponding nondimensional stream function $f(\eta)$ in the following manner:

$$\eta = \frac{(Gr_z)^{1/4}}{z}\int_0^x \frac{\rho}{\rho_\infty}dx, \qquad (8.5.13)$$

$$\psi = 2\mu_\infty(Gr_z)^{1/4}f(\eta), \qquad (8.5.14)$$

where $\psi(x,z)$ is the stream function defined by

$$\frac{\partial\psi}{\partial x} = \frac{\rho w}{\rho_w}, \quad \frac{\partial\psi}{\partial z} = \frac{\rho u}{\rho_w},$$

and $Gr_z$ is the local *Grashof number* defined by

$$Gr_z = h_L z^3(g\cos\varphi)/4c_p T_\infty v_\infty^2.$$

Substituting the above variables into the $x$-momentum equation (8.5.3), the following nondimensional equation results:

$$\frac{d^3f}{d\eta^3} + 3f\frac{d^2f}{d\eta^2} - 2\left(\frac{df}{d\eta}\right)^2 = -B + \frac{h_w}{h_L} - B\theta, \quad \text{for } \eta \le \eta_f, \qquad (8.5.15a)$$

or

$$\frac{d^3f}{d\eta^3} + 3f\frac{d^2f}{d\eta^2} - 2\left(\frac{df}{d\eta}\right)^2 = -\left(B + \frac{h_w}{h_L}\right)\frac{\theta}{\theta_f} - B\theta, \quad \text{for } \eta > \eta_f, \qquad (8.5.15b)$$

where $\eta_f$ is the flame position. The mass transfer driving potential $B$ is defined in the usual manner by

$$B = \frac{Y_{O\infty}Q}{Y_O n_O' L} - \frac{h_w}{h_L}, \qquad (8.5.16)$$

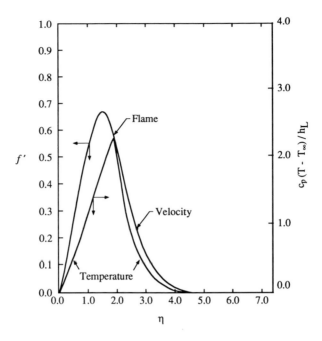

**Figure 8.12** Temperature and velocity distribution in the vicinity of the flame as a function of $\eta$, from [93].

and $\theta$ is the normalized *Schvab-Zeldovich variable* given by

$$\theta = -\frac{1}{B}\left(\frac{h}{h_L} + \frac{Y_O - Y_{O\infty}}{Y_O n'_O h_L}\right), \qquad (8.5.17)$$

where $\theta_f$ is the value of $\theta$ at $\eta_f$.

For a value of the Lewis number $Le = 1.0$, the energy and the species conservation equations become identical in form. The equations take the following form:

$$\frac{d^2\theta}{d\eta^2} + 3Pr\frac{d\theta}{d\eta}f = 0. \qquad (8.5.18)$$

The boundary conditions become in nondimensional form:

$$\theta(0) = 1, \qquad \theta(\infty) = \frac{df}{d\eta}(0) = \frac{df}{d\eta}(\infty) = 0, \qquad f(0) = \frac{B}{3Pr}\frac{d\theta}{d\eta}(0). \qquad (8.5.19)$$

The solution for the coupled ordinary differential equations, (8.5.15) and (8.5.18), was obtained numerically in [93]. Figure 8.12 shows the computed temperature and velocity profiles for burning methanol in air. It shows that the maximum temperature occurs in the flame. In similarity flows, the temperature peak coincides with the position of chemical energy release. The maximum velocity occurs inside the flame. The position of the flame is indicated by $Y_F = Y_O = 0$ and may be determined from the concentration profiles shown in Figure 8.13.

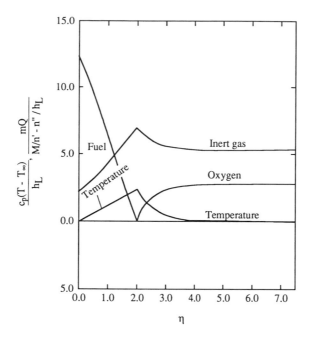

**Figure 8.13**   Concentration distributions for both the fuel and the oxidizer, from [93].

The mass transfer rate is given by the parameter $f(0)$. Figure 8.14 shows the calculated streamlines for the flow in terms of the physical coordinates. A demarcation, termed the *converging line*, separates the flow originating at the surface from that entrained from the ambient fluid. Therefore, convection aids the outward diffusion of fuel vapor between the wall and the converging streamline and opposes it beyond that line. The parameter $B$ was found to be the dominant chemical parameter in determining burning rates. Good agreement between theoretical and experimental results were obtained for lower molecular weight fuels, as shown in Figure 8.15, where the vertical burning rate is shown as a function of the Grashof number in the form $4c_p T_\infty Gr/h_L$. However, Lewis number effects become important for higher molecular weights.

## 8.6 FLAME SPREAD IN LOW GRAVITY

Understanding the mechanisms of fire spread in spacecrafts is an extremely important fire safety issue that has lead to a large body of research in low gravity combustion. The main difference between terrestrial and low-gravity opposed flame propagation lies in the value for the ratio of the oxidizer velocity with respect to the flame $U_a$ to the flame spread rate $U_f$. In flames for which $U_a/U_f \gg 1$, there exists sufficient understanding of the physics of the flame spread process. However, this ratio in low-gravity environment may be of order one under some circumstances. For the case in which $U_a/U_f \approx 1$, the gas and fuel surface radiative processes become important. Radiative effects are important due to the

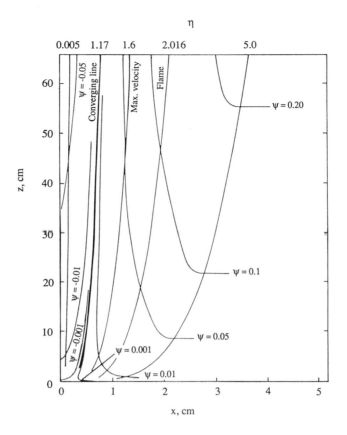

**Figure 8.14**   Concentration distributions for both the fuel and the oxidizer, from [93].

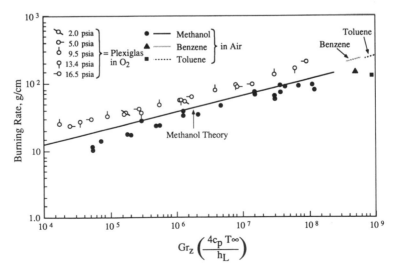

**Figure 8.15**   Comparison between experimental and theoretical burning rates as a function of the Grashof number, from [93].

decrease in the flow velocities as a consequence of the elimination or reduction of buoyancy-driven flows. Reduced convection will also limit the rate of oxygen transport to the flame.

To fully understand the effects of reduced gravity on flame spread, detailed solutions of the combustion equations must be developed. These solutions must then be validated with low-gravity combustion experiments for different regimes of the parameters of the problem. The validation of these solutions can then lead to a better understanding of the physics involved in low-gravity combustion. This understanding will lead to enhanced design practices, and ultimately result in enhanced fire safety on board spacecrafts.

In the analysis on flame spread in Section 8.4, the spread rate was assumed to be a known constant, while in Section 8.5, the effects of buoyancy on the spread rate were investigated using boundary layer theory. In these models the gas flow speeds were either forced or buoyancy induced, and their values were much greater than the flame spread rate. In a quiescent, low-gravity environment, the gas and the solid both approach the flame at comparable speeds, which is the approximate spread rate. The spread rate in this case is an unknown that must be determined as part of the solution for the full combustion problem. Such a solution can be obtained only through a complete numerical solution of the equation.

Among several numerical solution schemes available in the literature, the following method of [90] for the low-gravity flame spread problem is outlined here. This solution correlates reasonably well with data from low-gravity experiments. The equations governing the problem in this case are the conservation equations of mass, momentum, energy, and species, respectively, with the value for gravity set at $g = 0$. These equations in a steady-state, two-dimensional domain can be written in the following nondimensional form:

$$\frac{\partial(\rho u)}{\partial x} + \frac{\partial(\rho w)}{\partial z} = 0, \tag{8.6.1}$$

$$\frac{\partial(\rho u^2)}{\partial x} + \frac{\partial(\rho uw)}{\partial z} = -\frac{\partial p}{\partial x}\left[\frac{\partial}{\partial x}\left(\mu Pr\frac{\partial u}{\partial x}\right) + \frac{\partial}{\partial z}\left(\mu Pr\frac{\partial u}{\partial z}\right)\right] + S_x, \tag{8.6.2}$$

$$\frac{\partial(\rho uw)}{\partial x} + \frac{\partial(\rho w^2)}{\partial z} = -\frac{\partial p}{\partial x} + \left[\frac{\partial}{\partial x}\left(\mu Pr\frac{\partial w}{\partial x}\right) + \frac{\partial}{\partial z}\left(\mu Pr\frac{\partial w}{\partial z}\right)\right] + S_z, \tag{8.6.3}$$

$$\frac{\partial(\rho uT)}{\partial x} + \frac{\partial(\rho wT)}{\partial z} = \left[\frac{\partial}{\partial x}\left(\mu\frac{\partial T}{\partial x}\right) + \frac{\partial}{\partial z}\left(\mu\frac{\partial T}{\partial z}\right)\right] + \Delta H_c\Gamma, \tag{8.6.4}$$

$$\frac{\partial(\rho uY_F)}{\partial x} + \frac{\partial(\rho wY_F)}{\partial z} = \left[\frac{\partial}{\partial x}\left(\mu\frac{\partial Y_F}{\partial x}\right) + \frac{\partial}{\partial z}\left(\mu\frac{\partial Y_F}{\partial z}\right)\right] - \Gamma, \tag{8.6.5}$$

$$\frac{\partial(\rho uY_O)}{\partial x} + \frac{\partial(\rho wY_O)}{\partial z} = \left[\frac{\partial}{\partial x}\left(\mu\frac{\partial Y_O}{\partial x}\right) + \frac{\partial}{\partial z}\left(\mu\frac{\partial Y_O}{\partial z}\right)\right] - n\Gamma, \tag{8.6.6}$$

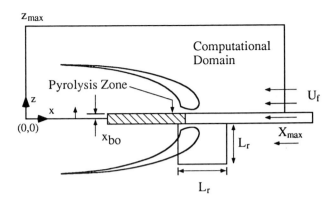

**Figure 8.16**  Finite difference solution domain for the flame problem, from [90].

where

$$S_x = \frac{1}{3}\frac{\partial}{\partial x}\left(\mu Pr\frac{\partial u}{\partial x}\right) + \frac{\partial}{\partial z}\left(\mu Pr\frac{\partial w}{\partial x}\right) - \frac{2}{3}\frac{\partial}{\partial x}\left(\mu Pr\frac{\partial w}{\partial z}\right), \quad \text{or } S_x = 0, \quad (8.6.7a)$$

$$S_z = \frac{1}{3}\frac{\partial}{\partial z}\left(\mu Pr\frac{\partial w}{\partial z}\right) + \frac{\partial}{\partial x}\left(\mu Pr\frac{\partial u}{\partial z}\right) - \frac{2}{3}\frac{\partial}{\partial z}\left(\mu Pr\frac{\partial u}{\partial x}\right), \quad \text{or } S_z = 0, \quad (8.6.7b)$$

and

$$\Gamma = Da_r \rho^2 Y_F Y_O e^{-E_g/T},$$

where $Da_r$ is a reference *Damkohler number* in the gas phase defined by $Da_r = \rho_r^* B^* L / U_a$, $B^*$ is a frequency factor whose value is $B^* = 3.125 \times 10^7$ m³/kg sec, and $E_g$ is the dimensionless activation energy in the gas phase. $n$ in equation (8.6.6) is the stoichiometric ratio of the oxidizer to the fuel, and $\Delta H_c$ in equation (8.6.4) is the nondimensional heat of combustion (see [90]). Note that the Lewis number has been assumed to be $Le = 1$ in the derivation of equations (8.6.1)–(8.6.7).

The field variables, namely the velocity components $u$ and $w$, the temperature $T$, and the pressure $p$, have been made dimensionless in the following manner:

$$u, w = (u^* w^*)/U_f, \quad T = T^*/T_\infty, \quad p = (p^* - p_\infty^*)/\rho^* U_f^2. \quad (8.6.8)$$

All of the remaining variables have been made dimensionless using specified reference values (e.g., ambient conditions) and a length scale given by $L = \kappa/U_f$. Figure 8.16 shows a schematic diagram of the solution domain for this problem from [90].

Reference [90] uses a simple equation of state for an ideal gas given by

$$\rho T = \rho_\infty^*/\rho_r^*. \quad (8.6.9)$$

The boundary conditions applicable for this problem are the following (see [90]):

$$u = -U_f, \quad w = 0, \quad Y_F = Y_{O\infty}, \quad T = 1, \quad \text{at } x = X \qquad (8.6.10)$$

$$p = 0, \quad \frac{\partial u}{\partial x} = \frac{\partial w}{\partial x} = \frac{\partial Y_F}{\partial x} = \frac{\partial Y_O}{\partial x} = \frac{\partial T}{\partial x} = 0, \quad \text{at } x = 0 \qquad (8.6.11)$$

$$u = -U_f, \quad \frac{\partial w}{\partial z} = 0, \quad Y_F = 0, \quad Y_O = Y_{O\infty}, \quad T = 1, \quad p = 0, \quad \text{at } z = z_{\max} \quad (8.6.12)$$

where $U_f$ is the nondimensional fuel velocity. The conditions at $z = 0$ depend on $x$, whether $x < x_{b0}$ or $x > x_{b0}$, where $x_{b0}$ is the horizontal location of the fuel leading edge. These conditions are

$$\frac{\partial u}{\partial x} = \frac{\partial w}{\partial x} = \frac{\partial Y_F}{\partial x} = \frac{\partial Y_O}{\partial x} = \frac{\partial T}{\partial x} = 0 \text{ at } z = 0, \ 0 < x < x_{b0}, \qquad (8.6.13)$$

and

$$u = -U_f, \quad w = w_w, \quad T = T_s \text{ at } z = 0, \quad x_{b0} < x < X, \qquad (8.6.14)$$

where $U_f$, $w_w$, and $T_s$ are determined from the solution for the solid-phase problem, and $Y_F$ and $Y_O$ at $z = 0$ and $x_{b0} < x < X$ are given by the interfacial species mass balances in reference [90].

The solution for this system is very complex and must be determined through numerical means. The procedure is initiated by providing the solid temperature and density distributions $T_s$ and $\rho_s$ as well as the flow and thermodynamic variables. Also, a value for the flame propagation speed $U_f$ is guessed and set equal to the reference velocity. Many iterations on the solution are performed until a converged solution on $U_f$ is obtained. If $U_f \neq 1$, a new value for $U_f$ is chosen and the iteration procedure is repeated. The calculations are continued until a value close to $U_f = 1$ is determined. The solution appropriate for that value of $U_f$ is the correct solution.

The domain of the numerical solution of [90] for the thin fuel bed problem is shown schematically in Figure 8.16. Figure 8.17 shows the nondimensional isotherms in the $(x, z)$-plane resulting from the numerical solution in the gas phase for 30% oxygen at 1.5 atmospheres and finite rate chemistry at zero-$g$. Figure 8.18 shows the flow velocity vectors with respect to the stationary solid fuel for the same problem.

Figure 8.19 shows the calculated spread rate as a function of oxygen concentration for both finite and infinite rate chemistry and assuming zero-$g$. Also shown in Figure 8.19 are results from drop tower experiments together with the simplified analytical solution of reference [91], which was derived in Section 8.4. The figure shows that the finite rate chemistry results come closest to the experimental data. This is because with finite rate chemistry the flame spread is

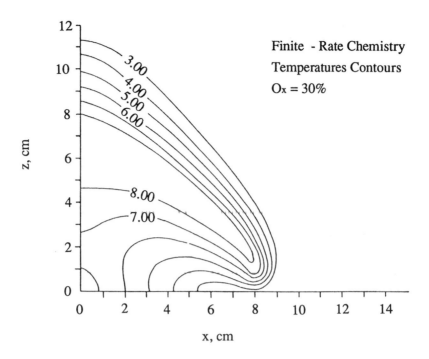

**Figure 8.17** Nondimensional isotherms for the combustion process of reference [90].

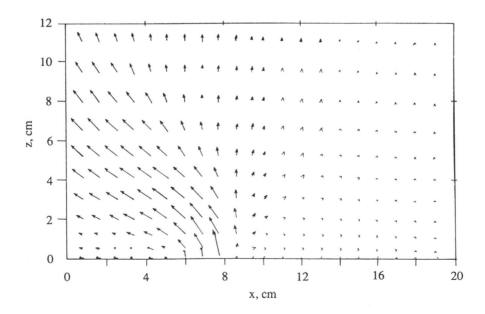

**Figure 8.18** Velocity vectors for the combustion problem of reference [90].

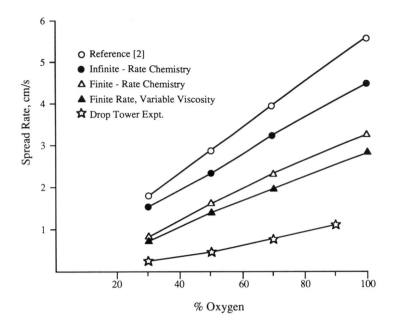

**Figure 8.19** Comparison of flame spread rates from different calculations and experiments, from [90].

depressed further, leading to a slower spread rate. The discrepancy between the calculated results and the drop tower experiments may be due to the neglect of the radiation effects in the numerical simulation.

Low-gravity experiments have been conducted in order to study flame spreading without the added complications due to the heat and mass transports associated with buoyancy-induced flows. Figure 8.20 shows the flame spread rate over a thin paper sample as a function of the ambient oxygen concentration in terrestrial and low-$g$ stagnant environments. It is seen that while the flame spread rate is basically unaffected by buoyancy in highly enriched oxygen environments, as the oxygen concentration is reduced the flame spreads considerably slower in low gravity. Figure 8.21 correlates the flame spread velocity over thin paper samples as a function of the characteristic relative velocity experienced by the flame. It is seen that the spreading response is not monotonic but that the spread rate exhibits a peak value at certain characteristic relative velocity.

The above results demonstrate that suppression of convective flows does not necessarily lead to the most favorable situation to retard flame spreading. In fact, it is seen that flame spreading is significantly promoted in low-gravity environment with a gentle breeze, a situation typical of the working environment in space. It is also shown that there exists an absolute minimum oxygen concentration of about 15%, below which flame spreading is not possible. These are very significant results for fire prevention in low-gravity environment.

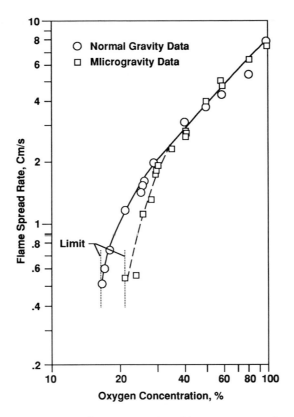

**Figure 8.20** Flame spread rates for terrestrial and low gravity conditions, from [95]. Ambient oxygen concentrations.

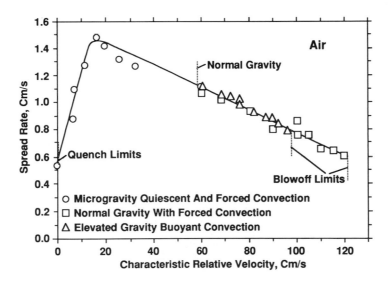

**Figure 8.21** Flame spread rates for terrestrial and low gravity conditions, from [95]. Characteristic relative velocity.

# Chapter 9

# Fluid Management in Space

$F$ luid management in space has been central to the development of space-based technologies since the beginning. The term implies the ability to position and control both liquids and gases in specifically identified locations within various system mechanisms in orbit. In terrestrial applications this does not pose undue problems, because gravity helps to keep order by sorting fluids into layers by density; this is the stratification property. Liquids, normally heavier than vapor, tend to be located below vapor. The absence or diminution of gravity in space environment therefore necessitates the use of other means for these purposes.

The earliest application of fluid management in space arose from the need to extend the life of the satellite cooling systems. Most of them employ liquid cryogens, which depend on phase change heat release to lower temperature.

With the advent of the Apollo program, another application arose for fluid management. In Apollo missions, liquid propellants were required to restart the engines of earth-orbiting rockets for injection into lunar orbit, and they had to be positioned at the engine intake port. Cryogens were also involved in this application and in substantially larger quantities, as the fuels were liquid hydrogen and liquid oxygen. The basic fluid management technologies required for these early applications, including positioning and stability, were specifically developed for each case, i.e., each system was analyzed individually as the need arose.

In the new era of expanded space activities, many space-based operations are required to carry out and support new space-based transportation missions. These missions include the Space Station Freedom, the Orbital Fuel Depot, the Space Transfer Vehicle, the Resupply Tanker, and, in the future, such human exploration missions as lunar basing and piloted Mars expeditions. These endeavors are technically challenging and will require strong sustained technology development efforts. A key aspect of these missions is their dependence on high-energy cryogenic propellants. Long-lived space systems that use subcritical cryogens present low-gravity fluid management challenges, including special storage utilization problems due to their low liquid temperatures. Cryogenic fluid management is clearly a critical development area for these future space missions.

Space-based systems developers are continually being challenged with immediate needs for engineering data bases, validated performance models, and brassboards or prototypes that do not exist at the present time. Only a handful of key features of these requirements have been demonstrated in the appropriate environment and at the appropriate systems level. In-space testing will be required to validate much of the technology needed for future space missions. Such testing must address systems-level behavior and control for realistic operations under low-gravity conditions.

## 9.1 *FLUID POSITIONING*

Liquid positioning and expulsion are the primary problems that faced technology planners in the early days of space flight. The need originally arose during the Apollo mission for liquid propellant positioning at the engine intake while in orbit. If surface tension and gravity are the only forces acting on a liquid volume, then its location can be determined with some of the methods presented in Chapter 3. Normally, a spacecraft in Earth's orbit is subjected to a number of forces and accelerations to render calculations based on those forces alone meaningless. Under these circumstances means had to be devised to actively control the location of the liquid even when different forces of similar magnitude act on the spacecraft.

The two methods for positioning liquids in low gravity are the active and the passive method. The active method requires generating and maintaining accelerations in the direction opposite to the desired position. Such a force can be created by strategically placed thrusters attached to the spacecraft. However, jet thrust is usually reserved for trajectory attitude control and cannot be used for other purposes.

The passive method for positioning liquids in tanks consists of providing strategically located surfaces for the liquid to wet. We remember that under low-gravity conditions, liquids that were wetting tended to wet more. Due to these enhanced effects, extensive use of baffles has become an important method for locating liquids at the desired locations.

Another passive method for liquid positioning is the use of parallel meniscus systems. The preferred configurations for each component of this system may be determined by using the principle of minimum capillary area or, alternatively, a force balance consideration. This method is illustrated, with the following example, from reference [107].

Consider two capillary surfaces formed by two concentric tubes, as shown in Figure 9.1. Such a capillary system can serve, for instance, as a propellant positioning device in large booster fuel tanks. If the radii of both tubes are properly chosen, the liquid can be made to flow into the inner tube as the vehicle enters low gravity. Reference [107] presents the following simple analysis for determining the preferred configuration of the system.

Assuming perfect wetting (i.e., zero contact angle), and neglecting the thickness of the inner tube, the pressure in the liquid in each tube—$p_a$ and $p_b$—

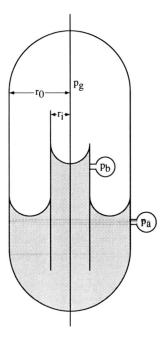

**Figure 9.1**

respectively, may be calculated from Laplace's condition:

$$p_g - p_b = \frac{2\sigma}{r_i} \, , \tag{9.1.1}$$

$$p_g - p_a = \frac{2\sigma}{r_o - r_i} \, , \tag{9.1.2}$$

where $r_i$ and $r_o$ are the inner tube and outer tube radii, respectively, and $p_g$ is the gas phase pressure. Normally, fluids subjected to a pressure gradient can be made to flow in such a way that the motion is from the higher pressure to the lower. Consequently, the liquid will flow into the inner tube if $p_b < p_a$. Hence, the liquid will move up the annulus whenever

$$\frac{1}{r_i} < \frac{1}{r_o - r_i}. \tag{9.1.3}$$

The liquid will remain stationary in any position when

$$\frac{1}{r_i} = \frac{1}{r_o - r_i}. \tag{9.1.4}$$

Finally, the liquid will move up the central tube whenever

$$\frac{1}{r_i} > \frac{1}{r_o - r_i}. \tag{9.1.5}$$

The pressure difference across liquid/gas interfaces of different curvatures can be employed to pump liquid from one tank to another. The pumping rates

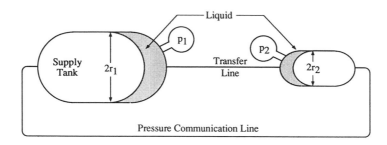

**Figure 9.2**

under these conditions are very small due to the minute pressure differences available with such a method. An example from [107] is shown in Figure 9.2. Laplace's condition applied to this system indicates that the pressure differential across the gas/liquid interface in the larger tank is less than that across the interface in the smaller tank. This is because the curvature of the interface in the larger tank is smaller. In this case, $p_1 > p_2$, and the liquid will flow to the smaller tank. Assuming laminar flow and zero losses due to plumbing, the flow rate can be calculated easily for a circular connecting pipe of radius $r_0$. For example, if $r_1 = 4.5$ m, $r_2 = 1.5$ m, and $r_0 = 0.15$ m, then a flow rate on the order of 0.005 kg/hr can be sustained for liquid hydrogen. This flow rate could be increased considerably, of course, by baffling the smaller tank and thus creating more highly curved free surfaces and consequently larger driving pressures.

Applications of capillary pumping of this type are not readily suited for larger tank systems for the following reasons. First, the time required to transfer large quantities of liquids from one tank to another, or from one end of a tank to the other, can be prohibitively large for operational convenience. Second, the pumping action will occur only as long as no adverse forces act on the tank, and forces of any size at all would probably stop the pumping action. However, capillary pumping on a small scale can be used as means for providing small quantities of needed liquid for certain feed devices.

The balance of forces acting on a liquid mass can also be used to achieve liquid control in low gravity. As an example of such a configuration is the central standpipe for liquid positioning that has been tested in early spacecraft designs. The actual design of such a device is discussed in [107]. The device consisted of an 8.9-cm ID plastic sphere with a 2.79-cm diameter standpipe fixed to one end, as shown in Figure 9.3. The lower edge of the standpipe was penetrated by 4 semicircular holes each 0.79 cm in radius to provide communication between the central standpipe and the tank. The tank was designed in such a way that the gravity body force acted in the direction indicated by the arrow. The resultant liquid level, when only Earth's gravity is acting on the sphere, is shown by the dotted line. Using the Bond number as the principal parameter for modeling purposes, with known spacecraft acceleration data, shows that the maximum transverse $g$-loading with such a system can retain control of the liquid, given by

$$g_{\max} \approx 67 \frac{\sigma}{\rho D} \tag{9.1.6}$$

where $D$ is the baffle diameter.

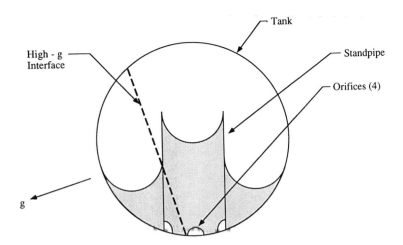

**Figure 9.3**

An estimate on the time required for a baffle of this type to gain control of the liquid can be made by considering the different time scales involved in such a design. When the system is in a gravity-dominated environment, the effective dynamic time scale according to dimensional analysis is given by

$$t_g \propto \sqrt{\frac{L}{g}}\,.$$

(9.1.7)

On the other hand, if the system is in a surface tension-dominated environment, then the effective time scale is

$$t_{st} \propto \sqrt{\frac{\rho L^3}{\sigma}}\,,$$

(9.1.8)

where $L$ is the appropriate length scale for the apparatus. For the tank in Figure 9.3, the surface tension time scale $t_{st}$ is the more relevant one during space flight. This time scale can be calculated to yield the required control time

$$t \propto \sqrt{\frac{\rho L^3}{\sigma}} \approx 1.5\sqrt{\frac{\rho D^3}{\sigma}}\,.$$

(9.1.9)

When the body forces are directed parallel to the axis of the tank and toward the end where the baffle is located, the height of the capillary rise can be estimated using the theory of equilibrium surfaces.

An important use of equilibrium surface stability analysis is the enhancement of meniscus stability with screens. A typical application is the placement of a plane screen mesh or perforated sheet metal baffle across a tank as shown in Figure 9.4. When the liquid in the tank is subjected to body forces acting to propel the liquid in the $z$-direction, the liquid ahead of the baffle will do so if the meniscus $A$ is unstable. The liquid behind the mesh, however, will keep its position at the capillary surfaces due to the effects of the screen. If the liquid in the tank is subjected to a transverse body force, the liquid will be kept from

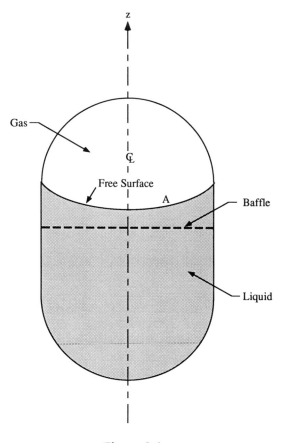

**Figure 9.4**

running out through the screen by the capillary pressure across the liquid-gas interfaces in the screen mesh. This method of containment and positioning is being used extensively in spacecraft designs.

The use of perforated material such as screening has become prevalent in the design of devices for liquid positioning and is discussed in the next section. Such material affords useful weight saving, since the liquid wets it almost as readily as a solid sheet metal.

## *9.2 LIQUID ACQUISITION DEVICES: LADs*

Some of the passive techniques for repositioning liquids in low-gravity tankage systems were discussed in the previous section. These techniques are not always possible, and when they are, they are not always dependable. For some time, industry has relied exclusively on a specifically designed instrument that uses surface tension forces for liquid positioning. Such an instrument is commonly known as a *propellant management device* (PMD) or alternatively a *liquid acquisition device* (LAD). These devices are used to ensure that gas-free liquid or liquid-free gas is available when and where it is needed.

The primary objective of a LAD is to keep the tank outlet covered with liquid whenever liquid outflow is desired. Secondary objectives include control of the center of gravity of the liquid, minimization of liquid sloshing, and the ability to vent liquid-free gas. There are many types of liquid management devices that make use of surface tension to maintain the liquid in a tank at a specific location. Reference [104] classifies LADs into one of three main categories: *partial communication*, *total control*, and *total communication*. These devices can also be classified in terms of the range of acceleration levels, or effective gravity, for which they can maintain control over the liquid position.

Partial communication LADs hold only a fraction of the liquid over a specific location and leave the remaining liquid free. They are very useful when a spacecraft must maneuver considerably. If thrusting is always sufficient to settle the liquid, the LAD can be refilled during the settling operation and the size of the LAD can be made just large enough to provide the gas-free liquid needed. These devices can be either refillable or nonrefillable. Both types are usually constructed with walls made of fine mesh screens or a similar porous material. Capillary forces in the pores of the screen effectively prevent external gas from entering the LAD during spacecraft maneuvers.

Figure 9.5 shows schematics of an idealized refillable and a nonrefillable LAD. In the latter device each successively lower volume of liquid is emptied in turn where the exposed screen prevents gas entry from the empty volume just above. This type of LAD is sometimes called a *sponge*. Due to their small size, partial communication LADs can be designed to be stable against large acceleration disturbances.

Total control LADs hold all the liquid over the desired surface. In effect, this type of device is just a nonrefillable LAD that occupies the whole of the tank volume. These devices are mainly used in spacecrafts for which slosh control is needed.

Total communication LADs are designed to establish a flow path from the bulk liquid to the tank outlet at all times. Due to the fact that the liquid tends to remain bound to a wall, generic forms of total communication LAD include a liner and a series of galleries, both made of fine-mesh screens and located near the tank wall. Figure 9.6 shows a schematic of a gallery type LAD. The flow channels, or galleries, in that device are all connected to a manifold at the tank outlet. As long as one of the channels remains in contact with the liquid bulk, tank pressurization will drive the bulk liquid into the channel and then along the channel into the manifold.

Because of their large size, total communication LADs are stable against moderate forces but will admit gas when the forces are large. Another type of total communication LAD uses a set of central vanes connected to a standpipe to position the liquid. If any vane is in contact with the liquid, liquid is said to be *wicked* to the standpipe. By tapering the open area of the vanes properly, a vaned LAD can also position gas over the vent as shown in Figure 9.7. Because the interfaces are large, the liquid position in vaned LADs is relatively unstable and as such they can only be used when the disturbing forces are small, such as in deep space probes and synchronous spacecraft.

There are a number of design considerations for the screens used in these devices that must be addressed. Fine-mesh metal screens are used extensively in LADs, but not exclusively. The most desirable characteristic of a screen is that

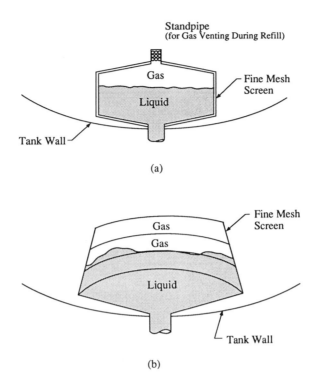

(a)

(b)

**Figure 9.5** Schematic of partial communication LADs, from [104]: (a) refillable and (b) nonrefillable.

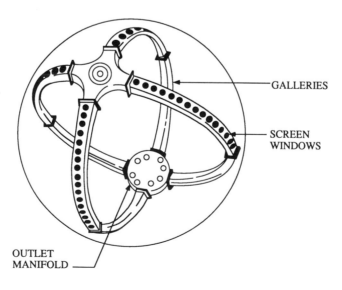

**Figure 9.6** Sketch of a total commthe unication LAD, from [104].

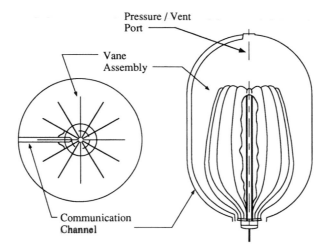

**Figure 9.7** Sketch of a vaned type total communication LAD, from [104].

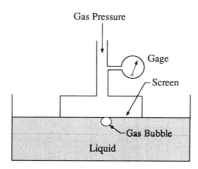

**Figure 9.8** Definition sketch of bubble point pressure, from [104].

when it is wet it can withstand a pressure differential from the gas to the liquid side. This maximum possible pressure differential is a function of the screen weave and the liquid properties. This pressure is commonly called the *bubble point pressure*, $p_{bp}$. Such designation is due to the type of test used to determine this pressure (see [104]). The test is performed by increasing the pressure differential across a screen in contact with liquid on one side and gas on the other until a bubble is forced through the screen, as shown in Figure 9.8. If the screen pores are exactly circular with, say, a diameter $D_0$, then the bubble point pressure can be evaluated from

$$p_{bp} = \frac{4\sigma}{D_0} . \qquad (9.2.1)$$

Expression (9.2.1) is derived by assuming a small contact angle $\alpha$ such that $\cos \alpha \approx 1$.

In general, the screen ports are not circular and hence the measured value of $p_{bp}$ is interpreted as defining an effective circular pore diameter $D_{bp}$ for each

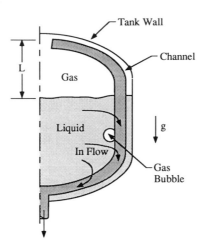

**Figure 9.9**

type of screen. Accordingly, $D_{bp}$ is defined in the following manner:

$$D_{bp} = \frac{4\sigma}{p_{bp}}.$$   (9.2.2)

The effective diameter for most screens is related to the pore size and is not necessarily identical to it. Screens are available commercially with $D_{bp}$ as small as 10 $\mu$m.

Knowledge of the bubble point pressure is sufficient to determine the stability of a LAD. Following reference [104], consider the total communication channel sketched in Figure 9.9. The pressure in the channel liquid, at the intersection of the liquid bulk interface with the channel wall, is equal to the gas pressure. This is due to the fact that there is no capillary force at any point where the screen is immersed in liquid on both sides. Thus the liquid at the top of the channel must support a negative hydrostatic pressure relative to the gas pressure in the approximate amount of $\rho g L$. To prevent gas from being withdrawn into the top of the channel to relieve the negative pressure, the bubble point must have the value

$$p_{bp} > \rho g L,$$   (9.2.3)

where $g$ is the component of the linear acceleration aligned with the channel axis. In other words, the bubble point diameter of the screen must be chosen such that

$$D_{bp} < \frac{4\sigma}{\rho g L}.$$   (9.2.4)

Many space-based systems are launched with tanks that are not completely full. According to expression (9.2.4), the value of $g$ at launch imposes a critical design requirement on the LAD, which may be very large.

As an example, the active liquid volume in the simplified partial communication LAD shown in Figure 9.10 will remain free of gas under lateral accelerations if

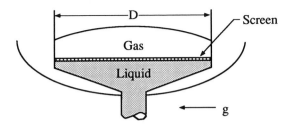

**Figure 9.10**

the bubble point pressure of the bottom-most screen exposed to the gas satisfies the requirement

$$p_{bp} > \frac{4\sigma}{\rho g D} \, . \tag{9.2.5}$$

Note that if gas enters on the right side of the screen where the pressure is, most negative liquid must leave on the left. According to the remarks made earlier concerning the bubble point pressure, the pressure required to force liquid out is negligible. In this case the stability of the LAD is a function only of the pressure differential required to allow gas in it.

   Another characteristic parameter of fine-mesh screens that must be considered in designing a LAD channel is the pressure difference $\Delta p$, required to establish a flow across a wet screen from the liquid bulk to the interior of the device. According to reference [104], extensive testing has shown that the following empirical correlation for the pressure difference holds:

$$\Delta p = C_1 \left( \frac{Ba^2}{\epsilon^2} \right) \mu v_{sf} + C_2 \left( \frac{B}{D_0 \epsilon^2} \right) \rho v_{sf}^2 \tag{9.2.6}$$

where $v_{sf}$ is the superficial liquid velocity defined as the volume flow rate per unit surface area of the screen, $\mu$ is the liquid viscosity, $a$ is the ratio of surface area to screen volume, $B$ is the screen thickness, $D_0$ is the screen pore diameter, and $\epsilon$ is the volume void fraction of screen. The constants $C_1$ and $C_2$ have the universal values of 8.61 and 0.52, respectively. The values of $C_1$ and $C_2$ have been shown through experiments to depend on the specific screen design.

   The pressure drop in expression (9.2.6) is the sum of two parts where one part is due to viscous effects and the other to inertia effects. The actual value for the superficial velocity can be calculated from the total volume flow rate divided by the screen area in contact with the bulk liquid.

   The screen area must be large enough to yield a flow pressure drop that will not sweep any gas bubbles from the bulk liquid into the LAD. For the example of Figure 9.9 the following relation must be satisfied among the various pressures:

$$p_{bp} > \rho g L + \Delta p. \tag{9.2.7}$$

Expression (9.2.7) can be solved for the screen area, from which the screen width is calculated for a given channel length. If the flow rate is constant (i.e., independent of the tank fill level), then $(\Delta p + \rho g L)$ will eventually exceed $p_{bp}$. This is due to the fact that the screen flow area in contact with bulk liquid becomes small when the

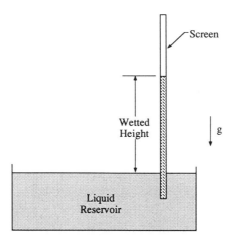

**Figure 9.11**   Suction pressure measurements of screens.

tank is nearly empty. When this occurs, gas-free liquid is no longer available. Thus the combination of bubble point pressure and flow pressure drop may determine the expulsion efficiency of the tank for some systems. Both of these pressures are functions of the screen weave.

The third design consideration is the ability of the screen to wick liquid along its length. The wicking ability of any screen is a function of the screen weave. A plain square weave screen will not wick. More complex weaves such as *dutch twill* will permit capillary surfaces to be established across the screen thickness and will wick liquids from wet areas to dry areas. The effective suction pressure for producing the wicking action can be determined by measuring the height to which liquid rises along a vertical sample of the screen immersed in a liquid reservoir. A schematic of such a test is shown in Figure 9.11. The maximum capillary pressure, $p_w$, available to cause wicking, is the hydrostatic head corresponding to the height of the liquid in the test. This may be correlated by

$$p_w = \frac{4\lambda\sigma}{D_{bp}} \, , \tag{9.2.8}$$

where $\lambda$ is a constant that depends on the screen weave. For a square-weave screen, $\lambda = 0$, and for a dutch twill screen $\lambda$ ranges from 0.2 to over 1.0 depending on the density of the weave.

The liquid wicking velocity $U_w$, from a wet to a dry area of the screen is determined by the wicking pressure. Assuming $g = 0$, or for a horizontal screen, the wicking velocity can be correlated as

$$U_w = \frac{p_w B^2}{C_w \mu L} \, , \tag{9.2.9}$$

where $L$ is the length along the screen between the wicking front and the reservoir. $C_w$ is a nondimensional resistance factor that is a function of the screen weave. Typical values for $C_w$ range from $\sim$300 to several thousand.

When the liquid evaporates from a wet screen it must be replaced by wicking from the bulk liquid, $L$ in this case is interpreted as the total length of the screen. Equation (9.2.9) implies that there is a maximum possible value of evaporation rate per unit screen surface area corresponding to the maximum possible $p_w$ that can be compensated by wicking. For nonvolatile liquids wicking is not usually a quantitative consideration, so long as screens are chosen that ensure some wicking occurs to keep wet the exposed parts of the LAD. When the liquid evaporates easily, such as for cryogens, wicking characteristics become more important.

Most liquids being proposed for use in space systems at the present are cryogens. The LADs discussed above must be modified to deal with the fast evaporation rates associated with cryogenic transfer. When a cryogen is evaporating from a screen and is being replenished by wicking, the bubble point is lowered from its normal value. This phenomenon is understood physically as a thinning or withdrawal of the liquid to the interior of the screen and a subsequent increase of the curvature of the capillary surfaces from their minimum value. Hence, the bubble point pressure available to prevent gas ingestion decreases as the wicking pressure, or evaporation rate, increases. To a first approximation, the available bubble point pressure $p_{bp}$ and the available wicking pressure $p_w$ can be related linearly in the following manner:

$$p_{bp} = \frac{4\sigma}{D_{bp}} - C\left(\frac{p_w}{\lambda}\right), \tag{9.2.10}$$

where $C$ is a parameter that must be determined experimentally as a function of screen weave (a typical value is $\approx 0.5$).

When the screen is not wicking, $p_w = 0$, and equation (9.2.10) shows that the available $p_{bp}$ has its maximum value of $4\sigma/D_{bp}$. This is the pressure measured in a conventional bubble point test. Similarly, when the available $p_w$ is at its maximum value of $4\lambda\sigma/D_{bp}$ corresponding to the maximum possible wicking flow, the bubble point pressure is reduced to 50% of its maximum value for $C = 0.50$. Due to the decrease in the bubble point when wicking occurs, the ability of the LAD to keep out vapor can be substantially degraded when the tank liquid is a cryogen.

## 9.3 LIQUID DRAINING

Liquid expulsion from a container comprises another important task necessary for fluid management in low gravity. Liquid expulsion processes are necessary for feeding rocket engines and also for transferring liquids from one tank to another. The limited data available at the present time suggests that draining rates at low gravity are considerably less than those at one-$g$ for comparable systems. Visual studies of liquid draining at zero-$g$ clearly illustrate the relevant problems as demonstrated by reference [107].

In draining to a center hole a dip in the interface is always formed which is caused by the lower central pressure due to the inward acceleration of the liquid. Body forces in terrestrial environment act to smooth out the dip, but under low-$g$ conditions only the relatively weak surface tension forces act to prevent the liquid from being removed in a central core. A schematic illustration of draining

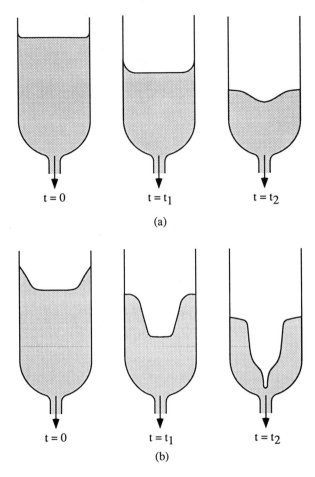

**Figure 9.12**   The free surface configuration during draining at $g/g_0 = 1$ in (a), and in low-$g$ in (b).

at identical rates in low- and high-$g$ conditions is shown in Figure 9.12. It is evident that a considerable amount of liquid could remain in the tank when draining at low-$g$, if adequate precautions are not taken.

If the draining rate is sufficiently low, the equilibrium surface could essentially retain its initial shape and a maximum amount of liquid will be withdrawn. An estimate of this maximum draining rate is very important. If the fluid is assumed to be inviscid and irrotational then it is possible to formulate a mathematical model for this problem. However, the large changes in the shape of the equilibrium surface with time means that nonlinear effects must be considered. A solution to such a problem normally requires an extensive computational effort.

Reference [107] shows that it is possible to estimate the maximum draining rate, under which the equilibrium surface will retain its initial shape, by using correlations based on the reorientation time. For the draining rate to be sufficiently slow such that the free surface has time to adjust, the reorientation period must be shorter than the time required for the surface to travel its length. The time required for an equilibrium surface to form its shape in zero-$g$, following the removal of a strong body force, is given by the surface tension time scale in (9.1.8). This time scale takes the value

$$t = CD^{3/2}\sqrt{\rho/\sigma},\tag{9.3.1}$$

where $C$ is a constant depending on the geometry and the fill level of the tank, and $D$ is the tank diameter. Expression (9.3.1) is found to agree well with experiments when using wetting liquids in which $C$ varies from 0.042 to 0.15. The time needed for the free surface to travel its depth can be calculated using one-dimensional mass conservation considerations given by

$$t_f = \frac{D/2}{U(d/D)^2},\tag{9.3.2}$$

where $U$ in (9.3.2) is the liquid velocity in the drain and $d$ is the drain diameter. Using (9.3.2) for the response time in the definition for the Weber number $We$, reference [107] arrives at the following estimate:

$$We_{\mathrm{max}} \approx \left(\frac{\rho U^2 D}{\sigma}\right)_{max} \approx 10(D/d)^4.\tag{9.3.3}$$

Thus, as long as $We < We_{\mathrm{max}}$, the equilibrium surface should retain its general shape during draining, resulting in optimum draining conditions. Reference [107] points out, however, that this criteria is only an approximation and there is no assurance that it will always hold.

When the tank Weber number exceeds the critical value estimated by expression (9.3.3), the diameter of the equilibrium surface can be estimated with the assumption that the Weber number based on the free surface diameter is equal to the critical value. This is equivalent to assuming that the equilibrium surface diameter gets as large as possible in the period of time it has to adjust to changes in its position. Expression (9.3.3) can be employed for such an estimate to yield

$$D_{\mathrm{max}}U^2\rho/\sigma \approx 10D^4/d^4,\tag{9.3.4}$$

from which the ratio $D_{\mathrm{max}}/D$ can be estimated.

Using estimate (9.3.4), the retained liquid volume can be calculated as

$$V_{liq} = \pi(D^2 - D_{\mathrm{max}}^2)H/4 + \pi D_{\mathrm{max}}^3/24,\tag{9.3.5}$$

where $H$ is the initial mean depth of the liquid. This volume is estimated by assuming a liquid annulus on the wall, with liquid in the volume beneath the spherical free surface of diameter $D_{\mathrm{max}}$ when its nose just reaches the drain. The liquid fraction $f_l$, remaining in the tank after draining has ended, is therefore estimated to be

$$f_l = \frac{V_0}{V_f} = [1 - (D_{\mathrm{max}}/D)^2] + D_{\mathrm{max}}/(8H).\tag{9.3.6}$$

At very low discharge rates, in which the meniscus spans the tank, the volume of liquid beneath the meniscus will be trapped when it reaches the drain. For the simple model used above this fraction is given by $D/(8H)$. More accurate calculations can be made using the equilibrium free surface theory of Chapter 3.

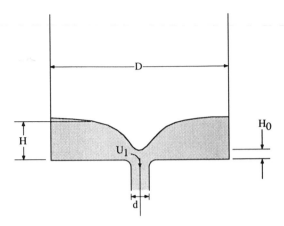

**Figure 9.13**   Liquid draining at high Bond number.

Reference [107] cites an experimental effort for validating expression (9.3.6) with $H/D = 1.50$. For such a configuration the Weber number is calculated to be $We = 2.5 \times 10^5$, with $d/D = 0.10$; expression (9.3.6) gives $f_l \approx 0.88$. In a corresponding low-gravity experiment $f_l$ was measured to be $f_l \approx 0.76$.

Reference [107] shows that when the Bond number is sufficiently large, an entirely different type of retention mechanism becomes important. Suppose a pump is sucking liquid from a tank, as shown in Figure 9.13. The pumping capacity must be less than the critical flow rate for the specific drain. The flow rate can be approximately determined by requiring the flow over the lip of the drain to be critical. This condition is reached when the Froude number is one. The maximum flow rate can therefore be estimated by the following relationship:

$$Q_{\max} \approx \rho H_0 \pi d \sqrt{gH_0} \leq Q_{pump}. \tag{9.3.7}$$

When the liquid depth at the lip, $H_0$, falls below the level implied by the above estimate, the required fluid cannot be delivered and thus the draining process is terminated. This condition is called *chocking*. In low-gravity environment, the critical depth can be large making the chocking criteria easily attainable.

Assuming that the flow rate, for which the liquid level drops, is slow compared to the liquid velocity over the drain, a quasi-steady estimate of liquid height, $H$, can be made using one-dimensional theory in the following manner:

$$\sqrt{2g(H - H_0)} = U_1 = \sqrt{gH_0}. \tag{9.3.8}$$

An estimate for the trapped liquid can then be made using equation (9.3.8) in the following way:

$$\pi(D^2 - d^2)H/4 = 3\pi(D^2 - d^2)H_0/8 = (3/8)\pi(D^2 - d^2)\left(\frac{Q_{pump}}{\pi \rho d \sqrt{g}}\right)^{2/3}. \tag{9.3.9}$$

The Bond number involved in the draining of a large booster tank under low-$g$ conditions may be well over unity, in which case model experiments at one-$g$

in smaller tanks can give insight into the draining problem. Reference [107] describes an experimental study of draining of cylindrical tanks with flat bottoms in the Bond number range 100–1000. Such a range for the Bond number may be considered high for the case in which the surface tension effects are expected to be minor. This was the case for these experiments in which it was found that the point at which gas ingestion occurred depended only on the Froude number. The height $H$ from the drain at which injection occurred was correlated reasonably well using the expression (see [107]):

$$H = 0.43D \ \tanh\left[1.3\left(\frac{U^2}{gd}\right)^{0.29}\right], \qquad (9.3.10)$$

for a Froude number range of $0.05 < Fr < 100$ and a $D/d$ range of $3 < D/d < 20$. For this experiment injection at high rates of draining occurred when the height at the wall was 0.43 D.

In cases for which qualitative estimates are not sufficient, complex computational efforts can provide better quantitative values. Recent detailed calculation of the meniscus shape during draining of tanks in low gravity confirmed in general the qualitative calculations of [107] outlined above. Reference [106] gives details of numerical calculations in which the draining problem is compared under different levels of gravity. A simple tank geometry was considered in these calculations with the initial liquid free surface appropriate for equilibrium conditions. For a specific draining rate these calculations predicted the final free surface shape as well as the residual liquid remaining at the termination of draining. Figure 9.14 shows the final liquid configurations resulting from draining liquid hydrogen for values of $g/g_0$ of 1.0, $10^{-2}$, and $10^{-4}$. The initial liquid fill level for both cases is 70% with a steady draining rate of 120 cm$^3$/sec. The calculations of [106] shows that there is indeed a higher liquid residual for the low-gravity case than that for the terrestrial case.

Figure 9.15 shows a summary of the percent liquid residual as a function of liquid draining rate for three values of $g/g_0$. It is very clear that the lower the value of gravity, the higher the liquid residual after draining. Also, Figure 9.15 shows that higher draining rates do not affect the residual considerably after a specific value of draining is achieved.

Another aspect of liquid draining from tanks that must be carefully assessed according to [107] is the formation of vortex flows. Since axial body forces do not enter the equations for the angular momentum, the formation of vortex flow can be expected to develop at the same rate under low-$g$ conditions as under high-$g$ conditions. When the vortex does occur, its effects can be much more pronounced under low-$g$ conditions, in which case the vortex funnel penetrates deeper into the liquid than at terrestrial conditions.

Due to the fact that draining of tanks under low-gravity conditions results in high trapped residuals, the use of mechanical means for flattening the meniscus inside the tank is highly desirable. Reference [107] cites experiments using baffles at the outlet and at the gas inlet indicating that such devices can be used to allow for a substantial increase in the liquid drain rate from tanks with little increase in the trapped liquid fraction. The use of a plane baffle over the tank drain at high flow rates results in some reduction in the trapped quantity at fixed flow rate. Reference [107] points out that this may be due to the elimination of the

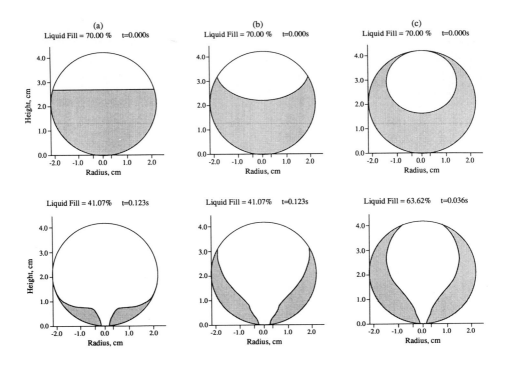

**Figure 9.14** Tank draining under different gravity levels, from [106]: (a) $g/g_0 = 1.0$, (b) $g/g_0 = 10^{-2}$, and (c) $g/g_0 = 10^{-4}$.

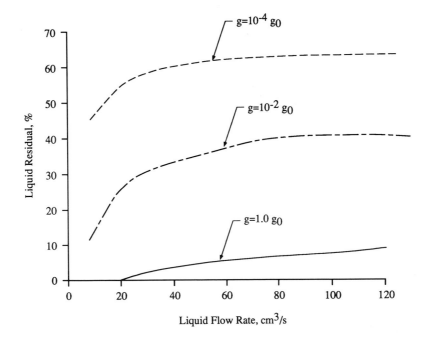

**Figure 9.15** Liquid residual as a function liquid draining rate, from [106].

central vortex. The use of baffles at the gas inlet in some experiments reduces the trapped residuals for a given flow.

When more positive control is necessary over the liquid during drainage, some sort of a flexible containment device such as a bladder or bellows must be used. Bladder use for such purposes is recommended but the bladder material must be chosen carefully. If the liquid being expelled is a cryogen, then a metal bladder or metal bellows must be used.

## *9.4* VAPOR VENTING

Vapor venting is another area of fluid management in low-gravity environment that is the subject of extensive study. The problem in here is the ability to vent vapor without simultaneously expelling liquid. This problem is especially critical in fluid management of cryogenic liquids. The storage and use of cryogenic liquids requires the need to periodically relieve pressure buildup in storage tanks due to liquid evaporation. For storable propellants that are normally used with pressure-fed devices it may be necessary to effect a pressure reduction in the storage tanks prior to use. The venting of pressurization gas and propellant vapors presents no difficulty whatsoever in a terrestrial environment in which case the location of the liquid is well defined. Under low-gravity operating conditions, however, the liquid may be located in a variety of places inside the tank depending on the level and direction of the body forces acting on the liquids and to some extent on the flight history. Some effort is therefore necessary to assure that when the tank vent is open, only gas is expelled and not liquid.

If the flight trajectory is well known, it may be possible to use passive means to hold the liquid away from the vent. Such a passive method may be positioning the vent at the opposite end of the tank along which there is an acceleration force. When this is not possible, some active means for removing the liquid from the vent must be used. Consider, for instance, the storage tank of Figure 9.16. The vent at the top is temporarily covered by liquid that had settled there under the action of forces encountered by the vehicle during a prior time. Application of a thrust in the acceleration direction will cause the liquid in the tank to move to the opposite end of the tank. For proper design considerations knowledge of both the minimum force required to bring about this change and the time required for its implementation is necessary.

The acceleration produced by the thrust must be sufficient to destabilize the equilibrium liquid surface whose configuration can be estimated by the methods of Chapter 3. Proper design would normally require the use of several times the theoretical minimum acceleration that may constitute a small thrust for a large fluid system. Care must be exercised, however, not to use too much thrust in order not to structurally damage the tank by fluid impact. It is also important to develop a good estimate on the rundown time and the velocity of the liquid mass on impact. Unfortunately, there are no empirical data for making such estimates and the designer must rely on educated guesses and model experiments.

Reference [107] shows that an estimate of the minimum time $t_{min}$ and the maximum impact velocity $U_{max}$ can be made, using liquid reorientation methods, by assuming a free-falling liquid. Under such an assumption these scales may be

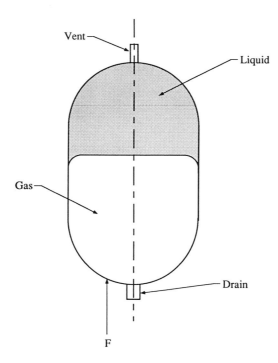

**Figure 9.16**

estimated in the following manner:

$$t_{min} = \sqrt{2L/g}, \qquad U_{max} = \sqrt{Lg}. \tag{9.4.1}$$

With drag, the actual impact velocity may be considerably less. In studies of the motion of bubbles in tubes, for modeling this phenomenon, velocities on the order of one third of that given above are found. Thus, a conservative design procedure would be to use reorientation acceleration somewhat in excess of the critical value determined from stability analysis, and design the structure to take the impact velocity given above. In addition, it is prudent to allow 5 to 10 times the reorientation time calculated above for the liquid to clear the vent.

Reference [107] cites an experimental study conducted on liquid behavior during reorientation. In that experiment it was found that, as the acceleration is applied, the liquid begins to flow down the wall in a relatively thin liquid sheet. If the Bond number is on the order of 10 or less, a large central bubble forms along the axis of the cylinder. If the Bond number is greater than 150, the formation of the wall sheet is accompanied by the delayed formation of a central plateau of liquid that grows down the axis of the cylinder, narrowing as it falls. The time trajectory of the wall sheets was approximated in [105]:

$$H/r_0 = 0.69\tau + 0.41\tau^2 \text{ for } H/r_0 < 1.2, \;\; 3.5 < Bo < 10, \tag{9.4.2}$$

$$H/r_0 = 0.65\tau^2 \text{ for } Bo \approx 150, \tag{9.4.3}$$

where $\tau = t\sqrt{g/2r_0}$ and $H$ is the liquid fall distance. For the lower $Bo$ case the large central bubble slows down as it approaches the top of the tank making a slight projection of the vent pipe into the tank help to insure that no liquid is removed.

In a partially filled tank it is possible to reduce the vapor pressure without vapor venting. This could be achieved if the vapor is made to condense by some means, including active or passive pressure reduction methods. In the active method, vapor condensation is achieved when the liquid surface adjacent to the vapor is made cooler to enhance condensation. One method for accomplishing this is to stir the liquid bulk either mechanically or by inducing convection in the liquid and thus moving fresh cold liquid into the liquid/vapor interface region.

Reference [103] describes a recent low gravity experiment on the control of cryogenic storage tank pressures by active mixing. In that experiment the possibility of jet-induced mixing in the liquid was explored as means for promoting vapor condensation at the liquid/vapor interface, to achieve vapor pressure reduction. The liquid was stirred at different speeds by means of a liquid jet initiated in the liquid bulk. Different liquid tank fills were also tested. In all of the tests of this experiment, precaution was taken so that the jet did not break the liquid/vapor surface. The main difficulty with this experiment is that only a single liquid/vapor surface was assumed. This is not always a realistic configuration for tanks in low-gravity environment.

One passive method for vapor reduction in a partially filled tank is to use what is commonly known as the *thermodynamic vent system* TVS. A TVS functions by using liquid withdrawn from the tank and passing it through a Joule-Thomson expander. The Joule-Thomson expander causes flashing of the liquid, and thus reduces the liquid's pressure and temperature. The two-phase fluid downstream of the expander is routed through a heat exchanger in contact with the bulk liquid. Since the bulk liquid is warmer than the two-phase fluid, heat is transferred from the bulk liquid to the two phase fluid. The heat flux causes boiling in the two phase side, and boiling will continue until all the liquid has evaporated. With this method the bulk liquid temperature is brought down, reducing the tank pressure while the gas from the two-phase side is vented overboard.

## *9.5* *FLUID MANAGEMENT EXPERIMENTS CONCEPT IN ORBIT*

In most applications of fluid management, the individual technology areas discussed above, such as liquid expulsion, positioning, draining, and vapor venting, can occur simultaneously in complex space-based system. In many instances the method that works best for one operation could be detrimental to another. The question must be answered of how a fluid acquisition and storage system can be made to operate in an optimal and efficient manner in low-gravity environment. To answer such a question, large-scale fluid system experiments in low-gravity have been proposed in the past. These remain a topic of great interest among the low gravity community.

Space-based fluid management technology requirements can be categorized into either enabling or enhancing technologies, depending on the amount of data collected to date on the specific technology. A certain technology is considered enabling when a data base is absolutely needed to accomplish a mission. Enhancing technology data base is not essential but could provide substantial mission benefits in the areas of performance, cost risk, and schedule. The enabling technologies identified to date are the following categories: tank pressure control, tank chilldown, tank no vent fill, LAD fill and refill, mass gauging, and slosh dynamics and control. The following enhancing technologies have been identified: tank thermal performance, pressurization, low-$g$ settling and out flow, LAD performance, transfer line chilldown, outflow subcooling, low-$g$ vented fill, fluid dumping, and advanced instrumentation.

Pressure control technology is needed due to the fact that little is known about long-term fluid conditions in a well-insulated cryogen storage tank in space-based, low-gravity environment. Techniques to control tank pressure without undue losses of fluid are required. Five methods for controlling pressure have been identified including the following: thermal stratification, passive thermodynamic vent systems to hold tank pressure constant, passive thermodynamic vent systems for pressure reduction, pressure control through mixing destratification, and active thermodynamic vent systems thermal conditioning. Thermal stratification, caused by heat leaks into the tank, can cause higher rates of pressure rise than if the fluid had remained in thermal equilibrium. Means must be developed to counteract the effects of the heat leaks or mix the liquid to induce uniform temperature distribution. Mixing can be achieved in tanks by natural convection which, in turn, depends on the degree of stratification and the direction of the body force vector. Active mixing may be accomplished by stirring the liquid using mechanical means, including liquid jet injection as discussed in the previous section.

When transferring cryogens to empty hot tanks, the receiving tank must be chilled down to temperatures closer to that of the liquid cryogen, in order to prevent wasting a great deal of the initial liquid used for that purpose. There are no data on liquid jet or spray cooling characteristics in low gravity, or even whether or not such characteristics can be a function of the acceleration level. Tests have been designed to provide a better understanding of tank chilldown in low gravity and to determine techniques for minimizing the cryogens required in such processes.

Filling a tank with liquid cryogen inevitably produces vapor, the pressure of which can prevent the tank from totally filling if not vented. The positioning of vents in low-gravity tanks can be very problematic due to the uncertain location of both the liquid and the vapor in the tank. Precious liquid can be lost during a vent process if the vent is not located at the appropriate position. Thus a no-vent fill process would be the method of choice for achieving low-gravity tank fill. It is not possible to depend on terrestrial no-vent fill characteristics since there is no corresponding analogy to liquid vapor separation in low-gravity environment. Data on low-gravity, no-vent fill are sorely needed for the proper application of such a technology.

Although liquid acquisition devices have been designed for low-gravity operations, some key characteristic operating parameters must be established before a final LAD design is adopted. Experiments in low-gravity environment need to be conducted with several LAD fill and refill cycles in order to address these issues. There are many questions concerning the filling of a total communication LAD in low-gravity environment. Some of the major outstanding issues concern the effect of the vapor pressure on the collapse of the vapor in the LAD, the mass flow effects on LAD successful filling, tank fill level effects on LAD filling, and TVS venting effects on LAD filling. The filling of the LAD may ultimately depend on the collapse of trapped vapor bubbles in the LAD.

In recognition of these technology needs the National Aeronautics and Space Administration (NASA) has been conducting extensive ground-based research while planning for a cryogenic flight experiment. In the early 1980s a scaled flight experiment system was defined by NASA through which data can be gathered on both the enabling and enhancing technologies. The Cryogenic Fluid Management Flight Experiment (CFMFE), as described by reference [100], was conceived to be a subcritical liquid hydrogen experiment to be performed in the cargo bay of the Space Shuttle orbiter. Plans for the CFMFE were discontinued after reassessment of payload safety criteria by NASA following the Space Shuttle *Challenger* accident.

Nevertheless, the data required for fluid management in space are still of critical importance to NASA and the investigations continues into these technologies. Studies were conducted on the feasibility of a Cryogenic On-Orbit Liquid Depot Storage, Acquisition, and Transfer Satellite (COLD-SAT). Such satellite would be totally dedicated to fluid management experiments and would be a free-flying orbital experiment to be launched using an expendable vehicle. The free flight requirement for COLD-SAT was fulfilled by the use of liquid hydrogen as the working fluid for the low-gravity tests. Feasibility studies were conducted and initial plans were made for the COLD-SAT experiments (see [102]). However, due to financial constraints, COLD-SAT was subsequently scrapped. Nevertheless, the technology needs, for which COLD-SAT was planned, remain.

Recently elements of the COLD-SAT experiment have been proposed to be performed onboard the Space Shuttle with liquid nitrogen serving as model for cryogenic liquid fuel. The experiment is named the Cryogenic Orbital Nitrogen Experiment (CONE); its purpose is to test needed critical components and technologies. Reference [101] gives a detailed description of the experiment and its operation. The CONE payload will provide an opportunity to demonstrate the feasibility of combining various methods of integrating pressure control, liquid acquisition, and tank fluid outflow in low-gravity environment. The experiment is designed to gather data on the following specific technologies: cryogenic liquid storage, liquid nitrogen resupply, pressurant bottle recharging, and LAD performance. The CONE experiment is scheduled at the present time to fly around mid-1997.